INCLUSION CHEMISTRY WITH ZEOLITES:

NANOSCALE MATERIALS BY DESIGN

TOPICS IN INCLUSION SCIENCE

VOLUME 6

INCLUSION CHEMISTRY WITH ZEOLITES: NANOSCALE MATERIALS BY DESIGN

Edited by

NORMAN HERRON and DAVID R. CORBIN
DuPont Company,
Central Research and Development Experimental Station,
Wilmington, DE, U.S.A.

Papers also appear in *Journal of Inclusion Phenomena and Molecular Recognition in Chemistry,* Vol. 21, Nos. 1–4, 1995.

SPRINGER-SCIENCE+BUSINESS MEDIA, B.V.

Library of Congress Cataloging-in-Publication Data

A C.I.P. Catalogue record for this book is available from the Library of Congress.

ISBN 978-0-7923-3606-8 ISBN 978-94-011-0119-6 (eBook)
DOI 10.1007/978-94-011-0119-6

Printed on acid-free paper

Contents

Chapter 2. Structure-direction in zeolite synthesis
R. F. Lobo, S. I. Zones, and M. E. Davis

Chapter 3. Structural case studies of inclusion phenomena in zeolites: Xe in RHO and stilbene in ZSM-5
John B. Parise

Chapter 4. Zeolite-included molecules studied by NMR
C. Dybowski

Preface

Zeolites, with their crystalline microporous structures, are cordial hosts to a wide variety of guests. However, it was the abrupt and unexpected departure of one of these *guests* (water) from a *host* (stilbite) on heating which led Cronstedt, in 1756, to coin the term "zeolite" (from the Greek meaning "boiling stone") to describe this material. Since that time, approximately 40 different naturally-occurring zeolites have been discovered on earth. Recent studies of meteorite compositions have shown that these guest-host materials (e.g., sodalite) occur in other parts of the universe as well. However, it wasn't until the twentieth century that synthetic routes to zeolites and other non-aluminosilicate molecular sieves were discovered. In addition, with the development of X-ray diffraction and the various spectroscopies, better understanding of the nature of the cavities, cages, and channels of these materials has led to the industrial exploitation of their guest-host properties. The world of zeolites has now expanded into a greater than 2 billion pound per year business, with major applications in detergent formulations, catalysis, and as adsorbents and desiccants. Their economic impact is difficult to determine; however, the improvement in gasoline yields alone (from catalytic cracking) must account for hundreds of billions of dollars in increased GDP.

In this volume, we have brought together a sampling of recent developments in various areas of guest-host or inclusion chemistry in zeolites. With the relevant chemical literature straining at the seams - there are over 44,000 references to "zeolite" or "molecular sieve" in Chemical Abstracts since 1967 - we have tried to include what we consider to be some of the more recent and leading-edge work in the field. Contributions on structure-direction in synthesis, characterizational techniques, computational aspects, coordination and organometallic chemistry, photochemistry, and electronic and optical materials illustrate the diversity and wide utility of this facet of inclusion chemistry. The breadth of information presented will, hopefully, inspire the creative reader to realize the title of this volume and to actually produce some novel *"nanoscale materials by design"*.

Finally, we would like to thank the authors for their thoughtful contributions (and their patience with us in assembling this volume) as well as the DuPont Company for providing the time and facilities for completion of this project. The assistance of Ms. Nina Paglia Urban and Mr. Jack Jensen in the technical assembly of the manuscripts is greatly appreciated.

David R. Corbin
Norman Herron
Wilmington, DE

March 1995

List of Contributors

L. Abrams,
The DuPont Company, Central Research and Development, Experimental Station, Wilmington, DE 19880, USA.

K. J. Balkus, Jr.,
University of Texas at Dallas, Department of Chemistry, Richardson, TX 75083-0688, USA.

N. P. Blake,
Department of Chemistry, University of California, Santa Barbara, CA 93106, USA.

A. K. Cheetham,
Materials Department, University of California, Santa Barbara, CA 93106, USA.

D. R. Corbin,
The DuPont Company, Central Research and Development, Experimental Station, Wilmington, DE 19880, USA.

M. E. Davis,
Chemical Engineering, California Institute of Technology, Pasadena, CA 91125, USA.

D. E. De Vos,
Centrum voor Oppervlaktechemie en Katalyse, K. U. Leuven, K. Mercierlaan, 92, 3001 Heverlee, Belgium.

P. K. Dutta,
Department of Chemistry, The Ohio State University, 120 West 18th Avenue, Columbus, OH 43210, USA.

C. Dybowski,
Department of Chemistry and Biochemistry, University of Delaware, Newark, DE 19716, USA.

A. G. Gabrielov,
University of Texas at Dallas, Department of Chemistry, Richardson, TX 75083-0688, USA.

N. J. Henson,
Materials Department, University of California, Santa Barbara, CA 93106, USA.

N. Herron,
The DuPont Company, Central Research and Development, Experimental Station, Wilmington, DE 19880, USA.

P. A. Jacobs,
Centrum voor Oppervlaktechemie en Katalyse, K. U. Leuven, K. Mercierlaan, 92, 3001 Heverlee, Belgium.

P. P. Knops-Gerrits,
Centrum voor Oppervlaktechemie en Katalyse, K. U. Leuven, K. Mercierlaan, 92, 3001 Heverlee, Belgium.

R. F. Lobo,
Chemical Engineering, California Institute of Technology, Pasadena, CA 91125, USA.

J. B. Parise,
Department of Earth and Space Sciences, State University of New York at Stony Brook, Stony Brook, NY 11794-2100, USA.

R. F. Parton,
Centrum voor Oppervlaktechemie en Katalyse, K. U. Leuven, K. Mercierlaan, 92, 3001 Heverlee, Belgium.

V. Ramamurthy,
Department of Chemistry, Tulane University, New Orleans, LA 70118, USA.

R. A. Schoonheydt,
Centrum voor Oppervlaktechemie en Katalyse, K. U. Leuven, K. Mercierlaan, 92, 3001 Heverlee, Belgium.

G. D. Stucky,
Department of Chemistry, University of California, Santa Barbara, CA 93106, USA.

N. J. Turro,
Department of Chemistry, Columbia University, New York, NY 10027, USA.

B. M. Weckhuysen,
Centrum voor Oppervlaktechemie en Katalyse, K. U. Leuven, K. Mercierlaan, 92, 3001 Heverlee, Belgium.

S. I. Zones,
Chevron Research and Technology Company, Richmond, CA 94802-0627, USA.

Journal of Inclusion Phenomena and Molecular Recognition in Chemistry **21**: 1–46, 1995.

Probing Intrazeolite Space

LLOYD ABRAMS and DAVID R. CORBIN

1. Introduction

Molecular sieves are crystalline materials with open framework structures. Of the almost two billion pounds of molecular sieves produced in 1990, 1.4 billion pounds were used in detergents, 160 million pounds as catalysts, and about 70 million pounds as adsorbents or desiccants [1]. Zeolites, composed of aluminosilicates, represent a large fraction of known molecular sieves. The primary building blocks of zeolites are $[SiO_4]^{4-}$ and $[AlO_4]^{5-}$ tetrahedra which are linked by their corners to form channels and cages or cavities of discrete size The pore openings to these channels and cages generally range from about 3 to 20 Å. As a result of the difference in charge between the $[SiO_4]^{4-}$ and $[AlO_4]^{5-}$ tetrahedra, the total framework charge of an aluminum-containing molecular sieve is negative and hence must be balanced by cations, typically protons, alkali, or alkaline earth metal ions.

Over 500 materials are listed in the *Handbook of Molecular Sieves* including at least 40 naturally occurring zeolites [2]. The numerous framework topologies of the molecular sieves offer various systems of channels and cavities resulting in one-, two-, or three-dimensional diffusion for included guest molecules. For example, so-called one-dimensional molecular sieves have channels parallel to one another and there are no connecting channels large enough for guest molecules to cross from one channel to the next.

The variety in structure types provides a range of pore dimensions from the small-pore materials consisting of ring openings containing eight of the tetrahedra, to the medium-pore zeolites with 10-membered ring openings, and the large-pore materials with 12-membered ring openings. Larger pore openings have been reported for the aluminum phosphate molecular sieves, such as $AlPO_4$-8, VPI-5 with 14- and 18-membered rings, respectively, and cloverite, a $GaPO_4$ with 20-membered rings [3]. For the same number of ring members, variations in pore geometry, for example, elliptical vs. circular, can have a marked effect on sorption properties especially when the openings are similar in size to the sorbing molecules.

Molecular sieves can be characterized not only by their channel systems but, in many cases, also by the polyhedral cavities or cages present. The aluminosilicate backbone of a zeolite molecular sieve can be represented in a number of ways which are illustrated in Figure 1 for the beta- or sodalite-cage.

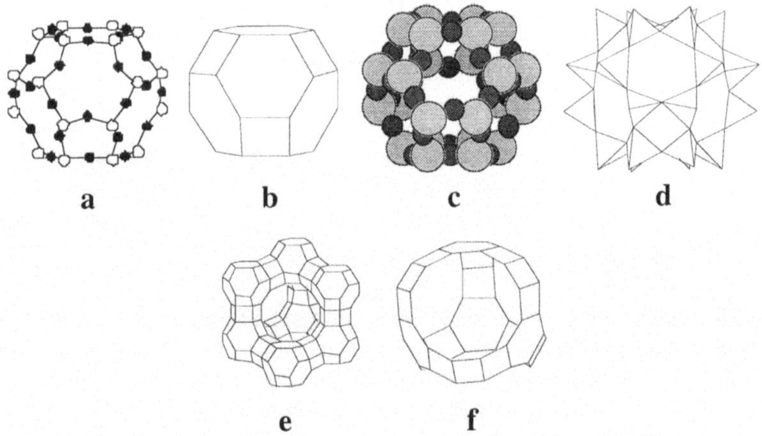

Figure 1. Representations (a-d) of the structures of the sodalite cage [using (a) a 'ball and stick' model; (b) a 'tetrahedral array' in which 'single bonds' join the framework T (T = Si, Al) atoms (The bridging oxygen atoms are not shown for simplicity); (c) 'space-filling' model; and (d) 'tetrahedral presentation' showing the arrangement of framework tetrahedra]; (e) faujasite ('tetrahedral array' with bridging oxygens removed); and (f) the supercage ('tetrahedral array' with bridging oxygens removed).

The sodalite cage structure is constructed of openings containing 4- and 6-membered rings of $[SiO_4]^{4-}$ and $[AlO_4]^{5-}$ tetrahedra. (In the literature, TO_2 (T = Si, Al) is also used to represent the framework tetrahedra [4].) The faujasite structure (see Figure 1) is constructed of sodalite cages connected through oxygen bridges at the 6-rings and the largest pore opening is a 12-membered ring. The sodalite cages surround an even larger cage, the supercage, which forms a three-dimensional network with each supercage connected tetrahedrally to four other supercages through the 12-membered ring opening.

In order to quantify the amount of molecular sieve in a sample and to understand and predict packing, sieving, and migration effects of absorbed molecules, the sizes of the framework constituents and internal void geometry must be known with a high degree of precision. Neutron diffraction provides excellent structural resolution but is not a common laboratory tool. Therefore, structures of molecular sieves are generally derived from X-ray diffraction patterns of powders because large single-crystals are not available (most molecular sieve preparations yield particles in the micron size range). The exact size and shape of the internal structures are, at best, only approximated by the X-ray technique because:

- different framework atoms (e.g., silicon and aluminum) are not readily distinguished nor is their siting necessarily identical from unit cell to unit cell.
- the location of charge compensating ions (especially protons) is difficult to identify.
- the size of the framework oxygen must be estimated.

Furthermore, the presence of small amounts of other phases and/or inclusions within the framework may not be detected or quantified. However, it is essential to have a good measure of the pore opening, channel, and cage dimensions of molecular sieves in order to describe:

- the ability to differentially absorb molecules on the basis of size (sieving ability).
- the rates of migration of molecules into and within the structures.
- the packing of molecules absorbed within the framework.

As noted above, the precise location of active sites, ions, and occlusions within the frameworks may not be known. Without detailed measurements of these factors, it is difficult to predict the chemical reaction selectivity imposed by framework structures. Sorption measurements using molecular probes of appropriate size, geometry, and chemistry can help provide such a description. Such measurements are relied upon to provide information about the framework dimensions, i.e., size and shape selectivity, as well as accessible framework volumes. For example, Na-A zeolite does not absorb molecules larger than 3.6 Å at 77 K, while at 300 K it will absorb molecules of 4.0 Å in diameter [5]. Such a change in channel dimensions with temperature is much larger than would be inferred from coefficient of expansion data but re-siting (migration) of the Na ion from the 8-ring to the cage would permit such absorption. As such, absorption measurements provide a good description of effective channel dimensions which is essential if inferences are going to be made concerning diffusion of molecules within framework structures and chemical reaction selectivity imposed by those structures.

This chapter will focus primarily on the authors' studies on zeolites A, rho, and ZSM-5 using physical sorption measurements in conjunction with X-ray diffraction. The authors will use the term "adsorption" to include all sorption phenomena involving sorption on external particle or crystal surfaces and will reserve the term "absorption" for use in describing sorption inside the frameworks. The amount adsorbed can be distinguished from the absorbed amount (*vide infra*). This chapter is not intended to be all-inclusive regarding sorption phenomena but to provide the reader with important concepts and the basis for characterizing other framework structures. Other techniques, such as chemisorption, IR, NMR, etc. will be discussed somewhat in other chapters or in other monographs [6]. Here, the utility of adsorption measurements will be demonstrated to show that a given framework dictates the equilibrium sorption values depending upon the sorbates' size and geometry. By using X-ray diffraction coupled with adsorption measurements, a good estimation of the accessible (available for absorption) framework can be made; in some cases, this capability permits the crystallinity of a sample to be determined.

2. Considerations

As noted above, when X-ray diffraction analysis is applied to framework structures there are some limitations compared to its application on dense crystalline solids. The unit cells of molecular sieves are often larger and the diffraction patterns have many more lines. Small amounts of impurity phases are virtually invisible but they can have pronounced effects on a sample's absorption properties. In order to fully understand the dimensions controlling absorption and filling of frameworks, the size of the framework oxygen must be defined. Once having established how oxygen is bound to other framework species, other factors become important. Filling of the framework and the estimation of pore volume, the packing of the absorbed molecules, estimation of the amount adsorbed on the external crystal surfaces versus the amount absorbed into the framework, the choice of molecules to probe the framework, and what experimental difficulties typically befall the experimentalist will be addressed in this section.

2.1 SIZE OF OXYGEN

A dilemma exists in the field concerning the usage of X-ray diffraction and absorption of molecules, that is, in making the "correct" choice of sizes that describe both the molecular sieve and sorbate dimensions. For the ideal planar configuration illustrated in Figure 2, the O-O interatomic distance across the 8-membered ring would be 6.9 Å. With the

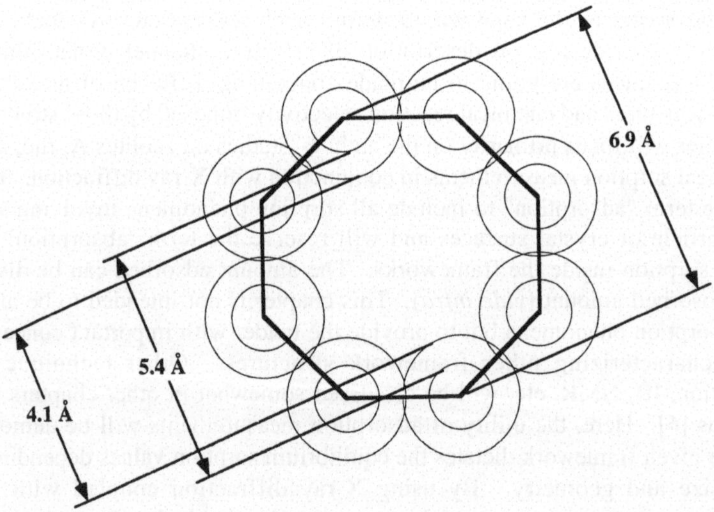

Figure 2. An ideal planar configuration of an 8-membered ring system showing the effect of oxygen ionicity on the pore opening dimension.

Table 1. The ideal planar pore: the dependency of size on oxygen ionicity.

Number of T Atoms in Ring# Å	Interatomic Distance Å	Circular Pore Opening Distance Ionic O† Å	Circular Pore Opening Distance Covalent O† Å
4	3.71	0.96	2.21
5	4.03	1.69	2.95
6	4.46	1.71	2.96
8	6.86	4.11	5.36
10	8.55	5.80	7.05
12	10.21	7.46	8.71
14	11.89	9.14	10.39
18	15.21	12.46	13.71
20	16.87	14.12	15.37

† A diameter of 2.7 Å is used for ionic oxygen [7]; a diameter of 1.5 Å for covalent oxygen [7].

\# Number of tetrahedra forming the pore.

framework T (Si, Al) and O atoms fixed, the size of the oxygen atoms will determine the size of the pore opening. If the framework oxygen were ionic in character, its diameter would be 2.7 Å [8] and the opening of the ring would be 4.1 Å. If the oxygen were purely covalent, the ring opening would be 5.4 Å. As shown in Table 1, the openings can vary significantly depending upon the ionicity of the framework oxygen.

The assumption that the framework oxygen is ionic (i.e., 2.7 Å diameter) has been made so often that the literature has many examples of apparent contradictions: for example, molecules almost 5 Å in size (e.g., isopropanol, trimethylamine) sorb into 8-ring structures such as zeolites A and rho, although the ring openings are cited as 3.6 Å and 4.1 Å, respectively [7]. There appears to be ample evidence that the framework oxygen is more covalently bound to silicon such that its size is closer to 1.5 Å (probably ~ 1.7 Å). Furthermore, ZSM-5 zeolites are generally hydrophobic which is difficult to rationalize in terms of highly ionic framework oxygen. The amount of water sorbed by ZSM-5 depends directly on the aluminum content of the framework [9] but the amount of hexane sorbed is independent. ZSM-5 is typical of the 10-membered ring systems which readily absorb benzene which is listed as having a width of 6.6 Å (Pauling) [5]; this dimension is larger than the ring opening of 5.8 Å based on the crystal structure and using the diameter of ionic oxygen. Other investigators have addressed the problem of assigning the appropriate molecular dimensions [10]. Dimensions used in this chapter were obtained from molecular models, Fisher-Hirschfelder-Taylor [11], which seem to provide a consistent basis for comparison of liquid density, molecular cross-sections for estimating kinetic diameters, and molecular sieving effects. From these models, Table 2 was developed and presents a summary of the dimensions of the probe molecules cited in this chapter.

Table 2. Approximate dimensions of probe molecules and abbreviations used in the text.

Molecule	Dimensions, Å [11]	Molecule	Dimensions, Å [11]
nitrogen, N_2	2.6x2.6x3.8	i-propanol, iPOH	4.6x5.0x6.1
water, H_2O	2.6x2.8x3.6	*m*-xylene, mxyl	3.6x7.2x8.6
methanol, MeOH	3.6x3.9x4.4	*p*-xylene, pxyl	3.6x6.5x8.9
ethanol, EtOH	3.9x4.2x6.0	*o*-xylene, oxyl	3.6x7.1x8.1
n-propanol, nPOH	3.9x4.3x7.5	3-methylpentane, 3MeP	5.1x5.9x8.0
n-hexane, nHex	3.9x4.3x9.7	toluene, tol	3.6x6.5x8.1
xenon, Xe	4.0	benzene, bz	2.5x6.5x7.2
carbon tetrachloride, CCl_4	5.4	cyclohexane, 0Hex	5.1x6.0x6.5
monomethylamine, MMA	3.7x3.9x4.4	2,2-dimethylpropane, neopentane	5.5x5.8x6.2
dimethylamine, DMA	3.9x4.7x6.0	trimethylamine, TMA	3.9x5.4x6.1

The different molecular sieve structures have slightly different sizes and shapes from the 'ideal' circular values listed in Table 1. For the various framework structures discussed in this chapter, Table 3a shows their pore and channel dimensions, "ionic" compared to "covalent" oxygen, and lists examples of those structures. Table 3b lists similar data for the more common cage structures.

2.2 ISOTHERMS AND THE LIQUID STATE

Molecular sieves yield Type I (BET) adsorption isotherms or the so-called Langmuir adsorption isotherm; characteristics of this isotherm are a steep rise in absorbed amount at low pressures and an almost constant absorbed value at high pressures. Recall that an isotherm is equilibrium adsorption data as a function of vapor pressure at a constant temperature. Since it is equilibrium data, it can and should provide thermodynamic values. At present, there is no good theoretical treatment developed that completely predicts the absorption isotherm in microporous materials such as molecular sieves.

For the most part, a molecular sieve will be filled in the relative pressure range of 0.1-0.5. In essence, filling of a molecular sieve stops the major sorption process and provides the distinctive isotherm shape. Generally, the Gurvitsch Rule governs the sorption of species into zeolites: "the uptake at saturation of different sorbates on the same adsorbent is nearly equal if expressed as the volume of liquid at the temperature of sorption" [12].

But the Gurvitsch Rule may not always be true, for example,

- when equilibrium has not been reached.
- in the temperature region near or above the homogeneous liquid's critical point.
- when steric constraints are operating.

Table 3a. Oxygen ionicity, its effect on pore and channel dimensions.

Framework Code	T atoms	Ionic Oxygen [7]	Covalent Oxygen [8]	Examples
RHO	8	3.6	4.8	rho
CHA	8	3.8	5.0	chabazite
KFI	8	3.9	5.1	ZK-5
LTA	8	4.1	5.3	A
ERI	8	3.6x5.1	4.8x6.3	erionite
MEL	10	5.3x5.4	6.5x6.6	ZSM-11, silicalite-2
MFI	10	5.3x5.6	6.5x6.8	ZSM-5, silicalite
	10	5.1x5.5	6.3x6.7	
FER	10	4.2x5.4	5.4x6.6	ferrierite
	8	3.5x4.8	4.7x6.0	
HEU	10	3.0x7.6	5.2x8.8	clinoptilolite
	8	3.3x4.6	4.5x5.8	
	6	2.6x4.7	3.8x5.9	
FAU	12	7.4	8.6	X, Y
BEA	12	7.6x6.4	8.8x7.6	beta
	10	5.5	6.7	
GME	12	7.0	8.2	gmelinite
	8	3.6x3.9	4.8x5.1	
LTL	12	7.1	8.3	L
MAZ	12	7.4	8.6	mazzite, omega
MOR	12	6.5x7.0	7.7x8.2	mordenite
	8	2.6x5.7	3.8x6.9	
OFF	12	6.7	7.9	offretite
	8	3.6x4.9	4.8x6.2	
AET	14	7.9x8.7	9.1x9.9	$AlPO_4$-8
VFI	18	12.1	13.3	VPI-5

The difficulty in dealing with absorption of vapors into microporous sorbents is that equilibration times are much longer, in general, than for adsorption on planar substrates. Instead of times on the order of seconds or minutes, the units are hours, days, and months. The problem is akin to the parking of cars in a large department store's parking lot on the day of a sale. At the early stage, there are many available spaces and parking is easy (equilibrium is reasonably rapid). During the peak of the sale, vacancy diffusion operates (a driver must wait for a space or diffusion path to open); equilibration times are long. Too often, researchers do not wait long enough to assure equilibrium, report rate phenomena on poorly characterized samples, or do not provide a large enough statistical base for interpretation of their measurements. In this chapter, we provide some

Table 3b. Oxygen ionicity, its effect on cage dimensions.

Cage or Polyhedral Building Unit[a] / Common Name(s)	Approximate Free Dimensions (Å)		Approximate Free Volume ($Å^3$)[b]	Zeolite Examples
	Ionic	Covalent		
$[4^6 6^2]$/D6R	1.1[c]	2.3[c]	5	FAU, KFI, CHA, ERI, OFF
$[4^8 8^2]$/D8R, Delta	3.3[d]	4.5[d]	31	RHO
$[4^6 6^5]$/Epsilon	2.3x3.5	3.5x4.7	24	ERI
$[4^6 6^8]$/Sodalite, Beta	5.4[e]	6.6[e]	151	SOD, FAU, LTA
$[4^9 6^2 8^3]$	4.8x6.2	6.0x7.4	157	GME, OFF, MAZ
$[4^{12} 8^6]$/Gamma	5.4x9.6	6.6x10.8	264	KFI
$[6^2 8^6]$	5.4x9.8	6.5x11	401	CHA
$[4^{12} 6^5 8^6]$	5.1x13.8	6.3x15	445	ERI
$[4^{12} 6^8 8^6]$Alpha	10.4[f]	11.6[f]	811[g]	LTA, KFI, RHO
$[4^{18} 6^4 12^4]$/Supercage	10.8[e]	12.0[e]	896	FAU

a. e.g., $[4^6 6^2]$ cage contains 6 four-membered rings and 2 six-membered rings.
b. Calculations of polyhedral volumes were performed using a modification of the POLYVOL program [D. K. Swanson and R. C. Peterson, "POLYVOL Program Documentation", Virginia Polytechnic Institute, Blacksburg, VA] assuming the radius of the TO_2 unit to be 2.08 Å (equivalent to that of quartz).
c. in plane of 6-ring.
d. in plane of 8-ring.
e. diameter of inscribed sphere.
f. diameter of inscribed sphere for LTA.
g. 811 $Å^3$ for LTA, 707 $Å^3$ for KFI, and 722 $Å^3$ for RHO.

guidance concerning these issues. Considerations of sorption phenomena in molecular sieves and microporous substrates have produced an extensive literature [13]. The references cited are not intended to provide a comprehensive bibliography on sorption in microporous substrates but were chosen to provide a reasonable starting point. The last words on sorption phenomena in molecular sieves have obviously not been written; a general treatment has not been developed to completely describe both sorption and diffusion. In order to completely describe sorption phenomena in micropores, it is essential that kinetic and equilibrium theories be compatible.

Experimental correlations are relied upon to describe molecular packing and diffusion within molecular sieves. For example, Moore and Katzer showed that a 0.4 Å difference (in molecular sieve pore versus sorbed molecule dimension) produces an order of magnitude change in effective diffusion coefficient [14]. Thus, for sorption rates of differently sized molecules into similar zeolites (for example, different preparations), rate differences relate to small differences (~0.1 Å) in channel dimensions. As such, sorption rate measurements are very sensitive to inclusions, ions, lattice defects, framework distortions, etc. which might not otherwise be detected by other analytical techniques.

The classical Kelvin Equation, used in physical adsorption to relate the vapor pressure of a solvent to the diameter of the filled pores, is not strictly valid for the region below ~20 Å. Below this size, the macroscopic properties used in the equation are not necessarily continuous nor well-defined. Typical of the sort of problems encountered when dealing with molecular sieve dimensions is the relationship of liquid density to critical temperature. Normally, the density of a liquid decreases as the temperature increases; an increase in pressure is needed to maintain the species in the liquid state. Near the critical point, the kinetic energy term overwhelms the interactive forces of the molecules and the species can no longer be liquefied.

But the situation is quite different inside molecular sieves. The absorbed molecules, in addition to their own interactive forces, are attracted by the framework - an interaction that might be much larger than the intermolecular forces of the liquid state. Well below the critical point, if there are no constraints upon packing, the sorbed state generally packs according to the liquid's density following the Gurvitsch Rule. Shown in Figure 3, the density of ammonia in the liquid state is compared to the density of ammonia absorbed by zeolite rho. The sorption data for ammonia were taken using a microbalance at an ammonia pressure of 0.3 atm. The weight of the ammonia absorbed was measured directly while the sorption volume for zeolite rho was determined from X-ray diffraction and other absorption measurements. For the liquid state, as the temperature is increased, the pressure on the ammonia must be increased to keep it liquefied. At the critical point ($T_c = 405.51$ K, $P_c = 111.3$ atm, $\rho_c = 0.235$ g/cm^3), the liquid density drops precipitously.

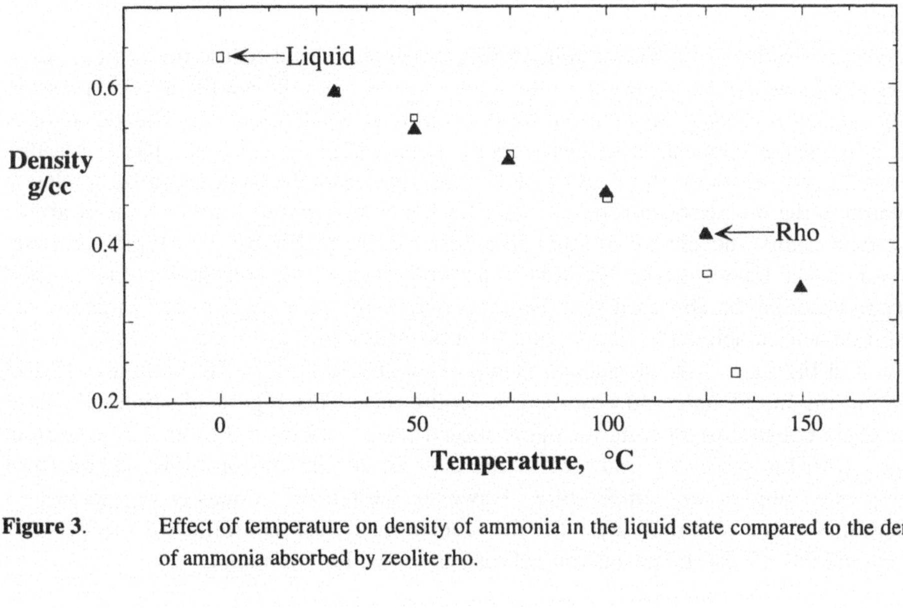

Figure 3. Effect of temperature on density of ammonia in the liquid state compared to the density of ammonia absorbed by zeolite rho.

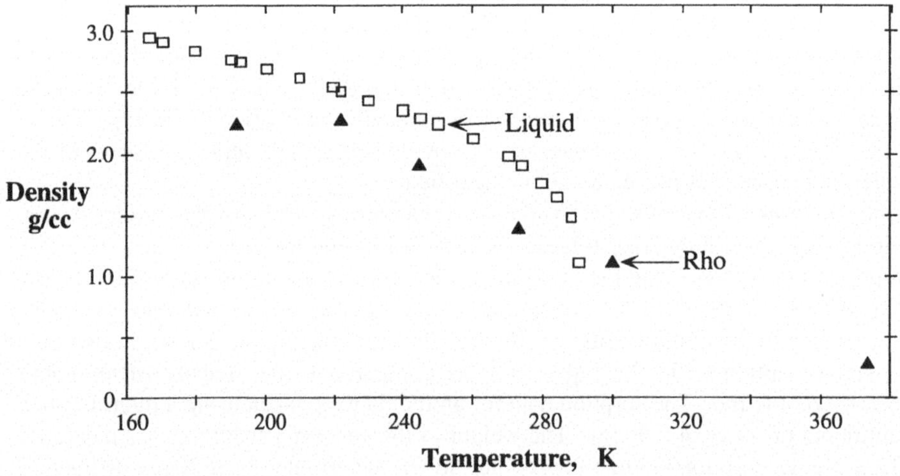

Figure 4. Effect of temperature on density of xenon in the liquid state compared to the density of xenon absorbed by zeolite rho.

However, the sorbed case shows quite different behavior: well-below the critical point, the absorbed state packs according to the liquid density even though the applied pressure may be a small fraction of that needed for the homogeneous liquid case. This behavior is termed, by us, the 'unconstrained case'. As the temperature increases and passes through the liquid's critical point, the density of the absorbed state continues to drop modestly compared to the precipitous drop in density for the homogeneous liquid. Near or above the critical point, the density-temperature plot shown in Figure 3 is typical for the behaviors of the homogeneous liquid vs. the absorbed state. At present, there is no good theory to describe the absorbed state behavior and so the packing density in micropores for a liquid near or above its critical point must be measured.

Shown in Figure 4 is the density of xenon (T_c = 289.75 K, P_c = 58.0 atm, ρ_c = 1.105 g/cm^3) in the liquid state and absorbed in zeolite rho. This Figure illustrates the case when steric constraints operate on the absorbed phase. Here, the large size of xenon coupled with the geometry and size of the cage of zeolite rho prohibits xenon from packing as a liquid at any temperature. However, the fall-off in density vs. increase in temperature of the absorbed state for the 'constrained case' is similar to the density-temperature fall-off for the unconstrained case as shown for ammonia.

2.3 EXPERIMENTAL DETAILS

Gravimetric measurements were used to follow the sorption of liquid vapors by molecular sieves. A sample cell employing a greaseless vacuum stopcock and an 'O-ring' seal was designed to contain the powders and fit on an analytical balance. Details of the activation procedure and measuring protocol were published previously [15]. Typically, samples were heated at 425° C until the pressure above the samples was < $2x10^{-5}$ torr. If the samples' outgassing rate was < 0.1 mtorr/min at 425° C, the samples were cooled under vacuum and weighed. If the outgassing criteria were not met, sorption weights were obtained 10-25% below expectations. It typically takes 2-4 days of heating at 425° C under vacuum to prepare small-pore zeolites for sorption studies whereas it takes < 1 day for ZSM-5 zeolites to meet the outgassing criteria. The weight of powder dried in this fashion was used to calculate the amount sorbed per gram of sample. Generally, the same samples were used for a series of absorptions and desorptions. Sample weights were monitored and an oxygen burn-off was used if there was any measurable build-up. X-Ray diffraction was used to detect gross crystallinity changes.

The initial weight loss upon activation at 425° C is recorded; water desorption generally accounts for this loss. If the sample has a low external surface area, then this weight loss may provide a measure of the framework sorption capacity.

In general, two weighings, at 3 and 20 hours, were taken routinely and the ratio of amounts is called the 'Rate of Filling Ratio'. When sorption was slow or the weight gain negligible, the duration of exposure was increased. Measurements at 3 and 20 h show:

- when the framework is completely filled. If the sorption amounts are > 10 wt % and virtually equal, then the absorbing molecules are smaller than the limiting channel dimension.
- that the molecules cannot enter the framework and only occupy the exterior surface. Again, the weight gains at 3 and 20 hours are equal but are generally < 2% for this case. Longer sorption times are then used to check this point.
- that migration of the sorbing molecule is restricted by a framework dimension. In this case, the 3 hour weight gain is approximately 10-75% of the 20 hour weight gain. By changing the size and/or shape of the sorbing molecule and using Moore and Katzer's correlation, an estimate of the limiting framework dimension(s) can be made. Again, if the weight gains are low, 2-5%, the duration of the sorption process can be extended.

Most of the sorptions were performed at room temperature, 20-25° C, because they were relatively easy to do and the possibility of reaction was reduced. Sorptions at higher temperatures could be done with some additional effort but, as the system could not be analyzed during sorption at higher temperatures and the possibility of reaction increased, such measurements were postponed until suitable instrumentation would be available.

2.4 ADSORPTION (EXTERNAL SURFACE) VS. ABSORPTION (INTERNAL FRAMEWORK)

Because many molecular sieves are synthesized as micron or submicron crystals, a significant contribution to the sorbed amount could be made by the external surface. Although the BET method is synonymous with the measurement of surface area, it provides an ambiguous result when applied to molecular sieves. In general, nitrogen is used as the BET probe molecule and, at liquid nitrogen temperature, it could condense within the sieve framework. For large pore zeolites, such condensation provides an estimate of the framework pore volume. This value would have to be corrected for adsorption on the exterior surface of the crystals. For small pore zeolites, nitrogen may not be able to enter the framework such that a real measure of the external surface area can be obtained. It is desirable to be able to measure both the external surface area as well as the internal pore volume of zeolite preparations. The measured external surface area can then be used to obtain the amount adsorbed on the external surface of a zeolite powder. Using mercury porosimetry, a method was developed to estimate the amount adsorbed on the surface of the particles [16]. The amount absorbed into the framework is obtained by subtracting this surface contribution from the total weight gain. All values reported in this chapter have been corrected in this manner.

2.5 ESTIMATION OF THE FRAMEWORK PORE VOLUME

Comparison of the absorption of molecules into molecular sieves versus their adsorption on planar substrates is similar to the filling of a pail versus pouring the same amount of water on a floor. A pail will hold a finite volume of liquid while the depth of water on your basement floor depends on how quickly a plumber can get to your residence. For molecular sieves, a method is needed to estimate the internal pore volume available for absorption. By using this volume in combination with sorption data, the framework crystallinity of a sample can be calculated. The steps to do this calculation are straight-forward:

- from Avogadro's Number and the molecular weight of the empty, dehydrated molecular sieve, the number of unit cells per gram can be calculated.
- X-ray diffraction provides the dimensions of a unit cell and indicates if other species (amorphous or crystalline) are present to any large extent.
- assume that each TO_2 (SiO_2 or AlO_2^-) group in a molecular sieve occupies the same volume as in quartz (~37.6Å^3). Subtracting the volume of the TO_2 groups from the unit cell volume yields the total pore volume in a unit cell. Note: the volume of charge compensating cations as well as the volume of framework sections that may be inaccessible for adsorption (for example, the ß- or sodalite cages) can be subtracted from the total pore volume to give the accessible pore volume. Dividing the accessible pore volume per unit cell by the unit cell volume yields the Void Fraction.

- finally, determine the volume of a gram of unit cells and multiply it by the Void Fraction to give the accessible framework pore volume. Typical values will range from 0.2-0.5 cc/g; zeolites are not only molecular sieves, they are also molecular sponges.

A listing of the calculated pore volumes for a variety of zeolites is given as Table 4. Some zeolites are listed twice to show how ions or inaccessible regions, for example the ß-cage, affect the pore volume calculation.

2.6 PACKING CONSIDERATIONS AND CHOICE OF PROBE MOLECULES

From equilibrium absorption data, we can get some idea of how molecules pack into a molecular sieve framework by using solvent molecules of different dimensions. For example, if two substances pack into a molecular sieve according to their liquid densities (at temperatures well-below their critical points), then the ratio of their absorbed weights

Table 4. Calculated sorption parameters from X-ray diffraction unit cell dimensions¥.

Zeolite	Formula	UC x10^{20}	UC_v Å^3	V_f cc/cc	V cc/g	ρ g/cc	V_p cc/g
H-Rho	$H_{12}Al_{12}Si_{36}O_{96}$	2.09	3375	0.465	0.705	1.418	0.328
H,Cs-Rho	$CsH_9Al_{10}Si_{38}O_{96}$	2.01	3375	0.465	0.675	1.481	0.308
H-ZK-5	$H_{30}Al_{30}Si_{66}O_{192}$	1.04	6539	0.448	0.681	1.468	0.305
H-Chabazite	$H_4Al_4Si_8O_{24}$	8.33	822	0.451	0.685	1.460	0.309
Ca-Chabazite	$Ca_2Al_4Si_8O_{24}$	7.54	822	0.442	0.620	1.614	0.274
Na-A	$Na_{12}Al_{12}Si_{12}O_{48}$	3.52	1870	0.491	0.659	1.519	0.323
Na-A	ß-cage blocked						0.270
Ca-A	$Ca_6Al_{12}Si_{12}O_{48}$	3.60	1870	0.504	0.673	1.487	0.339
Ca-A	ß-cage blocked						0.285
H-ZK-4	$H_9Al_9Si_{15}O_{48}$	4.17	1798	0.498	0.749	1.335	0.373
H-Erionite	$H_9Al_9Si_{27}O_{72}$	2.78	2300	0.411	0.639	1.565	0.263
Ca-Erionite	$Ca_{4.5}Al_9Si_{27}O_{72}$	2.58	2300	0.403	0.592	1.689	0.239
H-Ferrierite	$H_{5.5}Al_{5.5}Si_{30.5}O_{72}$	2.78	2027	0.332	0.563	1.776	0.187
H-ZSM-5	$H_2Al_2Si_{94}O_{192}$	1.04	5500	0.344	0.573	1.746	0.197
H-Clinoptilolite	$H_6Al_6Si_{30}O_{72}$	2.78	2100	0.355	0.583	1.714	0.207
H-Y	$H_{56}Al_{56}Si_{136}O_{384}$	0.521	15100	0.522	0.787	1.272	0.410
H-Y	ß-cage blocked						0.348

¥ Notes: UC (10^{20}) is the number of unit cells per g of dehydrated zeolite.
UC_v (Å^3) is the Unit Cell Volume.
V_f is the Void Fraction (cc/cc).
ρ (cc/g) is the density of the dehydrated framework.
V_p (cc/g) is the framework Pore Volume per g (dehydrated).

must be equal to the ratio of their liquid densities. Departure from the density ratio for differently sized absorbed molecules indicates that steric constraints are imposed. To provide a measure of the constraint imposed by a framework upon different guest molecules, the concept of 'packing ratio' is introduced. 'Packing ratio' is defined as the ratio of the equilibrium amounts absorbed of two solvents in a given molecular sieve sample; if the solvents are liquid-like in behavior within the framework, then the packing ratio should be identical to the density ratio. If, however, one molecule is more constrained within the framework than another, then the packing ratio will be less than the density ratio. The use of a ratio eliminates the need for an absolute measure of crystallinity for a given sample but serves to describe the packing of solvent molecules in the accessible framework.

For the ZSM-5 study, probe molecules were selected from the alkanes and aromatics to provide information concerning the internal framework volume accessible for absorption, cross-sectional dimensions, and migration limiting features. The interaction of these molecules with molecular sieves (at room temperature) is typical of physical adsorption processes with heats of sorption < 15 kcal/mole. Furthermore, the dimensions of these hydrocarbon molecules are sufficiently different such that slight changes in framework dimensions have profound effects on absorption data.

For the small-pore molecular sieve study, the choice of probe molecules was rather limited; methanol, ethanol, n-propanol, and linear alkanes were generally used. Branched alkanes could not enter such frameworks. For the present study, a compilation of probe molecules and their approximate dimensions is given in Table 2.

3. Characteristic Absorption Values for Zeolite Rho

In preparing a sample for sorption studies, the initial weight loss upon evacuation at 425° C is recorded and represents the desorption of water adsorbed on the surface and from within the framework. As such, the volume calculated from this weight loss from a saturated sample may provide an estimate (providing the sample is not hydrophobic) of the limit to the absorption amounts of other, larger solvent molecules. For zeolite rho, the subsequent absorption of water to measure the framework volume was impractical as the rate of sorption was very slow for all samples examined and the amount sorbed, even after prolonged exposures (~500 hours), did not equal the initial weight loss. However, the sorption of methanol is virtually complete within 3 hours. Shown in Table 5 is a compilation for several rho preparations of the initial weight loss upon evacuation at 425° C and the corresponding pore volume calculated from methanol, MeOH, absorption.

For each sample, the calculated pore volume from methanol absorption agrees very well with the accessible framework volume calculated from the initial weight loss (using a density of 0.998 for water). Thus, methanol absorption seems to provide an accurate measure of the accessible framework pore volume of a rho sample. This measure of

Table 5. Water and methanol absorption# data for rho samples.

Form @	Calc. Temp*	% Wt Loss#	Water # 3 h	Water # 66 h	Water # 100 h	MeOH# 20 h	V_p† cc/g
Ca,H	425/Vac	15.0	0.10	5.90		9.71	0.123
La,H	425/Vac	12.6	0.00	2.56		11.07	0.140
H	725/Air	17.4	2.59	12.14	12.49	14.12	0.178
H	600/Ar	24.3	7.25	18.99	18.62	19.53	0.247
H	600/Air	25.0	15.10	19.48	19.71	20.17	0.255
H	550/Air	26.9	15.83	18.62	19.47	21.11	0.267
Li,H	425/Vac	29.1	0.98	7.48		22.42	0.283
H	550/Air	31.4	4.47	21.99	22.02	24.58	0.310
H	500/Air	30.0	11.61	23.81	23.86	24.72	0.312

\# Dry sample weight basis, sample activated at 425° C and evacuated to $<2\times10^{-5}$ torr. Measurements were performed at 27° C and reported as grams absorbed per 100 grams bone-dry sample.

@ Charge compensating cations.

* Calcination conditions, temperature (° C)/ atmosphere.

† Pore Volume, V_p, calculated using a density of 0.792 g/cc for methanol.

accessible framework pore volume is the basis for calculating the amount of size and shape selective phase present in a given sample. It will be compared to the calculated framework pore volume of Table 4.

Almost 200 samples of rho zeolite were examined, about 20 separate preparations with different ion-exchanges and calcination conditions. Shown in Table 6a are the mean absorption values for 43 highly crystalline H-rho samples with a residual cesium content of <1/uc. The H-rho samples were generally prepared by exchanging the Na,Cs-form several times with a 10% NH_4NO_3 solution, washing and drying the material, and calcining in dry nitrogen at 550° C. Statistical analysis of the sorption data for different preparations, modifications, calcinations, etc. is needed to fully characterize and understand the effects of framework topography [16]. The sorption value for methanol, 24.00, corresponds to an accessible void volume of 0.303 cc/g which is in good agreement with the value calculated in Table 4. Thus, under the experimental conditions, methanol fills the available rho framework and packs according to its liquid density.

Table 6a. Characteristic absorption# values for zeolite rho - 43 samples, < 1Cs/uc.

Solvent	MeOH	EtOH		nPOH		nHex		iPOH
Time, h	20	3	20	3	20	3	20	20
Mean	24.00	21.12	21.70	14.26	20.20	9.00	10.85	0.89
Std.Dev.	0.89	0.96	0.97	3.88	0.88	2.72	1.66	0.39

\# g absorbed per 100 grams of dry sample.

Table 6b. Ratios of sorption data - 20 samples.

	Packing Ratios			Rate of Filling Ratios			
	EtOH/MeOH	nPOH/EtOH	nPOH/MeOH	MeOH 3/20	EtOH 3/20	nPOH 3/20	nHex 3/20
Mean	0.923	0.938	0.871	0.986	0.962	0.958	0.861
Std Dev	0.016	0.013	0.015	0.010	0.025	0.028	0.136
Liquid Density Ratio	0.996	1.019	1.015				

Lower values of methanol sorption indicate that the sample is <100% crystalline, a fraction of the framework pore structure is inaccessible, another phase may be present or, a combination of these possibilities. Use of other techniques, in some cases, provides the reason(s) for the lower framework absorption capacity.

Shown in Table 6b are data for 20 zeolite rho samples with the highest packing and rate of filling ratios. The rate of filling ratio for the solvents decreases as the chain length increases showing one manifestation of steric hindrance. For the C_1-C_3 linear alcohols, equilibrium was reached within 20 hours, longer exposures did not show any additional weight gain within experimental error. The packing ratios are 92-94% of their respective liquid density ratios, again showing the effect of steric constraints. As the guest molecules become longer, the packing ratio decreases. The difference in measured vs. calculated values (from liquid densities) of the packing ratios provides a measure of the constraint imposed on the ethanol and n-propanol molecules by the rho framework. These ratios serve to characterize the rho framework and provide a gauge of the crystallinity of a sample. For example, the samples shown in Table 5 had approximately the same packing ratios as shown in Table 6b. Therefore, the crystallinity of each sample could be estimated by dividing the measured methanol absorption value by the pore volume calculated in Table 4.

Once the sorption parameters for a molecular sieve are well-defined, the constraint

Table 7. Molecular constraint of packing in the zeolite rho framework, < 1Cs/uc.

Molecule	Density g/cc	g absorbed 100 g sample	Volume# cc/g	Constraint† Index	Molecular Dimensions, Å [11]
Methanol	0.792	24.00	0.3030	1.6	3.6x3.9x4.4
Ethanol	0.789	21.70	0.2750	10.7	3.9x4.2x6.0
n-Propanol	0.804	20.20	0.2512	18.4	3.9x4.3x7.5
n-Hexane	0.659	10.85	0.1646	46.5	3.9x4.3x9.7

\# Calculated assuming packing as a liquid at 27° C.

† Constraint Index is defined as $100*(V_p - \text{Volume}^{\#})/V_p$, where $V_p = 0.3078$ cc/g; the measured packing vs. the liquid density.

placed upon guest molecules can be calculated as shown in Table 7. In this case, the measure of constraint is the liquid density. The inability to assume all configurations of the liquid state is manifested by a low packing ratio or a high 'Constraint Index' and the order of constraint for the rho framework is:

n-hexane >> n-propanol > ethanol > methanol ~ H_2O ~ liquid.

The order of constraint follows the decrease in dimensions of the probe molecules.

4. Distortion of the Zeolite Rho Lattice

As illustrated in Figure 5, empty, the rings of H-rho are nearly circular and the lattice parameter, a, of this cubic zeolite is ~15 Å. However, upon ion-exchange, absorption, temperature, or with the absence or presence of other species in the rho framework, the lattice distorts [17]. The distortion of the double 8-ring in effect serves as a way to modify the aperture or pore dimension. A parameter, Δ, one-half the difference in the length of the major and minor axes of the elliptical 8-ring, was defined. The variation in lattice parameter vs. Δ for rho shown in Figure 6 was caused by the presence of different absorbates as well as different counter ions; in essence, the unit cell could be shrunk almost 25% in volume. In the filled state, with absorbed solvents such as methanol, water, or TMA, the structure is in the unstrained state and the double 8-rings are circular with Δ close to, if not, zero. Empty, the ion-exchanged lattice appears strained and the double 8-rings become orthogonal ellipsoids.

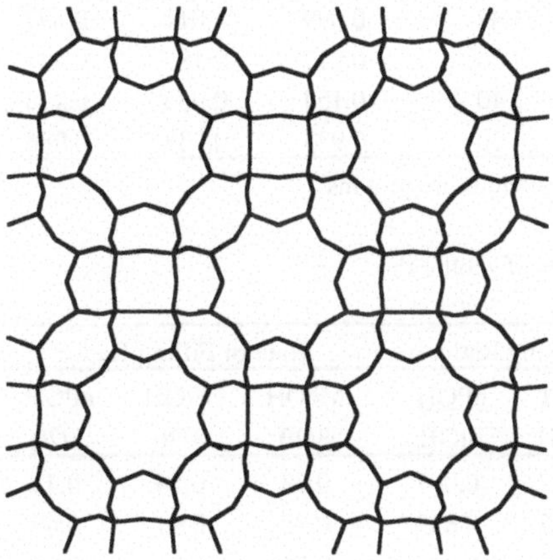

a

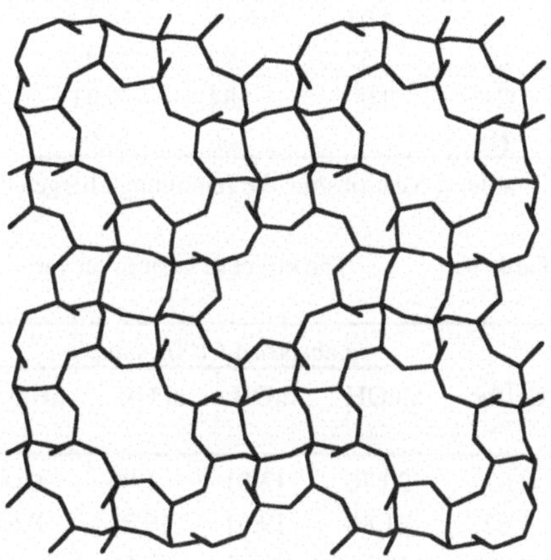

b

Figure 5. Distortion of the zeolite framework: a) dehydrated H-rho and b) dehydrated Ca,D-rho.

The effect of the size of the charge compensating cations on the sorption properties of zeolite rho can be significant as shown by the data in Table 8. At a nominal loading of six Cs ions per unit cell, absorption into the rho framework is virtually eliminated for methanol and larger probe molecules. Although the Δ is small, Cs sites itself in the double 8-rings and because of its large diameter compared to the nominal 5 Å opening of the rho pores, Cs is able to inhibit the absorption process. Sodium, a somewhat smaller ion, permits methanol to readily enter the framework and provides a fairly sharp size cut-off as shown by the elimination of the absorption of ethanol and larger molecules. For Li and Ca ions, one might expect the sorption process to be faster due to their smaller diameters, but that is not the case and X-ray diffraction provides the rationale. Referring to Figure 6, it is seen that the Δ parameter is quite large for these ions, that is, the double 8-ring is lensed down to a small aperture and produces the observed effect on the sorption process.

Progressive substitution for Cs ions by protons provides the data shown in Table 9. The large diameter of the Cs ion coupled with the relatively small pore diameter of zeolite rho also inhibited the ion-exchange process. Several exchanges with a hot 10% NH_4NO_3 solution were required to reduce the framework Cs content below the 1% level.

Table 8. The effect of charge compensating ions* on the sorption properties of rho.

		g absorbed /100 g sample				Packing Ratios		
Ion*	Diameter Å	MeOH 3 h	MeOH 20 h	EtOH 20 h	nPOH 20 h	EtOH/MeOH	nPOH/EtOH	MeOH 3/20
Cs	3.33	1.79	2.12	1.66	0.80	0.783	0.482	0.844
Na	2.04	16.26	18.13	1.80	1.82	0.099	1.011	0.897
Li	1.52	10.92	22.42	20.56	3.80	0.917	0.185	0.487
Ca	2.0	4.84	9.71	1.50	1.40	0.154	0.933	0.498
H		23.66	24.00	21.70	20.20	0.923	0.938	0.986

* at least 6 cations/uc, the remaining charge compensating cations are protons.

Table 9. The effect of cesium on sorption properties of zeolite rho.

	g absorbed / 100g sample				Packing Ratios		Rate of Filling Ratios		
Cs†/uc	MeOH	EtOH	nPOH	nHex	EtOH/MeOH	nPOH/EtOH	MeOH 3/20	EtOH 3/20	nPOH 3/20
2.67	20.76	13.01	1.00	0.00	0.627	0.077	0.73	0.54	0.42
1.62	21.86	19.61	16.33	0.64	0.897	0.833	0.95	0.91	0.69
0.87	22.96	20.65	17.86	11.14	0.899	0.865	0.95	0.92	0.96
0.50	22.81	20.57	17.84	11.71	0.902	0.867	0.96	0.95	0.97
H	24.00	21.70	20.20	10.85	0.923	0.938	0.99	0.96	0.96

Samples calcined for 4 h at 600° C under dry, flowing nitrogen.

† Cesium ions per unit cell; the remainder of the charge compensating cations are protons.

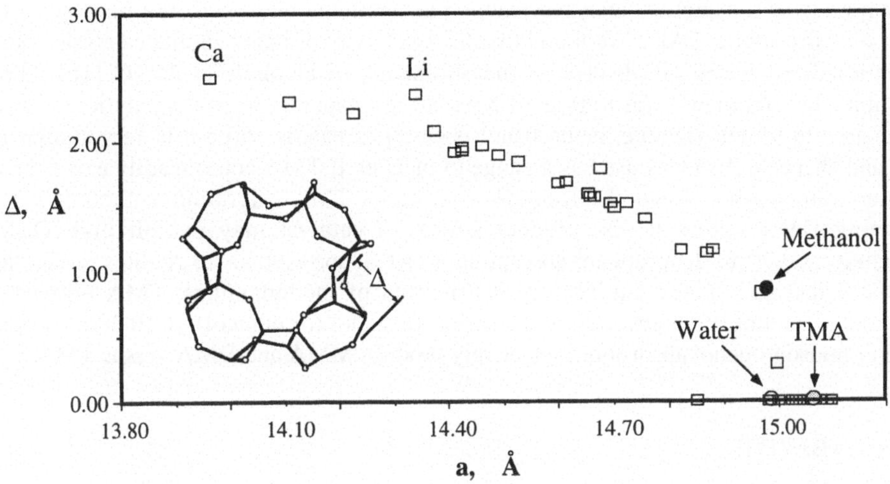

Figure 6. Plot of cell edge (a) vs Δ. The inset illustrates the definition of Δ which is 1/2 the difference in the length of the major and minor axes of the elliptical pore opening.

By carefully manipulating the amount of Cs in the rho framework, the sorption process can be well-controlled and molecules larger than specific sizes can be inhibited from entering the zeolite rho framework; this point is shown graphically in Figure 7.

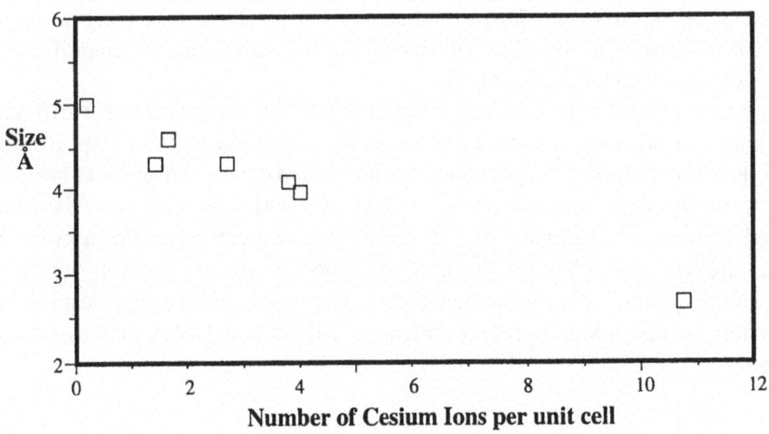

Figure 7. Effect of Cs content on molecular probe exclusion in zeolite rho.

5. Comparison of Rho with Other 8-Ring Zeolites

In our study of the zeolite catalyzed reaction of ammonia with methanol to selectively form dimethylamine, DMA, we found that the selectivity of many zeolite catalysts could be rationalized based on absorption measurements of alcohols at 25° C [16]. For example, low methanol absorption (3-5 wt %) corresponds to low catalytic activity. Compared to trimethylamine, isopropanol is easier to handle while it is very similar in size and shape, and so was used as a gauge to indicate if TMA could readily exit from a zeolite framework. When the framework restricts TMA migration without greatly affecting DMA egress to the product stream, a catalyst may exhibit high DMA selectivity, 60-70%. Isopropanol absorption >3 wt % corresponds to zeolites producing almost a thermodynamic equilibrium distribution of methylamines, TMA ~60-70%. Generally, zeolites, with absorption values for methanol (or ethanol) of 10-30 wt % and little or no isopropanol absorption, selectively produce MMA and DMA versus TMA.

5.1 CHABAZITE

Surprisingly, mineral chabazites provide a wide range of product selectivities towards DMA, although they have almost identical X-ray diffraction patterns and exhibit similar activities. Methanol absorption indicated that the samples' pore volumes were not 100% accessible and suggested the presence of occluded material within the framework. A Geometric Selectivity Index , GSI, was defined as the ratio of methanol absorption to the sorption of n-propanol. Methanol measures the accessible framework capacity while n-propanol sorption (at room temperature) seems to mimic the migration, under reaction conditions, of TMA from the chabazite framework into the product stream. For the variety of chabazite minerals studied, an increase in GSI was found to be directly proportional to an increase in DMA selectivity [16]. This was truly a surprising discovery and underscores the utility of absorption measurements in magnifying small differences in otherwise identical zeolites.

The GSI, however, failed to correctly predict the DMA selectivities for different rho zeolite samples. In this case, we used a microbalance coupled to a mass spectrometer to understand how the different rho samples discharge TMA into the product stream [18]. Mass spectrometry data indicate that, at 200° C, TMA is both physisorbed and chemisorbed with relative amounts of ~10 and ~90%, respectively. For a given zeolite rho sample, the yield of TMA in the methylamines synthesis reaction was found to correlate with the quantity of physisorbed TMA. The physisorbed entity desorbs intactly while the chemisorbed TMA pyrolyzes forming MMA and DMA at normal reaction temperatures.

5.2 ZEOLITE A

The effects of ion-exchange in zeolite A on ethanol packing are summarized in Table 10. By replacing six calcium ions with 12 sodium ions (almost the same size), the internal pore volume is decreased ~5% but the packing of ethanol is affected considerably, ~35% reduction. Exchange with a larger ion, potassium, further reduces the pore volume and virtually prevents ethanol from entering the framework which is indicates that K^+ resides in or near the 8-ring system. Most monovalent ions seem to site in the 8-ring opening so that the pore opening can be controlled somewhat. For the calcium form, no constriction of the pore is observed. The packing and rate of filling ratios are slightly higher for Ca-A than for H-rho especially for n-propanol. Calcium ions site in the 6-ring pores leaving the 8-rings open, and n-propanol easily enters the internal pores of the zeolite. Since the α-cages in the two zeolites are identical, the slight difference in sorption properties is probably caused by the slight constraint imposed by the double 8-ring of rho compared to the single 8-ring in Ca-A.

Table 10a. Effect of ion-exchange on absorption† by zeolite A.

	Ion		Calculated Values		Measured Values		
Type	No./UC	Diameter Å	MeOH	V_p# cc/g	MeOH 20 h	V_p cc/g	% Xtal#
5A	Ca/6	2.00	22.6	0.285	21.7	0.274	96
4A	Na/12	2.04	21.4	0.270	19.4	0.245	91
3A	K/12	2.76	17.2	0.217	16.9	0.213	98

\# The pore volume and crystallinity values are based on methanol absorption and the inaccessibility of the ß-cages to absorption.

† grams absorbed/100 grams dry sample.

Table 10b. Rate of filling and packing ratios for zeolite A.

Zeolite	Methanol 3/20	Ethanol 3/20	n-Propanol 3/20	EtOH/MeOH	nPOH/EtOH	nPOH/MeOH
5A	0.981	0.95	0.93	0.97	0.94	0.90
4A	0.90	0.4	0.2	0.62		
3A	0.85	0.1		0.24		

5.3 ZK-5

ZK-5 has α-cages like zeolites rho and A but the cages are joined by bridged double 8-rings (see Figure 8). As shown in Table 11, the EtOH:MeOH packing ratio approaches

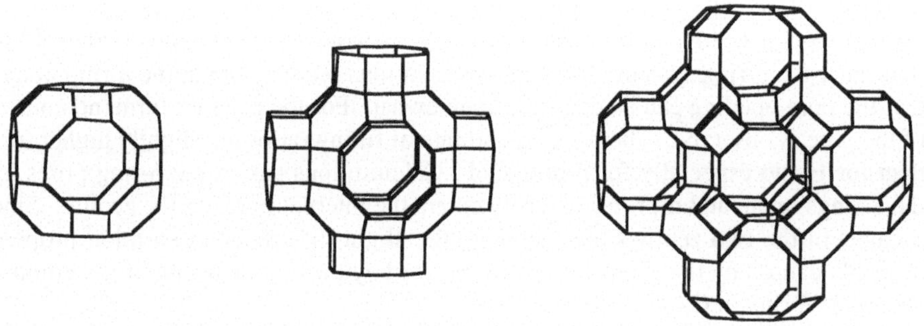

a b c

Figure 8. Structures of the α-cages and interconnections between α-cages for zeolites a) A, b) rho, and c) ZK-5.

Table 11. Absorption properties of H-ZK-5.

Measured Values			Packing Ratios		
MeOH† 20 h	Vp cc/g	% Xtal	EtOH/MeOH	nPOH/EtOH	nPOH/MeOH
20.4	0.257	84.2	0.955	0.839	0.801
19.8	0.250	81.9	0.975	0.762	0.743
19.5	0.246	80.6	0.977	0.708	0.692
19.9	0.251	82.3	0.953	0.677	0.645
20.0	0.253	83	0.883	0.628	0.554
20.3	0.257	84.2	0.945	0.604	0.571
19.1	0.241	79.1	0.993	0.573	0.569
19.9	0.251	82.2	0.955	0.358	0.342
16.7	0.210	68.9	0.994	0.357	0.355
19.4	0.244	80.1	0.889	0.303	0.269
18.6	0.235	76.9	0.992	0.282	0.280
22.0	0.278	91.2	0.876	0.115	0.101
19.8	0.250	82	0.987	0.108	0.106
19.5	0.246	80.7	0.865	0.046	0.040
17.2	0.217	71.3	0.766	0.010	0.008

† g methanol absorbed per 100 grams dry sample. The crystallinity is calculated on the basis of the methanol pore volume with 100% being equivalent to 0.305 cc/g.

1.0 and clearly demonstrates the effect of small changes in framework geometry on sorbate molecular packing. Another consequence of the structural differences shows up in the retention of framework crystallinity after calcination. Although calcined under almost identical conditions as zeolite rho to convert the ammonium form to the H-form, ZK-5 has lost significant amounts of crystallinity. Most of the resulting silica-alumina species remain within the framework as evidenced by the drastic lowering of the other packing ratios. This observation underscores the need to use molecules of different sizes (and/or shapes) to characterize accessible framework volumes. Reliance solely on the EtOH:MeOH packing ratio would be quite misleading especially if these samples were used for measuring the kinetics of absorption of n-propanol, hexane, etc.

5.4 EFFECT OF Si/Al RATIO

The Si/Al ratios for zeolites rho, ZK-5, A are 3, 2.2, 1, respectively. The packing ratios suggest that zeolite A has less constraint on nPOH than does zeolite rho; the slight increase in packing ratio is probably caused by the slightly larger bond and cage dimensions due to the higher Al content. The increase in pore volume is much less than anticipated by consideration of normal Al-O bond lengths because the bond length in zeolites is closer to the Si-O bond length. No unambiguous conclusion can be made for zeolite ZK-5 because of the extensive damage suffered during the calcination process to convert it to the H^+ form. Small occlusions within the framework can drastically affect packing ratios as shown in Table 11. For zeolite ZK-5, although most samples had greater than 80% crystallinity, there is a large variation in the nPOH:EtOH and nPOH:MeOH packing ratios. Appropriate calcination conditions were not found which would convert the ammonium form to the proton form while retaining all of the framework crystallinity. The instability of some frameworks in the proton form is not surprising. Zeolite ammonium-A will completely decompose if heated around 300°C.

5.5 CALCINATION OF MOLECULAR SIEVES

Calcination of molecular sieves was used to provide some understanding of their thermal stability as well as to provide samples with controlled changes in amounts of crystallinity and to determine what the effects on sorption properties would be. Such knowledge is important because:

- NH_4^+ containing zeolites are calcined to provide the acid H^+ form.
- molecular sieves are heated in some uses as well as during regeneration cycles to drive off absorbed species.
- if decomposition of the framework occurs, what is the fate of the decomposed material and what does it do with respect to the sieve's absorption properties?

Some of the effects of calcining a sample of zeolite rho are listed in Table 12. For comparison purposes, the expectation values for a typical sample of rho zeolite are included. The increase in sorption values upon calcining at 600° C rather than 500° C is attributed to the removal of small amounts of ammonium ion that had not dissociated during the 500° C calcination process. IR spectroscopy confirmed the presence of NH_4^+ in the lower calcination temperature sample. The small residual amount had a large effect on the amount of n-propanol absorbed as well as its 'rate of filling' ratio. At 600°C, ammonium ion was no longer detected. The crystallinity of the 600° C sample is around 88% (21.13/24) and one is prompted to ask 'what effect does the 12% lack of crystallinity have on the sorption properties?'. Since the packing ratios are the same as found for a typical rho zeolite, this suggests that the non-absorbing material either blocks off a portion of the framework or resides on the exterior surface of the crystals. If it were to reside within an accessible part of the framework, the packing ratios would be lower. The fact that the rate of filling ratio for n-propanol is much lower than that of a typical sample suggests that the non-absorbing material is on the external surface of the crystals near the pore openings such that it slows down the sorption process.

Upon raising the calcination temperature to 700° C and using undried nitrogen (probably 5-10 torr of water vapor was present), the changes produced by calcination are much more dramatic. In this case, 'house' nitrogen (the supply to our entire building) was used rather than a special tank of dry, high purity grade used to obtain BET surface areas. The sample's crystallinity has dropped to 51% and the propanol and hexane sorption was almost eliminated. The reduction in packing ratios reflects sizeable migration of the decomposed material within the framework. By contrast, calcination of other zeolite rho preparations in a very dry environment does not produce equivalent or larger effects even at 800° C as shown in Table 13. The calcining of zeolite rho and the effects produced vary somewhat depending upon the preparation. The amount of framework decomposed certainly increases as the calcination temperature increases but the trends in packing ratio vary erratically while the rate of filling ratios clearly decrease. For the most part, calcination under dry conditions does not provide a mechanism for material migration; the decomposed framework seems to remain in place. By contrast, as shown in Table 12, small amounts of water vapor facilitate the decomposition and migration processes.

Table 12. Effect of calcination on the sorption values of a H-rho zeolite sample.

Calc.	g absorbed/100g sample				Packing Ratios		Rate of Filling	
	MeOH	EtOH	nPOH	nHex	EtOH/MeOH	nPOH/EtOH	EtOH 3/20	nPOH 3/20
Temp., °C								
Typical	24.00	21.70	20.20	10.85	0.923	0.938	0.96	0.96
500†	20.20	18.18	12.27	4.77	0.900	0.675	0.97	0.58
600†	21.13	19.51	17.95	9.58	0.923	0.920	0.88	0.71
700*	12.66	9.16	1.99	0.37	0.723	0.218	0.81	0.44

† Calcined in a shallow bed for 4 hours under dry, flowing nitrogen.

* Calcined under flowing nitrogen (not dried) for 4 h.

Table 13. Calcination of different preparations of zeolite rho..

Prep	Calc. Temp °C†	g absorbed/100 g sample			Packing Ratios		Rate of Filling Ratios	
		MeOH	EtOH	nPOH	EtOH/MeOH	nPOH/EtOH	EtOH 3/20	nPOH 3/20
1	600	19.37	18.35	15.67	0.947	0.854	0.89	0.54
1	700	18.62	17.71	15.36	0.951	0.867	0.88	0.31
2	600	22.23	20.93	18.71	0.942	0.894	0.99	0.98
2	700	19.77	18.01	16.93	0.911	0.940	0.96	0.95
2	800	17.64	15.76	15.25	0.894	0.968	0.95	0.94
3	600	20.85	19.95	17.10	0.957	0.857	0.96	0.97
3	800	19.11	17.35	15.32	0.908	0.883	0.94	0.94
Typical		24.00	21.70	20.20	0.923	0.938	0.96	0.96

† Samples were calcined under dry flowing nitrogen for 4 h at the temperatures noted.

6. Characterization of ZSM-5 by Sorption Measurements

The pentasil family of zeolites, which includes ZSM-5, has exhibited very valuable catalytic properties including high chemical selectivity, low coking (aging) tendency, and high turnover activity for a variety of chemical reactions [19]. The impressive catalytic selectivity and in-use stability of the pentasil family of zeolites are attributed to a unique combination of properties, such as size, shape, and site sorption selectivities, chemistry, and substrate diffusion on the internal surface. X-Ray diffraction analyses show that ZSM-5 possesses two types of pores, both of which are composed of 10 tetrahedral-membered rings: one pore system is sinusoidal with a nearly circular cross section, while the other is straight and perpendicular to the sinusoidal system with elliptical pores. A representation of the internal surface of the channels (void space) of the ZSM-5 internal framework is shown schematically in Figure 9.

For the sorption of n-hexane into the H-ZSM-5 framework, there is a range of values 10-14 g/100 g reported in the literature [19,20]. There are several causes for the variety of values for the amount of n-hexane sorbed. As noted above, the duration of exposure is very important because of the slow absorption rates that are common during the final stages of filling the framework. Too short exposures to solvent vapors yield low 'equilibrium' absorption values. Furthermore, as noted earlier, a rigorous vacuum heat treatment is needed to prepare a molecular sieve for sorption studies. If some residue were contained in the framework, absorption values would tend to be lower. Another reason for low absorption values is a poorly crystalline sample or a sample with a non-zeolitic component. For high surface area solids, the external surface can make a

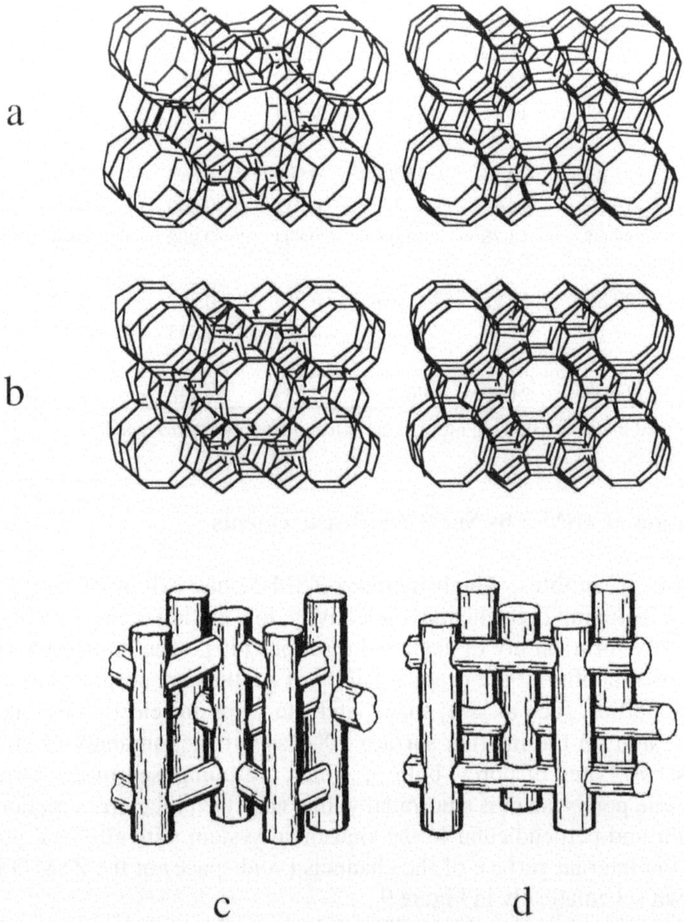

Figure 9. Stereographic drawings of the frameworks of a) ZSM-5 and b) ZSM-11 as well as representations of the void space (channel) structures of c) ZSM-5 and d) ZSM-11.

significant contribution to the sorption isotherm, generally in the range 5 to 20% of the total amount. Adsorption, if unaccounted for, would provide higher than expected absorption values. Interestingly, most investigators neglect surface contributions.

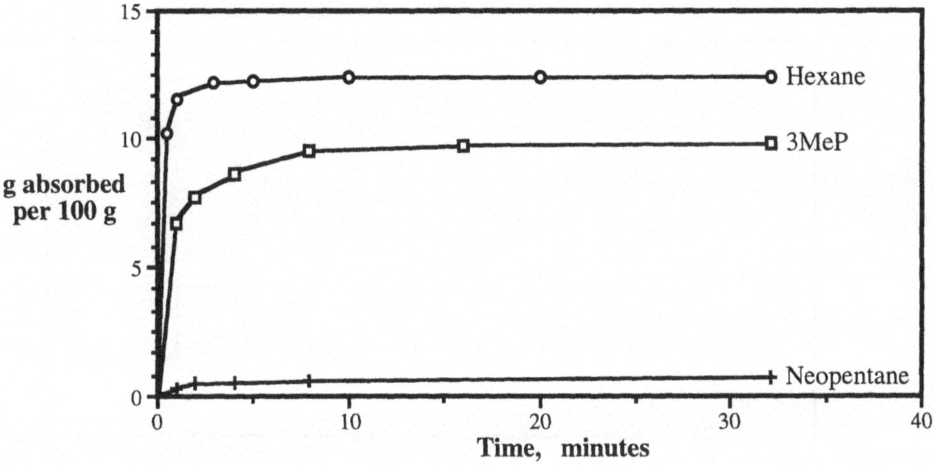

Figure 10. Absorption data for hexane, 3-methylpentane (3MeP), and neopentane into a sample of H-ZSM-5.

Shown in Figure 10 are the data for the absorption of three alkanes into a sample of H-ZSM-5. Hexane absorbs quickly into this sample of H-ZSM-5 and reaches about 95% of the final absorbed amount (0.125 g/g at 20 hours for this sample) in five minutes. The absorption of 3-methylpentane, 3MeP, a branched isomer of hexane, is somewhat slower as shown by the rounded 'knee' at a longer time in the sorption-time plot. The sorption of 2,2-dimethylpropane, neopentane, is very slow and continues for weeks without equilibrating. The relative diffusion rates of these sorbates can be estimated from the data [21] and decrease with an increase in cross-sectional dimensions of the molecules:

hexane:3MeP:neopentane = 0.92:0.25:0.001 = 3.9x4.3:5.1x5.9:5.5x5.8 Å

If a value of 6.8 Å is chosen as the 'critical' (smallest) pore dimension of ZSM-5 common to all the data, then the relative absorption rate-size data above correspond reasonably well to the correlation of Moore and Katzer [14]. The value for the pore dimension is close to the value in Table 1 estimated for covalent oxygen in a 10-membered ring.

As noted, the absorption of hexane into H-ZSM-5 is fairly rapid and reaches an apparent 'end point' or equilibrium value; the 3 and 20 hour sorption readings are the same within experimental error. A mean value of 13.07 g/100 g (std dev 0.27) was determined (for 29 highly crystalline samples prepared using tetrapropylammonium ion, TPA, as the templating ion) which corresponds to a framework volume of 0.198 (std dev 0.004) cc/g using a value of 0.659 g/cc as the density for hexane. Thus, hexane appears to fill the H-ZSM-5 framework in a 'liquid-like' manner as the volume of hexane absorbed agrees well, within experimental error, with the calculated framework volume

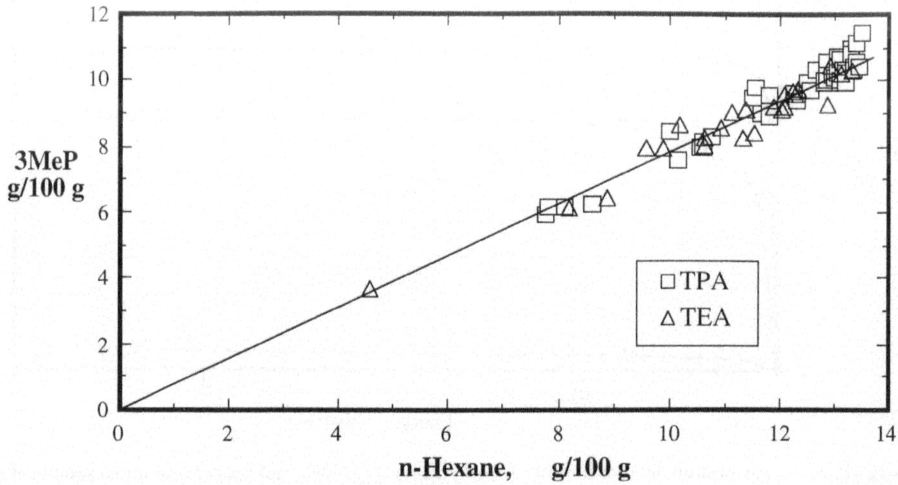

Figure 11. Absorption data for 3-methylpentane (3MeP) vs. n-hexane into samples of ZSM-5 prepared using TPA and TEA templates.

capacity (Table 4). Samples of H-ZSM-5 had Si/Al ratios ranging from 11 to >1000; the amount of aluminum in the framework did not appear to affect the value of hexane absorbed.

As shown in Figure 10, 3MeP is also absorbed rapidly into the H-ZSM-5 framework. Its liquid density is sufficiently close to that of hexane such that, if steric constraints were not imposed by the framework on the packing of 3MeP, the same sorption weight gain would be obtained. But, H-ZSM-5 samples absorb a significantly smaller amount of 3MeP compared to hexane. Shown in Figure 11 is the sorption data for 3MeP and hexane for a variety of H-ZSM-5 samples (included in this figure are data using two different synthesis template molecules; this will be discussed further). For a given sample, the amount of 3MeP absorbed is plotted versus its hexane absorption value. From the Figure, it is seen that the absorption ratio 3MeP:hexane is independent of the amount of hexane absorbed (that is, it is independent of the crystallinity of the sample). The packing ratio, 3MeP/Hex, of 0.789 (std. dev. 0.023 for 39 samples) is characteristic of the H-ZSM-5 framework. Similarly, the absorption of *p*-xylene is directly related to the hexane absorption amount with a packing ratio of 1.218 (std. dev. 0.045 for 30 samples); this is shown graphically as Figure 12. Again, the amount of aluminum in the framework did not seem to affect the observed ratio. Samples of H-ZSM-5 displaying these characteristic ratios have their crystallinity determined by dividing the hexane sorption by 13.07.

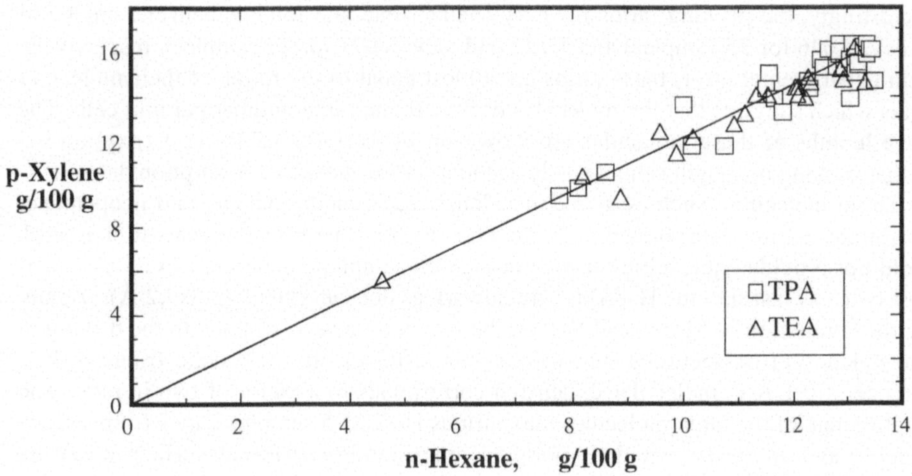

Figure 12. Absorption data for *p*-xylene vs. n-hexane into samples of ZSM-5 prepared using TPA and TEA templates.

Table 14. Characteristic absorption values for H-ZSM-5 (TPA template, 58 samples).

Solvent	g absorbed/ 100 g sample	Std Dev	No. #	Packing Ratio†	Density Ratio∂	Constraint Index®	Molecular Dimensions, Å [11]
n-Hexane	13.07	0.03	29	1.000	1.00	0	3.9x4.3x9.7
p-Xylene	15.78	0.43	22	1.218	1.31	7.5	3.6x6.5x8.9
3-Methylpentane	10.36	0.32	27	0.789	1.01	22.4	5.1x5.9x8.0
Toluene	12.91	1.03	14	0.974	1.31	25.5	3.6x6.5x8.1
Benzene	11.29	0.53	20	0.888	1.33	35.8	2.5x6.5x7.2
Carbon Tetrachloride	15.23	0.64	20	1.177	2.42	52.2	5.4
Cyclohexane	7.38	0.98	29	0.638	1.18	52.6	5.1x6.0x6.5

The number of values used to obtained the reported data.

† The ratio of the amount absorbed to the amount of hexane absorbed.

∂ The ratio of the solvent density to the density of hexane.

® The (density ratio minus the packing ratio) divided by the density ratio x 100.

Shown in Table 14 are the equilibrium sorption data, at 20 hours, for a variety of solvents and the packing ratios are characteristic values for H-ZSM-5. Hexane and smaller molecules (for example, butane, pentane, CO_2) pack according to their liquid densities; the same framework volume is accessible to all of these molecules. The other, larger, molecules listed in Table 14 are all constrained in some fashion.

Interestingly, the packing ratios for *p*-xylene:benzene and toluene:benzene are 1.384 (std. dev. 0.046 for 25 samples) and 1.122 (std. dev. 0.097 for 24 samples), respectively. Within experimental error, these ratios are almost equal to the ratios of their molecular weights which suggests that the molecules pack with the same number per unit cell. The relative lengths of these molecules are, however, in the ratio 1:24:1.12:1 (pxyl:tol:bz) such that packing by length cannot solely account for the measured absorption data.

For larger molecules, such as *o*- and *m*-xylenes, 2,2-dimethylbutane, and neopentane, the sorptions are not equilibrated at 20 hours; extending the measurements to one week still did not provide equilibrium data. For a given sample, *o*-xylene (3.6x7.1 Å) sorbs almost twice as fast into the H-ZSM-5 framework as does *m*-xylene (3.6x7.2 Å). Again, using the correlation of Moore and Katzer, the sorption rates correspond to the difference in the xylenes' cross-sectional dimensions; for diffusion into the same framework, a difference of 0.1 Å in molecular dimension corresponds to a factor of two in rate. The absorption rate of the large molecules into various H-ZSM-5 samples varies from sample to sample and suggests that their pore openings (and/or external surface area) are different. The different rates of sorption of *o*-xylene into ZSM-5 samples can be used to distinguish the various samples from each other even though they have similar hexane sorptions (similar crystallinities); *o*-xylene absorption values at 20 hours ranging from 0.37 to 9.7 g/100 G were obtained for 55 samples. Assuming that Moore and Katzer's correlation is correct, the overall variation in diffusion coefficients corresponds to a difference of 0.6 Å in pore dimension because the molecular probe, *o*-xylene, was kept constant. This difference in size is close to the size difference in pores for the straight and sinusoidal channels of the ZSM-5 framework. Values in between the extremes suggest a mix of the two channel systems.

6.1 COMPARISON OF TPA AND TEA PREPARATIONS OF ZSM-5

ZSM-5 can be prepared using a variety of template molecules; the data in Table 14 were obtained for samples prepared using tetrapropylammonium ion, TPA, as the template. For comparison, tetraethylammonium ion, TEA, was used as the template. The crystals have morphology (stubby, elliptical cylinders with flat ends) that is significantly different from the usual ZSM-5 crystal configuration of right-angle, prismatic, penetration twins obtained when TPA is used as the templating ion. However, the X-ray diffraction patterns of the TEA preparations are virtually indistinguishable from those using TPA. As shown in Figure 11, the absorption data for 3MeP vs. n-hexane does not distinguish the preparations on the basis of the template ion that was used. Further, as shown in Figure 12, absorption of a slightly larger molecule, *p*-xylene, doesn't clearly distinguish the different templated preparations. The error in the absorption measurements is larger than any difference in the framework structures. Table 15 summarizes the absorption data for the TEA templated preparations.

Table 15. Characteristic absorption values for H-ZSM-5 (TEA template, 24 samples).

Solvent	g absorbed/ 100 g sample	Std Dev	No. #	Packing Ratio†	Density Ratio∂	Constraint Index®
n-Hexane	12.56	0.49	10	1.000	1.00	0
p-Xylene	14.54	0.64	11	1.191	1.31	9
3-Methylpentane	9.68	0.47	10	0.777	1.01	23
Toluene	9.09	1.17	8	0.74	1.31	44
Benzene	8.09	0.73	13	0.66	1.33	50
Carbon Tetrachloride	11.58	1.07	9	0.94	2.42	61
Cyclohexane	4.43	1.30	13	0.34	1.18	71

\# The number of values used to obtained the reported data.

† The ratio of the amount absorbed to the amount of hexane absorbed.

∂ The ratio of the solvent density to the density of hexane.

® The (density ratio minus the packing ratio) divided by the density ratio x 100.

For molecules the size of hexane and smaller, the absorption data, within experimental error, is the same for the TPA and TEA template preparations. As with the X-ray diffraction patterns for the two different templated materials, the packing ratios of *p*-xylene:hexane and 3MeP:hexane also suggest that the zeolites are very similar, if not identical. However, as the size of the probe molecule increases, the constraint presented by the TEA framework is larger than the constraint exerted by TPA system. This observation suggests that the TEA preparations have slightly smaller channel dimensions than the TPA preparations and this difference is magnified as the absorbing molecule gets larger.

As noted above, ZSM-5 crystals have two slightly different channel systems capable of absorbing molecules. For the TPA template, conditions during the synthesis of a sample cause one channel system to dominate over the other, thereby providing a spectrum of sorption rates for large molecules. Using TEA as the template molecule seems to alter the crystal morphology and let the smaller channels dominate. The difference in channel dimension and relative channel population imposes greater constraint on the packing of probe molecules as they increase in size; this is consistent with the data shown in Tables 14 and 15. Thus, the choice of template molecule seems to control the external morphology of the ZSM-5 crystals, which channel system dominates, and the rates of absorption and amount absorbed into the framework.

6.2 CHARACTERIZATION OF ZSM-11 BY SORPTION MEASUREMENTS

The zeolite ZSM-11 also belongs to the pentasil family; it has two channels which, unlike ZSM-5, are straight and nearly circular (see Figure 9). Although we didn't prepare and characterize as many ZSM-11 samples as we did for ZSM-5, the few absorption values seem to confirm that ZSM-11 channels are slightly larger and constrain the probe molecules somewhat less than ZSM-5 as shown in Table 16. The rates of absorption of *o*- and *m*-xylenes into the framework seem to be higher for ZSM-11 than for ZSM-5 but the statistics may be faulty as there were not that many samples prepared and tested. Further, as noted for ZSM-5, the absorption of large molecules such as *o*- and *m*-xylenes did not equilibrate even after one week of exposure.

Table 16. Characteristic absorption values for H-ZSM-11 (9 samples).

Solvent	g absorbed/ 100 g sample	Std Dev	No. #	Packing Ratio†	Density Ratio∂	Constraint Index®	ZSM-5 Constraint
n-Hexane	13.52	0.33	5	1.000	1.00	0	0
3-Methylpentane	10.79	0.24	5	0.797	1.01	21	22
Carbon Tetrachloride	15.55		2	1.150	2.42	52	52
Cyclohexane	8.42	0.51	9	0.623	1.18	47	53

\# The number of values used to obtained the reported data.

† The ratio of the amount absorbed to the amount of hexane absorbed.

∂ The ratio of the solvent density to the density of hexane.

® The (density ratio minus the packing ratio) divided by the density ratio x 100.

7. Quantitative Aspects of Adsorption Measurements

Although there are numerous papers reporting on the characterization of zeolites, few deal with the problem of quantifying the amount of molecular sieve present in a sample. Gabelica et al. used X-ray diffraction, DTA, and NMR to characterize ZSM-5 preparations [22]. Derouane et al. used thermogravimetric analysis to follow the decomposition of the organic template ion in ZSM-5 preparations and calculated crystallinities from that data [23]. Jacobs et al. used IR spectroscopy to quantify the presence of crystalline ZSM-5 in X-ray amorphous samples [24]. Whether examining laboratory synthesized preparations as these investigators have done or examining natural minerals, it is desirable to know how much of the material present is zeolitic vis-a-vis, for example, an amorphous aluminosilicate. The use of absorption can provide a means to quantify the amount of molecular sieve in a given sample. For a given framework, by using the absorption of appropriately sized probe molecules, a set of packing ratios and specific absorption amounts can be determined. For any sample of that framework, X-ray diffraction is used to make sure that there is no other framework species present to any large extent. Then, the samples' absorption values are used to quantify the amount of molecular sieve phase present (crystallinity) or accessible.

7.1 COMPARISON OF CRYSTALLINITY MEASUREMENTS: X-RAY DIFFRACTION VS. ABSORPTION

It is common practice to use the intensity of some X-ray diffraction peaks as the basis for declaring molecular sieves to be 100% crystalline especially in the absence of other phases. However, X-ray diffraction is not a sufficiently sensitive tool for this task because the unit cells of many molecular sieves are on the order of 15-20 Å in size while large crystals, that is, much longer range ordering, are needed to clearly describe the resulting frameworks in XRD. In the synthesis process, small groupings (2-3 unit cells wide) can form and they may not be visible to X-ray diffraction - in fact, such preparations have been named 'X-ray amorphous'.

Shown in Figure 13 are the X-ray patterns of a highly crystalline ZSM-5 sample and one that appears to be ~25% crystalline; it has an average crystallite size of 300 Å estimated from line broadening (Debye-Scherer). Absorption of hexane, 3MeP, and *p*-xylene show typical ZSM-5 packing ratios from which it was determined that the sample is about 60% crystalline. The apparent discrepancy in crystallinity was resolved by using high resolution electron microscopy which revealed small domains of 60-80 Å within many particles of the preparation. These small domains account for the X-ray line broadening, the poor quality of the X-ray diffraction pattern, and the fact that crystallinity via absorption was significantly higher than via X-ray diffraction. What is so fascinating is that the domains, although small and disoriented, are almost completely accessible for absorption and that they show typical ZSM-5 packing fractions.

During the synthesis of molecular sieves, it is essential to know when the appropriate moment arrives to stop the synthesis and recover the product. If the synthesis were allowed to proceed unchecked, many zeolites would transform to other unwanted phases. This is the so-called 'art' of molecular sieve synthesis. However, by carefully controlling synthesis procedures and characterizing the products at various times during the synthesis, some of the 'art' can be eliminated. Shown in Table 17 is a comparison of the X-ray diffraction data, surface area measurements, and sorption data for samples of rho zeolite extracted from the synthesis autoclave. Following normal work-up procedures, the samples were ion-exchanged with NH_4NO_3 and calcined at 550° C to convert them to the H^+ form. As the synthesis proceeds, the X-ray counts remain quite low, the surface area decreases rapidly and, prior to collecting the 48 hour sample, the sorption values for the alcohols are almost identical, independent of molecular size. By not correcting for surface adsorption, the sorption values, shown in Table 17, can be quite misleading. The packing ratios are close to the density ratios, especially the iPOH:MeOH value, which is typical of surface adsorption. At 48 hours in the synthesis, the iPOH:MeOH packing ratio drops significantly towards the value 0.04, the value typical of a highly crystalline rho sample.

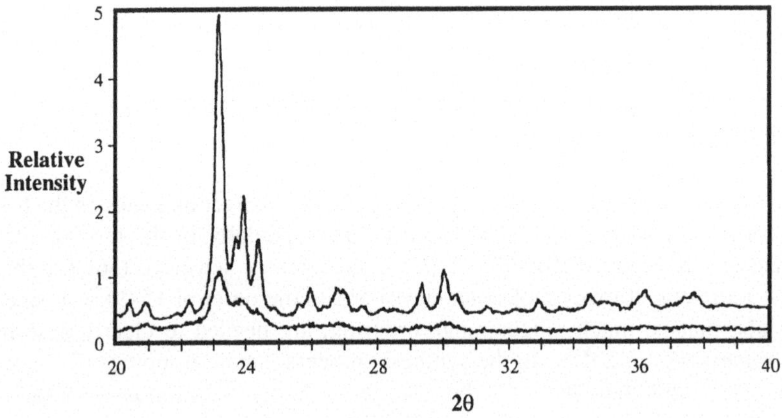

Figure 13. X-Ray powder diffraction patterns for a highly crystalline sample of ZSM-5 and one that is approximately 25% crystalline by XRD.

Correction of the sorption data for adsorption effects in the 32 h and earlier samples reveals that there is no framework absorption. Shown as Table 18 are the data for the 48 h and later samples corrected for adsorption. Both the 56 and 72 h samples are highly crystalline via X-ray diffraction. But absorption shows that something occurred during the synthesis, exchange, or calcination because, for the 72 h sample, the rate of

Table 17. The synthesis of zeolite rho: the first 48 hours.

							Packing Ratios		
Prep#	APD\$	SA†	g sorbed®/ 100 g dry sample				EtOH	nPOH	iPOH
Time	cps	m^2/g	MeOH	EtOH	nPOH	iPOH	MeOH	MeOH	MeOH
0	41	218	3.75	4.52	5.16	4.86	1.205	1.376	1.296
8	225	208	4.56	4.31	4.80	4.60	0.945	1.053	1.009
24	53	131	2.83	2.69	3.06	2.96	0.951	1.081	1.046
32	32	112	2.55	2.25	2.60	2.52	0.882	1.020	0.988
48	346	87	6.76	5.58	5.50	1.97	0.825	0.814	0.291

\# Hours in synthesis reactor, temperature 100° C.

\$ Automatic Powder Diffraction camera, counts per second of major X-ray diffraction peak; the expectation for a highly crystalline sample is 1700-1800 cps.

† Surface areas, ±2 m^2/g, determined by adsorption and mercury porosimetry; 30-40 m^2/g is typical of a highly crystalline sample.

® Total sorbed amounts: no correction for adsorption on external surfaces.

Table 18. Corrected absorption data for zeolite rho sampled during synthesis.

Prep.	g absorbed/100 g sample			Xtal†	Packing Ratios		Rate of Filling Ratios		
					EtOH	nPOH	MeOH	EtOH	nPOH
Time	MeOH	EtOH	nPOH	%	MeOH	EtOH	3/20	3/20	3/20
48	4.86	4.69	3.38	20.2	0.964	0.721	0.92	0.89	0.53
56	17.50	16.30	13.94	72.9	0.931	0.855	0.96	0.96	0.51
72	23.22	21.60	5.93	96.7	0.930	0.275	0.34	0.14	0.04
Typical	24.00	21.70	20.20	100	0.923	0.938	0.99	0.96	0.96

† per cent crystallinity - basis: 24 g/100 g of methanol is equivalent to 100% crystallinity.

adsorption of the linear alcohols and nPOH:EtOH packing ratio are much lower than usual. The presence of a sharp X-ray diffraction pattern is no guarantee for the synthesis of a well-behaved molecular sieve. This example clearly demonstrates the fact that absorption and X-ray diffraction measurements are complementary techniques and both are needed in characterization of molecular sieves.

8. Catalysis and Reactions Using Molecular Sieves as Supports

As noted above, molecular sieves are used as absorbents where a probe molecule's size and shape play important roles in the sorption process. The situation is similar when sieves are used as catalysts or supports for a reaction. As such, reactions and their product distributions can be used to characterize the framework topology, connectivity, and dimensions of molecular sieves. At least three types of reaction selectivity over zeolite catalysts have been described [25].

- Product selectivity. Molecular sieving selectivity of the zeolite provides the basis for determining the product spectrum: molecules smaller than the pore openings can exit from the framework into the product stream while larger molecules are retained within the framework.
- Transition state selectivity. Certain reactions are prevented because the corresponding transition state requires more space than is available within the framework of the zeolites.
- Reactant selectivity. Only those components of the feed stream that are small enough to enter the pores can react while larger ones are excluded.

As a practical application, zeolite H-ZSM-5 has been used as a catalyst for the conversion of pseudocumene and methanol to durene [26]. Different preparations of ZSM-5 gave different durene yields without any rational basis. Consider that the cross-section of durene is almost identical to *o*-xylene; sorption of *o*-xylene was shown to correlate with the catalytic selectivity towards durene. Thus, very slight differences in the accessible channel dimensions may have a profound affect on the catalytic selectivity.

8.1 USE OF ABSORPTION IN CATALYST DESIGN AND MODIFICATION

In the production of ethylene from the cracking of natural gas, small amounts of acetylene and butadiene are produced. Acetylene can present a hazard while non-selective hydrogenation removes valuable ethylene and butadiene as well as the target acetylene. A process improvement would eliminate acetylene from the product stream yet permit full recovery of butadiene and ethylene. The catalysts generally used are active metals supported on "inert" substrates but both acetylene and butadiene are hydrogenated to a substantial degree. Therefore, in the design of a new catalyst, these considerations were taken into account:

- selective hydrogenation of acetylene versus butadiene and ethylene.
- hydrogenation of acetylene to ethylene is preferable.
- minimal hydrogenation of butadiene and ethylene.
- the economics of plant investment.

Placing the active metal catalyst inside a small pore zeolite takes advantage of the differences in size and shape of the acetylene, butadiene, and ethylene molecules and establishes kinetic control over the hydrogenation [27]. Butadiene, because of its larger size, would be excluded from entering the zeolite framework which is accessible to the smaller acetylene and ethylene molecules. Then, acetylene could be hydrogenated within the framework at active metal sites within the pores while butadiene, because it cannot enter the framework, passes through the bed unreacted. Zeolite A was selected as the catalyst support because of its appropriate pore dimension, availability, cost, and the ability to be ion-exchanged which would control the transport processes. Absorption measurements could then be used to design and fine-tune the catalysts which were ion-exchanged with a small amount of Ni^{2+} and subsequently reduced.

The linear C_1-C_3 alcohols were used as probes for the study. If methanol could freely enter the zeolite, so could the smaller acetylene. Ethanol and propanol bracket butadiene in size. If propanol would have free access to the framework, so would butadiene. If the access of ethanol were restricted, then butadiene would be prevented from entering the framework (and hydrogenating). From the data presented in Table 19, it is apparent that the degree of ion-exchange and the size of the charge compensating cation, M, have profound influences on the amounts absorbed by the framework.

Reduction in methanol sorption capacity indicates decreased internal pore volume and relates to the exchanged ionic size and the amount of charge compensating cation. The Ca sample shows little differentiation for any alcohol which indicates that this sample should not be a selective catalyst (it wasn't). Calcium ions site in the 6-ring pores leaving the 8-rings open, and n-propanol easily enters the internal pores of the zeolite. Because it is smaller in size than n-propanol, butadiene should also access the internal framework of this sample. As in this case, if the active catalytic sites are within the framework and butadiene can access them, butadiene will also be hydrogenated.

Table 19. Absorption results for Zeolite Ni,M-A (Ni^0 is the active catalyst).

Ions# M	Absorption Data MeOH	EtOH	nPOH	Packing Ratios EtOH/MeOH	nPOH/EtOH	nPOH/MeOH	Volumes, cc/g Acc @	MeOH &	Occ % †
Ca	21.9	19.9	18.6	0.908	0.935	0.85	0.278	0.28	99.3
Na	19.5	16.3	9.6	0.836	0.586	0.49	0.281	0.25	87.5
K-1®	17	11.4	0.45	0.666	0.04	0.03	0.246	0.22	87.2
K-2®	16.5	1.7	0	0.106	0	0	0.236	0.21	88.1
Rb	14.6	8.6	0	0.589	0	0	0.224	0.18	82.2
Cs	10.6	0.19	0	0.018	0	0	0.202	0.13	66.4

\# M, the major cation: Ca^{+2}, K^+, Na^+, Rb^+, Cs^+.

@ the accessible framework volume taking into account the ionic volumes.

& the framework volume occupied by methanol assuming packing as a liquid.

† the percent occupancy of the accessible framework.

® K-2 has a higher percent of K^+ as charge compensating ion than K-1.

The Na sample exhibits some pore narrowing (as evidenced by reduced n-propanol sorption) which should translate to reduced sorption of the butadiene molecule. In the cases of the K^+, Rb^+, and Cs^+ exchanged forms, essentially no n-propanol is absorbed and ethanol sorption is significantly lowered for some samples; these catalysts should show a reduction in the hydrogenation of butadiene. Here, fine-tuning is essential since ethylene, produced by the hydrogenation of acetylene within the framework, should have difficulty in migrating out of the framework and entering the product stream (samples where the ethanol absorption is low). Thus, optimum catalyst design must take into account the effects of cation sizes, loadings, and the subtle changes in sorption behavior.

Although the absorption data were useful in demonstrating differences in molecular sieving from one exchanged form to the next, the data do not provide a measure of the catalytic activity of the external surface of the zeolite particles. A figure of merit, FM, was defined as the ratio of hydrogenated acetylene to hydrogenated butadiene under the condition that ethylene passed through the bed essentially unreacted. For the commercially available catalyst under plant conditions, about 60% of the butadiene and all of the acetylene in the feed is hydrogenated, yielding a FM ~1.6.

The initial zeolite catalysts that were prepared exhibited low FM values even though the absorption properties were fine-tuned. This apparent lack of selectivity suggested that the catalytic sites dominating the product spectrum must be located on the surface of the exchanged zeolites. Such metal sites on the external surface are more accessible in comparison to sites within the zeolite framework and can therefore override the contribution to the product spectrum made by the internal sites. The external sites were selectively poisoned by the addition of thiophene to the feed stream. Thiophene is larger than the pore opening of the zeolite; hence it selectively poisoned the metal catalyst on the external surface of the zeolite. Further improvements were realized by using tetraethylorthosilicate, TEOS, to deposit a silica coating on the external surface of the zeolite. For the samples shown in Table 20, the TEOS treatment improved the catalytic performance but did not alter the sorption data significantly. The TEOS treatment was mild enough such that the size of the zeolite pore openings was not affected. Apparently, the TEOS treatment is able to cover some of the metal catalyst on the external surface of the zeolite that would otherwise not react with thiophene.

Compared to a commercial catalyst which hydrogenates 60% of butadiene and all of the acetylene, the new catalysts totally hydrogenate acetylene with only 10-20% hydrogenation of the butadiene and almost no hydrogenation of ethylene. Thus, to achieve selective hydrogenation, that is, to obtain a product spectrum dominated by catalytic sites within the zeolite framework, poisoning of the metal sites on the external zeolite surface as well as fine-tuned diffusion control are essential.

Table 20. TEOS treatment of Ni,M-A zeolite catalysts.

Catalyst	g absorbed/100g sample			
M	MeOH	EtOH	nPOH	FM
K (Control)	17.0	3.0	0.4	5.9
K + TEOS	17.1	5.0	0.6	9.1
Rb (Control)	15.2	9.2	0.3	7.7
Rb + TEOS	15.1	8.6	0.3	10.0

8.2 COATING EFFECTS

In order to improve absorption selectivity or to inhibit surface reactivity when molecular sieves are used as catalysts or supports, the external surface is sometimes coated or chemically treated as noted above. Although changes in the framework caused by such treatments can be small or undetected, the framework absorption properties can be altered significantly.

In the methylation of toluene, H-ZSM-5 was used as the catalyst. As part of the improved process, an objective was to eliminate the unselective catalysis occurring on the external surface of the crystals. The zeolite was treated with TEOS and a silica coating deposited. The improvement in selectivity was much greater than anticipated but was reasonable after considering the absorption data shown in Table 21. The coating slightly reduced the amount of accessible framework for the readily absorbed hexane and *p*-xylene. But the larger cyclohexane molecule was virtually excluded from entering the framework as were the *o*- and *m*-xylene isomers. Thus, if toluene methylation were to produce either *o*- or *m*-xylenes inside the frameworks, they would be inhibited from exiting the framework because the coating slightly reduced the opening of the surface

Table 21. Effects of TEOS surface treatments of H-ZSM-5 on absorption.

Sample I.D.	g absorbed/100 g sample			p-Xyl / n-Hex	0-Hex / n-Hex	*m*-Xylene ®	*o*-Xylene ®
	n-Hexane	*p*-Xylene	cyclohexane				
1. Uncoated	9.2	11.9	4.4	1.293	0.478	1.2	1.4
1. Coated	8.8	11.3	0.1	1.284	0.011	0	0
2. Uncoated	11.3	14.3	2.4	1.265	0.212	0.6	0.1
2. Coated	10.6	13.4	0.2	1.264	0.019	0	0
Typical	13.07	15.78	7.38	1.218	0.638		

® absorption values at 20 hours; not equilibrium values.

pores. Thus, the silica coating appears to be bi-functional: it can reduce or eliminate the unselective surface activity while slightly reducing the dimensions of the pores on the surface of the crystals which controls product migration out of the framework.

In a similar fashion, silica coatings were deposited on the external surface of zeolite rho. The results are shown in Table 22; the sample designated EOS-2 had more silica deposited on it compared to EOS-1. The effect of the coatings on the absorption of methanol and ethanol seems to be slight, if any. However, the absorption of the slightly larger n-propanol molecule is significantly affected, both in the packing and rate of filling ratios. These experiments demonstrate that very fine tuning of the sorption process (as well as catalysis) can be achieved by controlled precipitation on molecular sieve surfaces.

Table 23 shows the effects of HCl and $SiCl_4$ treatment of chabazite and zeolite A. $SiCl_4$ is used to modify the silica:alumina ratio and to coat the zeolite with silica. It seems to be a very harsh treatment as the data for both treated zeolites show they lost considerable crystallinity (measured by methanol absorption) during the treatment and material migrated within the zeolite framework further constraining the packing of ethanol. Treatment by HCl seems to be somewhat milder considering that there is less crystallinity lost and the ethanol:methanol ratio is almost unchanged compared to the starting, untreated sample.

Table 22. TEOS treatment of zeolite H-rho.

Sample I.D.	g absorbed/100g sample			EtOH / MeOH	nPOH / EtOH	MeOH 3/20	EtOH 3/20	nPOH 3/20
	MeOH	EtOH	nPOH					
Uncoated	23.92	22.04	19.85	0.921	0.901	0.962	0.962	0.817
EOS-1	23.33	20.42	5.33	0.875	0.261	0.975	0.950	0.334
EOS-2	23.85	20.97	1.98	0.879	0.094	0.971	0.956	0.131
Typical	24.00	21.70	20.20	0.923	0.938	0.986	0.962	0.958

Table 23. The effects of HCl and $SiCl_4$ treatment of chabazite and zeolite A.

Sample I.D.	g absorbed/100 g sample			EtOH/MeOH	nPOH/EtOH
	MeOH	EtOH	nPOH		
H-Chabazite-1	13.47	12.52	2.65	0.929	0.212
$SiCl_4$-treated	10.71	8.18	1.22	0.764	0.149
H-Chabazite-2	17.88	15.73	1.92	0.880	0.122
HCl-treated	14.02	13.01	1.02	0.928	0.078
5A	20.89	20.30	10.74	0.972	0.529
$SiCl_4$-treated	14.89	11.44	3.67	0.768	0.321

8.3 MIGRATION OF SPECIES WITHIN THE FRAMEWORK

The concept of "molecular traffic control" was first suggested by Derouane and co-workers and was applied to describe absorption phenomena within pentasil zeolites [28]. Essentially, framework structures with intersecting channels can be thought of in macroscopic terms of avenues and streets. A stalled car can impede the flow of traffic just as a large chemisorbed molecule or charge compensating cation might hinder transport down a channel system. A bus making a sharp right turn into a narrow one way street has its equivalence, for example, on the molecular scale, of a dodecane molecule trying to go from the straight channel in ZSM-5 into a sinusoidal one. The effects of molecular traffic control are readily displayed as consequences of reaction selectivity imposed by dual channel framework structures.

In one study, the presence of co-absorbed water had a pronounced effect on migration and catalytic hydrogenation selectivity. The competitive hydrogenation of cyclopentene and 4-methylcyclohexene was catalyzed by zeolites containing rhodium ions [29]. For the nearly dry zeolite Rh-ZSM-11, there was little hydrogenation selectivity for the cyclopentene; this observation agreed with absorption data. However, when water was added to the zeolite prior to the reaction, the selectivity of cyclopentene (vs. cyclohexene) hydrogenation increased significantly as shown in Table 24. Water, which doesn't play a part in the reaction, is absorbed near the Rh-ions (and framework aluminum). At low co-absorbed water contents, the migration of the larger cyclohexene molecule towards the catalytic Rh-site is more affected than that of the smaller cyclopentene molecule; a consequence of Moore and Katzer's correlation of molecular size and diffusion kinetics. Thus, the hydrogenation selectivity increases until a maximum is reached at a specific water content. Above this level of co-absorbed water, migration of both molecules to the catalytic sites is quite impaired and, as a consequence, the hydrogenation selectivity falls off. As expected, the hydrogenation rates are significantly reduced because of the reduced migration (diffusion) rates. A reviewer of that study suggested that the water of hydration was not only surrounding the active Rh sites but also modifying their oxidation

Table 24. The effect of co-absorbed water on the selectivity of the Rh-ZSM-11 catalyzed hydrogenation of cyclopentene/4-methylcyclohexene mixtures.

Weight % Co-Absorbed Water	Hydrogenation Selectivity*
0.54	1.0
1.04	10.6
1.22	11.3
1.81	19.7
2.19	46.9
2.90	22.9
5.10	9.1
8.67	6.2

* Ratio of rate of hydrogenation of cyclopentene/rate of hydrogenation of 4-methylcyclohexene.

states. Several years later, it was unambiguously shown that co-adsorbed water could modify product selectivity without taking part in a reaction.

In this study, it was found that the photochemistry of *o*- and *p*-methylbenzyl benzyl ketones, *o*-A-C(O)-B and *p*-A-C(O)-B, in the presence of pentasil zeolites follows strikingly different pathways due to the location of the adsorbed ketone or the presence of co-absorbed water [30]. The product distribution demonstrated the effects of absorption and diffusion on the radical species produced by photolysis. *p*-A-C(O)-B is readily absorbed within the pentasil framework and produces *p*-AB as the primary product. In contrast, the photolysis product distributions of *o*-A-C(O)-B can be dramatically varied depending upon the extent of its absorption into the framework. By addition of a non-reactive titrant, water, after the ketone adsorption, the photolysis product distributions were systematically varied depending upon the aluminum content of the framework. For high aluminum samples, the addition of co-absorbed water forced the ketone to the external surface of the zeolite which the subsequent photolytic product distribution ably demonstrated. In some cases, the co-absorbed water forces the ketones into the hydrophobic regions of the ZSM-5 framework. The observed product distributions were completely described by considerations of

- the size and shape absorption of the pentasil zeolites.
- the sorption of water by the hydrophilic sites of the pentasil zeolites (which depends upon the framework aluminum content).
- the hydrophobic characteristics of the pentasil channels which do not contain framework aluminum.

Recently, Turro and co-workers showed how the product distributions of chlorination relate to the topology of the ZSM-5 framework and the geometry and location of the absorbed dodecane species [31]. It was observed that the photo-induced chlorination of absorbed n-dodecane selectively occurs at the ends of the chain rather than in the center. The observed selectivity is a function of the loading of the substrate, the nature of the charge compensating cation, and the framework aluminum content. The chlorinating radicals enter and diffuse into the framework while the target dodecane molecules are absorbed at specific regions within the ZSM-5 framework and are relatively immobile. The product distributions effectively characterize the locations of the absorbed dodecanes as well as the path the radical must have taken.

9. Summary and Conclusions

Absorption of probe molecules of varying geometries and sizes is used to characterize the framework dimensions and topography of molecular sieves. From unit cell dimensions derived from X-ray diffraction and assumptions about the size of the framework forming species, a pore volume can be calculated. The size and chemical behavior of the oxygen in the sieve suggests that the framework bonding is predominantly covalent with a radius 1.5-1.7 Å. The volumes of the absorbed probe molecules, using their liquid densities, are then compared to the calculated pore volume. The constraint on the packing of the absorbed molecules is quantified by comparing their packing density to their density in the liquid state. Further, the packing of probe molecules into the same pore volume is compared via a ratio technique called the 'packing ratio'. The effect of the lattice geometry and framework dimensions on the 'packing ratios' is to provide a set of characteristic values for a given molecular sieve. The combination of X-ray diffraction and absorption measurements permits the crystallinity of a molecular sieve to be quantified. The absorption values and packing ratios for molecular probes into zeolites rho and ZSM-5 are presented as expectation values for other scientists to use as bases of comparison.

Absorption is used to characterize zeolites such as chabazite, A, and ZK-5 as well as ion-exchange effects. The effect of different templates molecules on the absorption properties of ZSM-5 is discussed; the absorption and packing ratios provide sensitive indicators of very small changes in framework dimensions. Also discussed is the application of absorption measurements to understand the effects of surface treatments of zeolite catalysts as well as to provide a better understanding of migration of guest molecules within the pentasil zeolite frameworks.

Acknowledgments

Many of the chapters in this book are written by colleagues with whom the authors had the pleasure of working. In addition, the authors would like to cite these colleagues: Glover Jones for the in-situ X-ray diffraction studies, Ted Gier for the syntheses of rho, George Sonnichsen and Mike Keane for the methylamines reaction data, Gunter Teufer for discussions and X-ray diffraction analyses of the ZSM-5 system, Bob Shannon and Ed Moran for their ZSM-5 preparations, John Parise and Reinhard Fischer for neutron diffraction studies, Cris Bonifaz for acetylene hydrogenation data, Norman Herron and Bill Seidel for olefin hydrogenation data, and our technicians, Brian Arters, Bill Stevens,

Earl Jones, Bob Shiffer, Charlie Perry, Ron Nickle, Walt Weber, Penrose Hollins, and "Chunky" McCartney. DuPont Contribution No. 7102.

References

1. M. Smart, *Chemicals Economics Handbook*, SRI International, (1992).
2. R. Szostak, *Handbook of Molecular Sieves*, Van Nostrand Reinhold, New York (1992).
3. $AlPO_4$-8:
   R. M. Dessau, J. L. Schlenker, and J. B. Higgins, *Zeolites*, **10**, 522 (1990).
   J. W. Richardson, Jr. and E. T. C. Vogt, *Zeolites*, **12**, 13 (1992).
   VPI-5:
   L. B. McCusker, Ch. Baerlocher, E. Jahn, and M. Buelow, *Zeolites*, **11**, 308 (1991).
   M. E. Davis, C. Saldarriaga, C. Montes, J. Garces, and C. Crowder, *Nature* **331**, 698 (1988).
   Cloverite:
   M. Estermann, L. B. McCusker, Ch. Baerlocher, A. Merrouche, and H. Kessler, *Nature*, **352**, 320 (1991).
4. F. Liebau, *Structural Chemistry of Silicates*, Springer-Verlag, Würzburg, p 136 (1985).
5. D. W. Breck, *Zeolite Molecular Sieves*, J. Wiley and Sons, New York, pp 607, 633-41 (1974).
6. R. Szostak, *Molecular Sieves: Principles of Synthesis and Characterization*, Van Nostrand Reinhold, New York (1989).
   H. van Bekkum, E. M. Flanagen, and J. C. Jansen (Eds.), *Introduction to Zeolite Science and Practice*, Elsevier, Amsterdam (1991). (*Stud. Surf. Sci. Catal.*, **58** (1991)).
7. J. E. Huheey, *Inorganic Chemistry*, 2nd Edition, Harper and Row, New York, p 232 (1978).
8. W. M. Meier and D. H. Olson, *Atlas of Zeolite Structure Types*, Third Edition, Butterworth-Heinemann, Boston, p 9 (1992).
9. S. G. Hill and D. Seddon, *Zeolites*, **5**, 173 (1985).
10. W. W. Kaeding, C. Chu, L. B. Young, B. Weinstein, and S. A. Butter, *J. Catal.* **67**, 159 (1981).
    J. Koresh and A. Soffer, *J. Chem. Soc., Faraday I*, **76**, 2457 (1980).
    J. Koresh and A. Soffer, *J. Chem. Soc., Faraday I*, **76**, 2472 (1980).
    R. M. Moore and J. R. Katzer, A. *I. Ch. E. Journal*, **18**, 816 (1972).
11. Fisher Scientific Co., Cat. No. 12-824, *U.S. Patent 2,308,402* (1940).
12. L. Gurvitsch, *Russ. J. Phys. Chem.*, **47**, 805 (1915).
13. These references should provide a reasonable overview of sorption:
    See [5], Chap. 8, "Adsorption by Dehydrated Zeolite Crystals", p 593ff.

S. J. Gregg and K. S. W. Sing, Chap. 4, Adsorption of Gases on Porous Solids, *Surface & Colloid Science*, **19**, 231 (1976).
J. Koresh, *J. Colloid Interface Science*, **88**, 398 (1982).
J. L. Soto, P. W. Fisher, A. J. Glessner, and A. L. Myers, *J. Chem. Soc., Faraday I*, **77**, 157 (1981).
A. P. Vavlitis, D. M. Ruthven, and K. F. Loughlin, *J. Colloid Interface Science*, **84**, 526 (1981).
The titles of these papers are included to show the diversity of studies:
"Sorption and Diffusion of Gaseous Hydrocarbons in Synthetic Mordenite", C. N. Satterfield and A. J. Frabetti Jr., *A. I. Ch. E. Journal*, **13**, 731 (1967).
"Desorption and Counterdiffusion Behavior of Benzene and Cumene in H-Mordenite", C. N. Satterfield, J. R. Katzer, and W. R. Vieth, *Ind. Eng. Chem. Fundam.*, **10**, 478 (1971).
"Restricted Diffusion in Liquids within Fine Pores", C. N. Satterfield, C. K. Colton, and W. H. Pitcher, *A. I. Ch. E. Journal*, **19**, 628 (1973).
"Sorption and Diffusion Properties of Natural Zeolites", Y. H. Ma and T. Y. Lee, in *Natural Zeolites*, L. B. Sand and F. A. Mumpton (Eds), Pergamon Press, Elmsford, New York, p 373 (1976).
"Interpretation and Correlation of Zeolite Diffusivities Obtained from NMR and Sorption Experiments", J. Karger and J. Caro, *J. Chem. Soc., Faraday I*, **73**, 1363 (1977).
"Mass Transport in Heterogenenous Catalysts", J. C. Vedrine, in *Mass Transport in Solids*, F. Beniere and R. A. Catlow (Eds), Plenum, New York, Chap. 20 (1981).
"Kinetics of Nonisothermal Sorption: Systems with Bed Diffusion Control", D. M. Ruthven and L. -K. Lee, *A. I. Ch. E. Journal*, **27**, 654 (1981).
"Diffusion in Zeolites", J. Karger and D. M. Ruthven, *J. Chem. Soc., Faraday I*, **77**, 1485 (1981).
"Diffusion-Controlled Adsorption Kinetics. General Solution and Some Applications", K. J. Mysels, *J. Phys. Chem.*, **86**, 4648 (1982).
"Interpretation of Sorption and Diffusion Data in Porous Solids", R. Aris, *Ind. Eng. Chem. Fundam.*, **22**, 150 (1983).
"Adsorption and Diffusion of Gases in Zeolites", L. V. C. Rees, *Chemistry and Industry (London)*, 252 (1984).
"Nonisothermal Sorption Kinetics in Porous Adsorbents", R. Haul and H. Stremming, *J. Colloid Interface Science*, **97**, 348 (1984).
14. R. M. Moore and J. R. Katzer, *A. I. Ch. E. Journal*, **18**, 816 (1972).
15. L. Abrams and D. R. Corbin, *J. Catal.*, **127**, 9 (1991).
16. L. Abrams, M. Keane, and G. C. Sonnichsen, *J. Catal.*, **115**, 361 (1989).
17. J. B. Parise, L. Abrams, T. E. Gier, D. R. Corbin, J. D. Jorgensen, and E. Prince, *J. Phys. Chem.*, **88**, 2303 (1984).
18. L. Abrams, D. R. Corbin, and M. Keane, *J. Catal.*, **126**, 610 (1990).

19. D. H. Olson, G. T. Kokotailo, S. L. Lawton, and W. M. Meier, *J. Phys. Chem.*, **85**, 2238 (1981).
20. E. G. Derouane and Z. Gabelica, *J. Catal.*, **65**, 486 (1980).
E. M. Flanigen, J. M. Bennett, R. W. Grose, J. P. Cohen, R. L. Patton, R. M. Kirchner, and J. V. Smith, *Nature*, **271**, 512 (1978).
A. Auroux, H. Dexpert, C. LeClercq, and J. Vedrine, *Appl. Catal.*, **6**, 95 (1983). C. G. Pope, *J. Catal.*, **72**, 174 (1981).
H.-J. Doelle, J. Heering, L. Rickert, and L. Marosi, *J. Catal.*, **71**, 27 (1981).
R. LeVanMao, O. Pilati, A. Marzi, G. Leofanti, A. Villa, and V. Ragaini, *React. Kinet. Catal. Lett.*, **15**, 293 (1980).
H. Nakamoto and H. Takahashi, *Zeolites*, **2**, 67 (1982).
21. A comparison of the relative diffusion coefficients was made by using a modified form of the Crank equation: $(Q_t-Q_o)/(Q_{eq}-Q_o) = c^*(Dt)^{1/2}$, {[5] p. 673}where Q_t is the amount sorbed at time t, Q_{eq} is the equilibrium amount, c is a constant depending on the geometry of the crystals, and D is the diffusion coefficient. For the case where Q_o is zero, the ratio of Q_t to Q_{eq} is the occupancy of the crystals. Rearranging the equation yields: $D = (k/t)^*(Q_t/Q_{eq})^2$, which is valid only for short times.
22. Z. Gabelica, J. B. Nagy, and G. Debras, *J. Catal.*, **84**, 256 (1983).
23. E. G. Derouane, S. Detremmerie, Z. Gabelica, and N. Blom, *Appl. Catal.*, **1**, 201 (1981).
24. P. A. Jacobs, E. G. Derouane, and J. Weitkamp, *J. Chem. Soc., Chem. Commun.*, 592 (1981).
25. S. M. Csicsery, *Zeolites*, **4**, 202 (1984).
26. V. N. Ramannikov, A. S. Loktev, A. N. Spektor, G. L. Bitman, K. G. Ione, P. S. Chekrii, *Neftekhimiya*, **31**, 409-15 (1991).
27. D. R. Corbin, L. Abrams, and C. Bonifaz, *J. Catal.*, **115**, 420 (1989).
28. E. G. Derouane, P. Dejaifve, Z. Gabelica, and J. Vedrine, *J. Chem. Soc., Faraday Disc.*, **72**, 331 (1981).
E. G. Derouane, *J. Catal.*, **72**, 177 (1981).
29. D. R. Corbin, W. C. Seidel, L. Abrams, N. Herron, G. D. Stucky, and C. A. Tolman, *Inorg. Chem.*, **24**, 1800 (1985).
30. N. J. Turro, C. -C. Cheng, L. Abrams, and D. R. Corbin, *J. Am. Chem. Soc.*, **109**, 2449 (1987).
31. N. J. Turro, N. Han, X-G. Lei, J. R. Fehlner, and L. Abrams, *J. Am. Chem. Soc.*, (submitted).

Abstract. Molecular sieves are inorganic framework structures generally composed of crystalline aluminosilicate tetrahedra which are arranged to form channels of 2-10 Å diameters and cages with dimensions from 6-15 Å. Absorption of probe molecules of varying geometries and sizes is used to characterize the framework dimensions and topography in concert with X-ray diffraction identification of the specific structure. From unit cell dimensions and assumptions about the size of the framework forming species, a pore volume can be calculated. The volumes of the absorbed probe molecules, using their liquid densities, are then compared to the calculated pore volume. The constraint on the packing of the absorbed molecules is quantified by comparing their packing density to their density in the liquid state. Further, the packing of different probe molecules into the same pore volume is compared via a ratio technique called the 'packing ratio'. The effect of the lattice geometry and framework dimensions on the 'packing ratios' is to provide a set of characteristic values for a given molecular sieve. The packing ratios for zeolites rho and ZSM-5 are presented as expectation values for other scientists to use as bases of comparison.

Journal of Inclusion Phenomena and Molecular Recognition in Chemistry **21**: 47–78, 1995.

Structure-Direction in Zeolite Synthesis

RAUL F. LOBO, STACEY I. ZONES and MARK E. DAVIS*

1. Introduction

Zeolites and molecular sieves are important industrial materials that have a broad range of applications. Many zeolites with different structures and compositions are commercially available and used in refining and chemical processing as catalysts, absorption materials, and drying agents, as well as in ion-exchange operations [1,2]. Zeolites and molecular sieves show a strikingly close relationship between their microscopic structure and macroscopic properties. This relationship is particularly clear in their ability to sieve organic and inorganic molecules based on size. Also, the rigorous steric control that zeolite frameworks offer over the chemical reactions occurring inside their pores is directly related to their microscopic structure.

The relationships between the microscopic structure and macroscopic properties have been large driving forces for the synthesis of new materials with novel pore architectures and crystal compositions. It is expected that molecular sieves with new pore structures will show catalytic properties not observed before. Although over 89 different structures are known [3], this is just a small fraction of all the hypothetical structures that have been proposed [4,5].

In this review, we will focus on the relationships between the geometrical and chemical properties of the organic molecules used to direct the crystallization of a particular zeolite structure, and the geometry and characteristics of the cages and pores of the zeolites in which the organic molecules have been occluded. Additionally, we will look at the effects of alkali-metal ions and trivalent and divalent tetrahedral atoms on the zeolite structures formed and how these effects are coupled to the actions of the organic structure-directing agent.

We will emphasize high-silica materials because these molecular sieves are the ones that offer the greatest possibility of control through the use of organic molecules in the synthesis gel. Many new microporous aluminophosphates ($AlPO_4$) and metal-substituted aluminophosphates (MAPO) have been and are continuing to be reported. However, the connection between synthesis conditions and the structures of these molecular sieves is very poorly understood. In these materials, the crystal structure of the final product often depends on the presence of a particular organic molecule in the synthesis gel. However, the relationship between the organic molecules and the pores or cages is very loose, and sometimes the organic molecule is not even present inside the pores of the molecular sieve as is the case for VPI-5 [6,7]. Also, aluminosilicates and borosilicates with low ratios of SiO_2/M_2O_3 will not be covered because the structure of these hydrophilic and highly anionic framework materials is mostly dependent on interactions with the inorganic cations of the synthesis gel.

In addition to an enhanced interaction with organic structure-directing agents, materials formed mostly from tetrahedral silicon atoms (T) have the advantage over $AlPO_4$'s and aluminosilicates in that the T—O—T angle with only Si as the T atoms is much more flexible than other T—O—T angles [8,9]. This is an indication that, in principle, more frameworks should be attainable. Also, high-silica materials can have, among their structural building units, rings with odd and even numbers of tetrahedral atoms, which are not found in $AlPO_4$'s because of the rigorous aluminum-phosphorus alternation.

We begin our discussions with an introduction to the concepts of structure-direction by illustrating the synthesis of clathrasils. Next, we describe recent findings on the relative energetic stability of different pure-silica framework structures and discuss different factors that may affect the crystallization kinetics of zeolites. In the following section, we return to explore more complex zeolite syntheses in which the structure-directing effects not only depend upon the organic structure-directing agent, but also are influenced by the presence of different alkali-metal cations and trivalent and divalent T atoms - such as aluminum and zinc - in the synthesis gel.

The three-letter code of the International Zeolite Association will be used throughout this review [3] and some familiarity with zeolites and zeolite synthesis will be assumed on the part of the reader. Previous reviews and articles related to the material covered here can be found in the references [10-14] as well as in the general discussion of zeolite structure and properties in Chapter 1 of this volume by Abrams and Corbin.

2. Clathrasils

Clathrasils are a group of tectosilicates formed of polyhedra (cages) that enclose relatively small organic or inorganic molecules. The clathrasils can be considered as zeolites with pores of 0-dimension. The different cages found in the known clathrasils are illustrated in Table 1 and many of the structure-directing agents used for the synthesis of these tectosilicates can be found in [13].

Gies and Marler [15] have carried out the synthesis of clathrasils using the simple reaction system SiO_2/structure-directing agent/H_2O, by systematically changing the nature of the structure-directing agent. The advantage of this simple gel composition is that changes in the clathrate structure can be unambiguously attributed to the effect of the structure-directing agent. Gies and Marler found that sixty-one molecules varying from noble gases such as Kr and Xe to large amines such as 1-aminoadamantane can be used to control the structure of the clathrasil formed.

Table 1. Summary of the different cages found in the structures of clathrasils [15].

Clathrasil	Code	FD (T nm^{-1})	V of Largest Cage (nm^3)	Clathrasil Cages		
Melanophlogites	MEP	18.9	0.160	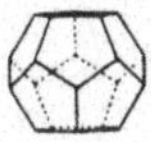 $[5^{12}]$	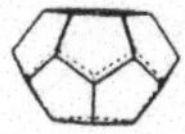 $[5^{12}6^{2}]$	—
Dodecasils 3C	MTN	18.7	0.250	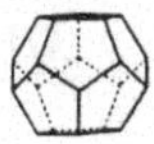 $[5^{12}]$	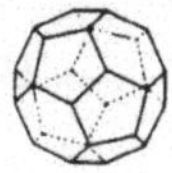 $[5^{12}6^{4}]$	—
Octa-decasil	AST	16.7	0.280	 $[4^{6}]$	—	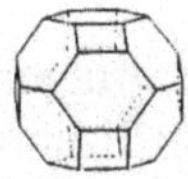 $[4^{6}6^{12}]$
Nonasil	NON	19.3	0.290	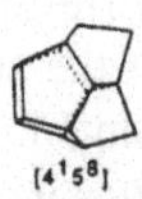 $[4^{1}5^{8}]$	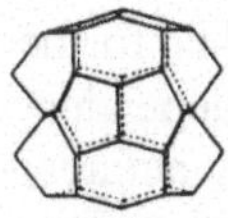 $[5^{8}6^{12}]$	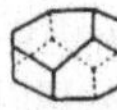 $[5^{5}6^{4}]$
Deca-dodecasil 3R	DDR	17.6	0.350	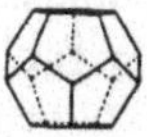 $[5^{12}]$	—	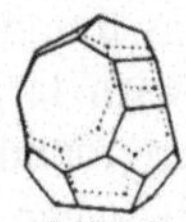 $[4^{3}5^{12}6^{1}8^{3}]$
Sigma-2	SGT	17.8	0.390	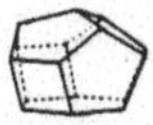 $[4^{3}5^{6}6^{1}]$	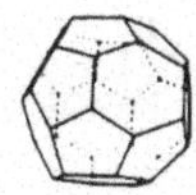 $[5^{12}6^{8}]$	—
Dodecasil 1H	DOH	18.5	0.430	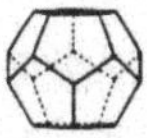 $[5^{12}]$	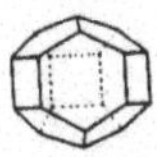 $[4^{3}5^{6}6^{3}]$	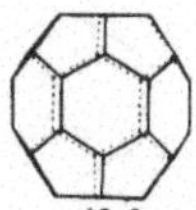 $[5^{12}6^{8}]$
Deca-dodecasil 3H	—	—	0.540	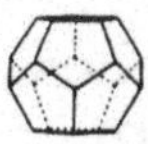 $[5^{12}]$	—	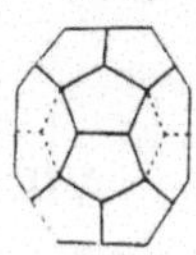 $[5^{18}6^{2}8^{3}]$
Sodalite	SOD	17.2	0.230	—	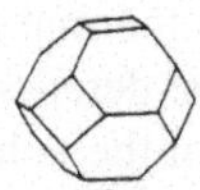 $[4^{6}6^{8}]$	—

Guest molecules of similar size and shape were found to direct the crystallization of the same product, regardless of differences in chemical properties. The following series of molecules formed the MTN structure in all cases:

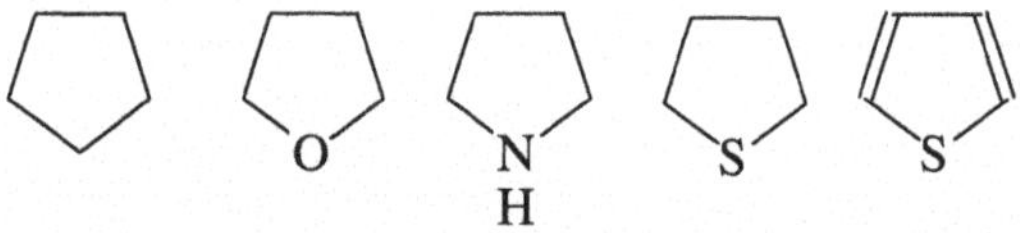

I **II** **III** **IV** **V**

Although the structure is independent of the chemical properties, they found that the kinetics of crystallization are very dependent on the basicity of the organic molecule [16]. The most basic molecule (**III**) gave the highest rate of crystallization and the least basic (**I**) gave the slowest.

Several guest molecules formed different frameworks under different synthesis conditions, and a strong dependence of the structure type on synthesis temperature was observed. For example, 1-aminoadamantane was found to form DDR, SGT, DOH, and deca-dodecasil 3H as the temperature was increased from 140 to 200° C. In general, the smaller cage was found at the lower synthesis temperatures. Although large molecules form crystals with larger cages, some relatively small molecules - like 2-methyl-pyrrolidine - were found in the large $[5^{12}6^{8}]$ cage of DOH.

The relation between the geometry of the organic molecule and the geometry of the cage is limited for clathrasils. For example, the following molecules can be used to synthesize DOH:

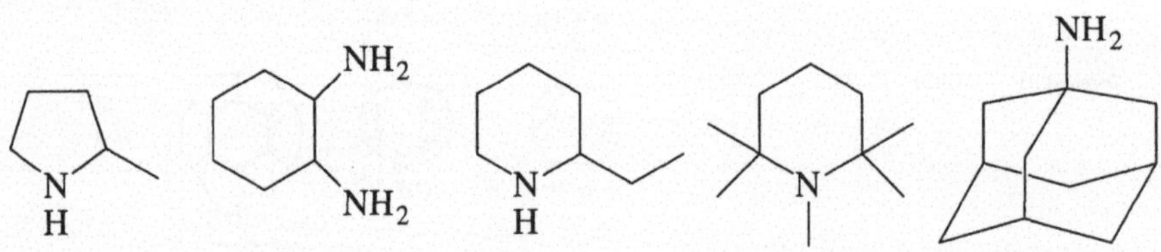

VI **VII** **VIII** **IX** **X**

These results when viewed in combination with the effect of the temperature on the size of the cage suggest that the thermal motion of the molecule in the crystallization gel may inhibit the imprinting of the detailed geometry of the molecule, and that the silicate gel 'senses' only the geometry of the volume of revolution of the structure-directing molecule.

We carried out a set of experiments similar to the ones reported by Gies and Marler in which the only components of the gel were the structure-directing agent, the silica source (Cab-O-Sil, M5) and water. The results of these experiments (Table 2) are roughly in agreement with those reported by Gies and Marler. Large molecules form clathrasils with large cages and smaller molecules form crystals with smaller cages.

Clathrasils can be also prepared using non-hydrothermal synthesis conditions, e.g., using an organic solvent. The synthesis of SOD was first reported in this manner by Bibby and Dale [17] and ethylene glycol was used as solvent and as the structure-directing agent. Originally, this result was intriguing because the sodalite structure is the only pure-silica clathrasil - besides AST - that contains only even-membered rings in the structure. The remainder of the clathrasils contain a large fraction of 5-rings as is generally observed for high-silica zeolites. The molecules of ethylene glycol are found inside the sodalite cages and are necessary for the synthesis of the material. However, recently, Post and co-workers [18] and Chen [19] reported the synthesis of pure-silica SOD under hydrothermal conditions using trioxane and ethylenediamine as structure-directing agents, respectively.

Although, there have been proposals suggesting that there is a relationship between the symmetry of the organic molecule and the symmetry of the cage [20,21], the results reported by Gies and Marler and those presented in Table 2 suggest that the relationships found between the point group of the molecule and the point group of the cage are probably accidental. Molecules with no symmetry - except the identity operation - are capable of directing the formation of cages with very high symmetry (compare Tables 1 and 2). As suggested above, it seems that the thermal motion of the organic molecule may average out the geometrical differences between the molecules. Definitive proof of the rapid motion of guest molecules inside the cages of clathrasils was recently provided by Gies [14] using static ^{13}C solid-state NMR. Gies showed that organic molecules inside the cages of MTN, DOH, and NON are rotating on a time-scale that is smaller than the time-scale of the NMR experiment. The rapid rotation of organic molecules inside the pores and cages of zeolites has also been reported by Hong and co-workers [22,23].

Using solid-state NMR with and without magic-angle spinning (MAS), they show that adamantane and naphthalene undergo rapid motion inside the pores of zeolite Na-Y. Although these molecules were absorbed inside the zeolite pores after the zeolite synthesis, similar results have been obtained for 15-crown-5 and 18-crown-6 that were used as structure-directing agents in the synthesis of FAU and EMT [24]. These results indicate that rapid motion of the structure-directing agents in the clathrasil cages is likely to be the 'normal' state of the organic molecule at synthesis temperatures.

Clathrasils have several types of cages in each framework structure (Table 1). How are these voids stabilized? Are any of them empty? Crystallographic studies of clathrasils show that the organic structure-directing molecules are found inside the large cages and small molecules are usually located inside the small cages. N_2 and Kr are occluded inside the [5^{12}] and in the [$5^{12}6^2$] cages of MEP [25], N_2 and ammonia are found inside

Table 2. Results from the synthesis of pure-silica clathrasils using different organic molecules as structure-directing agents.

Structure Directing Agent	Clathrasil
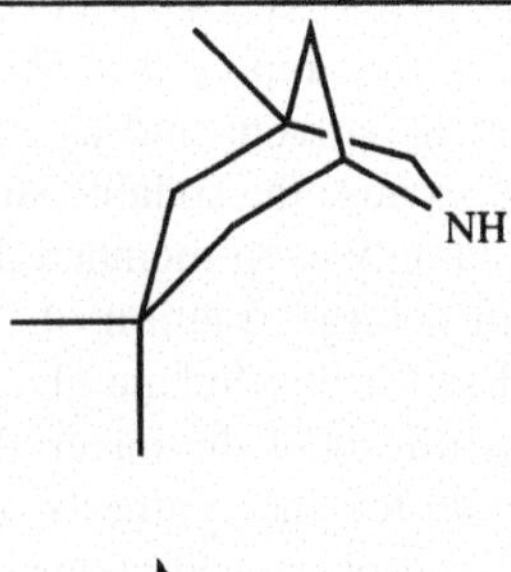	DOH
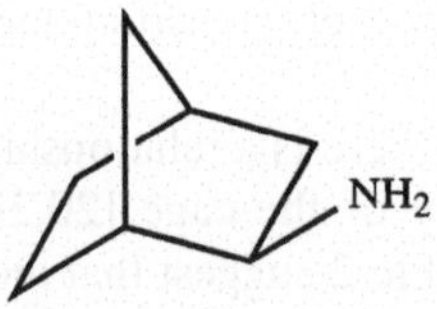	DOH
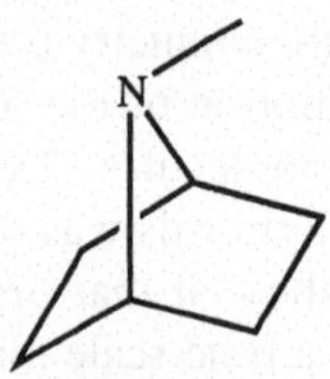	STG, DOH
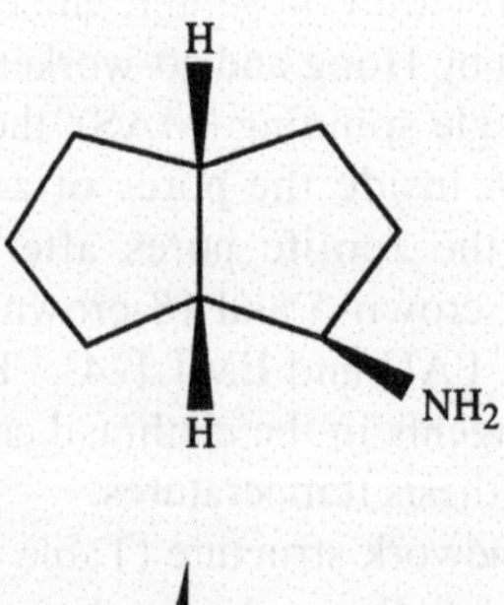	NON
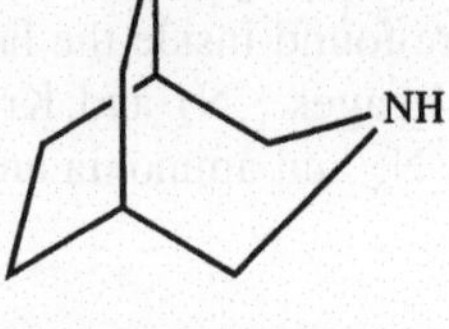	NON

the [5^{12}] and the [$4^3 5^6 6^3$] cages of DOH [26,27] and N_2 is also located in the [5^{12}] cages of MTN and DDR [28,29]. Note that the occluded N_2 was not purposely added to the synthesis, but was 'captured' from the small amount of atmospheric air trapped inside the silica-glass tubes used for the synthesis [26]. Cages smaller than the [5^{12}] cage are probably not filled with any molecules because the available void is too small to be occupied by guest species. It is interesting to speculate whether these materials could be synthesized in the absence of any small guest molecules that could fill these small cages, especially in the light of what is observed for other aluminous zeolite polymorphs. The small cages of many aluminous zeolites such as CAN, GME, OFF, SOD, etc., are invariably occupied by alkali-metal cations which stabilizes the cage. Therefore, it seems that in general, with zeolites and clathrasils, small cages are only formed when small guest molecules are present as stabilizing agents.

These crystallographic studies have shown that frequently the point group of the organic structure-directing agent is a subgroup of the point group of the cage. However, the axis of symmetry of the molecules and the cages often do not coincide with each other. 1-Aminoadamantane in the [$5^{12}6^8$] cage of SGT [30,31] is disordered and the axis of the molecule in the dominant conformation does not correspond to the axis of the cage. Also, the axis of 1-aminoadamantane in the [$5^{12}6^8$] cage of DOH [24], where the cage belongs to the same point group as the organic molecule, is rotated about 20° with respect to the six-fold axis of symmetry of the cage. On the other hand, the axis of 1-aminoadamantane was found to coincide with the three-fold axis of symmetry of the [$4^3 5^{12} 6^1 8^3$] cage of DDR [29]. These studies have been carried out at room temperature and as indicated above, it is likely that the organic molecule at synthesis temperatures (140° C – 200° C) is rapidly rotating. The results described above must be carefully interpreted when extrapolating to synthesis mechanisms.

Amines and quaternary ammonium ions are the organic molecules used most frequently in the synthesis of clathrasils and molecular sieves. The hydrophobicity of pure-silica frameworks suggests that the positive charge of the organic cation is balanced by a negative charge formed by a defect in the framework that could be illustrated as follows:

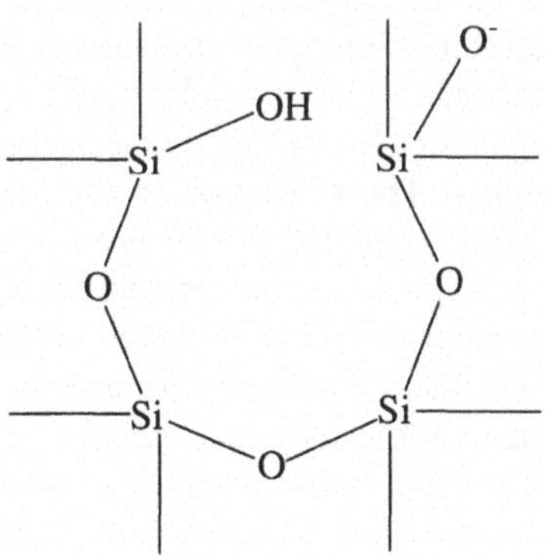

We investigated the possibility that defects such as these could balance the charge of the quaternary ammonium ions by obtaining ^{29}Si MAS-NMR on two pure-silica samples of nonasil (NON) synthesized with cyclohexylamine and a quaternary ammonium ion as structure-directing agents (Figure 1). The sample synthesized with a quaternary ammonium cation (Figure 1A) shows two peaks that can be assigned to the presence of

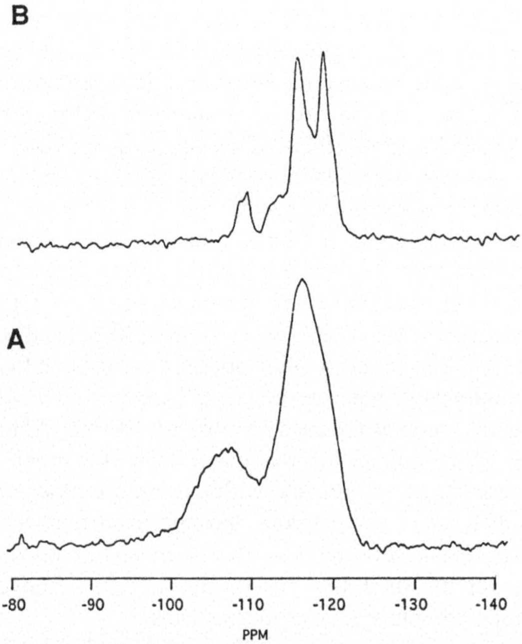

Figure 1. ^{29}Si CP/MAS-NMR spectra of samples of pure-silica nonasil (NON) synthesized using a quaternary ammonium cation (A) and an organic amine (B) as structure directing agent.

both Q_3 and Q_4 silicon atoms (Q_i: $Si(OSi)_iO_{4-i}^-$). The presence of Q_3 and Q_4 suggests that the charge of the structure-directing agent is balanced by a defect (deprotonated silanol group) in the zeolite structure. Similar results have been reported for a sample of pure-silica AFI [32] and for high-silica MFI [33] synthesized with quaternary ammonium ions as structure-directing agents. The resolution of the MAS-NMR spectrum of the sample synthesized with the amine (Fig. 1B) is enhanced to the point of recognition of most of the different tetrahedral environments present in NON [3]. The increase in resolution of ^{29}Si MAS-NMR spectrum has been linked to the absence of defects in the framework [34,35] and suggests that the amine is not protonated inside the NON cages. These results are in agreement with the idea that the charge of quaternary ammonium cations in high-silica zeolites is often balanced by deprotonated silanol groups in the structure. The infrared spectra of these two samples, only show weak evidence to suggest the presence of the O—H stretching band (expected at approximately 3300 cm^{-1}), indicating instead that the structure of the defects is closer to the following illustration:

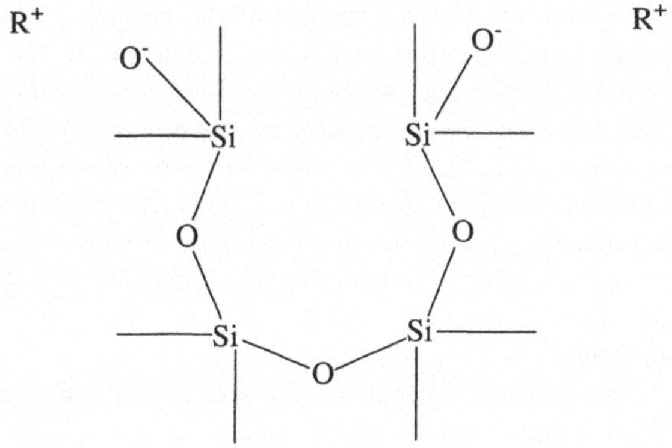

In this second alternative, two positive charges are balanced for each defect. If defects are energetically destabilizing, then it is expected that the structure will crystallize such that the least number of defects are formed to balance the positive charge of the organic guest molecule. This is accomplished by a mechanism best represented by the second illustration. However, this model does not necessarily exclude the formation of defects with protonated silanol groups. From the perspective of structure-direction, this is important because it suggests that molecules with large charge/volume ratio - tetramethylammonium, for example - are not good alternatives for the synthesis of clathrasils. It also indicates that materials with a large void fraction require organic guests with a very low charge/volume ratio, for the defects to have a small impact on the stability of the organic-inorganic composite.

The above discussions illustrate several important issues on the synthesis of clathrasils and pure-silica molecular sieves. The conditions at which an organic guest will best form a clathrasil have been summarized by Gerke, Gies and Liebau [cited in ref. 36] and are as follows:

1. The molecule should have sufficient room within a particular cage or pore.
2. The molecule must be stable under the synthesis conditions.
3. The molecule should fit the inner surface of the cage with as many van der Waals contacts as possible, but with the least deformation.
4. The guest molecule should have only a weak tendency to form complexes with the solvent.
5. More rigid molecules will tend to form clathrasils more easily than flexible molecules.
6. The tendency to form a clathrasil will increase with the basicity or polarizability of the guest molecule.

From the molecular point of view, the first condition indicates that if the organic molecule is bigger than the polyhedral cage, steric forces inhibit the formation of the clathrasil. The second point suggests that molecules that can react under basic, hydrothermal synthesis conditions will of course change during the synthesis and their decomposition products will affect the structure of the product. The third and fourth conditions are energetic considerations that indicate that during the formation of the clathrasil, the optimum of the van der Waals interactions between the organic molecule and the cage or zeolite pore depends on the organic guest, and that the energy employed in changing the conformation of the organic guest to fit the geometry of the pore or the cage destabilize the organic-inorganic composite. Finally, the basicity of the molecule is related to a) the effect that a higher pH has on the kinetics of silicate hydrolysis, and b) to the stabilizing effect that a higher polarizability of a molecule within the electric field created by the partial ionic character of the Si—O bond, has over the total energy of formation of the composite.

What is observed if we continue to increase the size of the structure-directing agent? This question was recently addressed by Nakagawa and Zones [37]. The following series of organic molecules were synthesized and used as structure-directing agents:

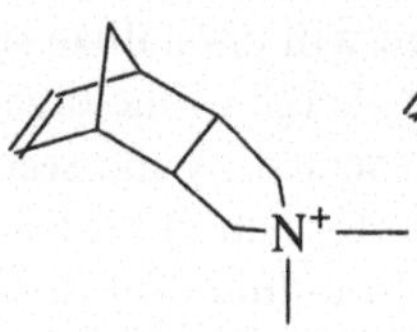

XI

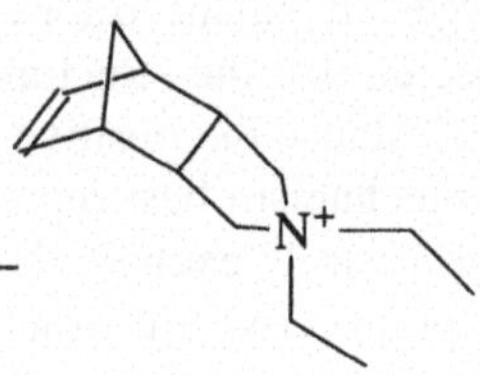

XII

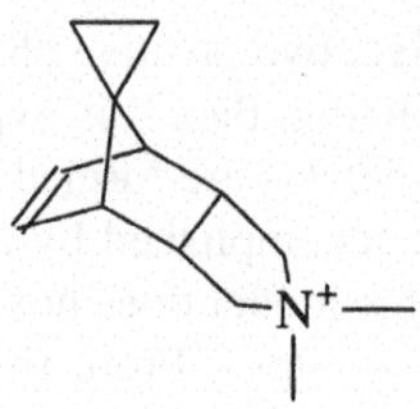

XIII

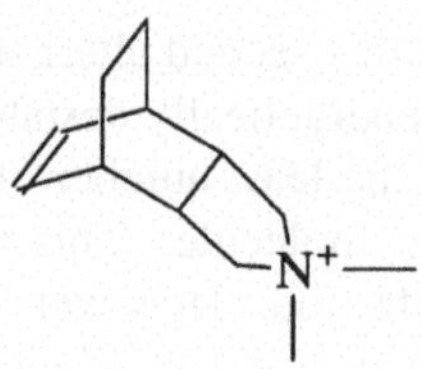

XIV

In the absence of any structure-directing agent, the product of the synthesis mixture was quartz. Using molecules **XI**, **XII**, **XIII** and **XIV**, the clathrasil nonasil and the one dimensional large-pore zeolites MTW, SSZ-31 and SSZ-31, were formed, respectively. Thus, Nakagawa and Zones found that when the size of the structure-directing agent is increased to a certain dimension, the product formed changes from a clathrasil to a microporous, large-pore molecular sieve.

If instead of the size, the geometry is changed from cyclic molecules to linear molecules, there is also a transition from clathrasils to microporous molecular sieves but now to microporous silicates with 10-ring pores [15]. For example, piperidine forms the NON clathrate, but 1-aminobutane forms TON and ZSM-48 at 160 °C and 180 °C, respectively. The size of the linear organic molecule is of relatively little importance compared to the effect of temperature on the structure of the zeolite formed and small molecules such as ethylenediamine (Table 2) and 1,5,9,13-tetraazatridecane [15] form ZSM-48 under approximately the same conditions.

There is another transition from 1-dimensional, 10-ring materials to 3-dimensional, 10-ring materials as the organic molecule geometry is changed from linear to branched. Diethylamine and dipropylamine form ZSM-48 (a linear 10-ring) but triethylamine and tripropylamine form the MFI structure. Note that in these organic-inorganic composites, the good geometrical fit reflects an optimization of the van der Waals contacts between guest and host [15].

3. Thermochemical Stability of High-Silica Zeolites

Zeolites are thermodynamically metastable phases accessible by kinetically favored pathways under hydrothermal synthesis conditions [39]. Recently, Petrovic and co-workers [40] shed some new light on the thermodynamics of high-silica zeolites. The enthalpy of formation of a series of high-silica zeolites (MFI, MEL, MTW, AFI, FAU and EMT) was determined using a high-temperature solution calorimetry method. The results were compared to the enthalpy of formation of dense polymorphs of silica. All the microporous frameworks are only 7-14 kJ mol^{-1} less stable than quartz. Additionally, no clear correlation between the enthalpy of formation of the zeolites and the framework density - or degree of *openness* - secondary building units such as large or small rings in the structure, or the mean Si–O–Si angle was found (however, see [41] for a different interpretation of the data).

In pure-silica zeolite structures, every silicon atom is bonded to four oxygen atoms in a relatively undistorted tetrahedral coordination and the oxygen atoms are connected to two silicon atoms with an average Si–O–Si angle close to 145°. Ab initio molecular orbital calculations have shown that there is a small barrier to variations in the Si–O–Si angle between 140° and 180°, but that there is a sharp increase in the energy for angles less than about 135°. Therefore, based on the above evidence and to a first degree of approximation, it seems now understandable that the energy of all these pure-silica frameworks should be very similar.

Petrovic et al. showed that there is a correlation between the fraction of Si–O–Si angles ≤ 140° and the enthalpy of formation of these pure-silica molecular sieves. This correlation is consistent with the shape of the potential energy curve of the Si–O–Si angle that suggests that small angles are much more destabilizing than large ones. The authors, however, note that these angles have been obtained mostly from analyses of X-ray powder diffraction data. These XRD data contain very short Si–O bond lengths (1.49 and 1.53 Å), very long Si–O bond lengths (1.67 to 1.70 Å), and a large range of O–Si–O angles (92° to 133°). By comparison, the bond lengths and O–Si–O angles of the single crystal structure determination of MFI are in very good agreement with the single crystal data from quartz. Therefore, these unexpected bond lengths and angles may be the result of the difficulty in obtaining a stable and acceptable solution with the Rietveld refinement methodology. Also, based on their work with the relatively simple structure of sodalite, Hu and Depmeir [42] have suggested that the symmetries of zeolites may have been overstated and that some zeolite structure solutions may need to be revised. In summary, to draw further conclusions, it may be necessary to wait until new single crystal X-ray data or other technique improvements confirm the anomalous bond lengths and angles.

The numerical similarity of the enthalpies of formation of the different pure-silica zeolites imply that the role of the organic molecule may be to select a particular structure among many possible of similar energy - i.e., a more kinetic rather than thermodynamic role. However, the incorporation of the structure-directing agent inside the zeolite pores reduces the total free energy of formation of the composite in the same way that a solution of ethanol and water has lower total free energy of formation than pure ethanol and water separately. Interactions that stabilize the organic-inorganic composite are the optimization of van der Waals contacts between the organic molecule and the surface of the pores or cages.

Is it energetically favorable for the structure-directing agent to be occluded inside the zeolite pores? Although most structure-directing agents are soluble in water, studies on the interactions between tetramethylammonium (TMA), tetraethylammonium (TEA), and tetrapropylammonium (TPA) [43,44] and water indicate that the interactions between TMA and water are mainly hydrophilic in character, but TPA, in spite of its positive charge, shows mostly a hydrophobic character due to its longer hydrocarbon chains. TPA orders water around itself with a concomitant entropic cost that balances the energetic gains obtained from the solvation of the positive charge. In an attempt to investigate these interactions with a greater variety of structure-directing agents, Zones has studied the transfer of different charged organic-structure directing agents from an aqueous solution to a chloroform phase as a function of the C/N^+ ratio. He found that the percentage of transfer is low (less than 10%) for molecules with a $C/N^+<11$ and that it is very large (more than 70%) for $C/N^+>15$ (see Figure 2). Interestingly, it is observed that organic molecules with C/N^+ between 11 and 15 work well as structure-directing agents yielding a great variety of new high-silica molecular sieves. This trend suggests that molecules that are moderately hydrophobic are the best candidates for structure-direction. This result is also in agreement with the ideas of Gerke, Gies, and Liebau [36] who indicated that the guest molecule should only have weak tendency to form complexes with the solvent. The percentage of transfer, however, does not correlate with the kinetics of zeolite crystallization, suggesting that geometrical properties also play an important role in the formation of crystal nuclei.

The size of the molecule (carbon+nitrogen atoms) correlates strongly with the selectivity of different structure-directing agents. As indicated previously, often one organic guest molecule can direct the crystallization of more than one zeolite structure. Zones has observed that as the number of carbon plus nitrogen atoms in the structure-directing molecule is increased, this tendency is reduced until finally for large molecules, only one structure is observed. Figure 3 illustrates this trend for a variety of molecules. The conclusion is that, in order to form new structures, larger molecules should be explored more carefully as structure-directing agents. This trend has to be balanced against the effect of the C/N^+ ratio as indicated above. If only one quaternary nitrogen is

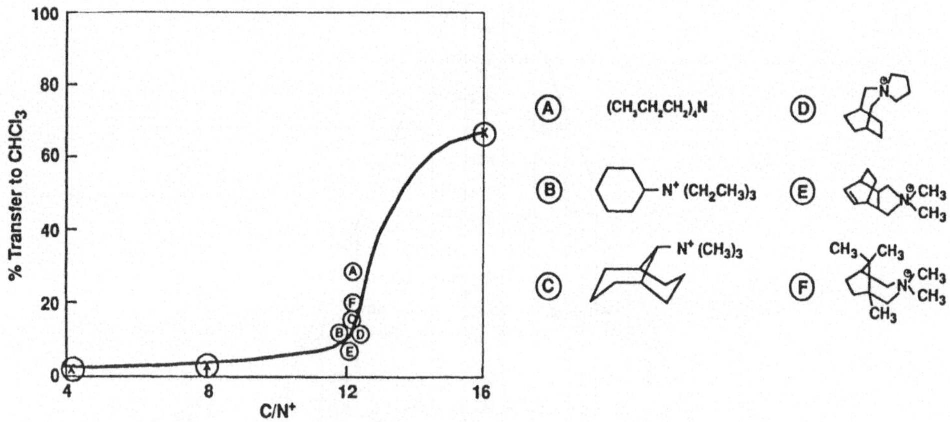

Figure 2. Partition of quaternary ammonium compounds between water and $CHCl_3$.

included in the molecule, eventually the hydrophobicity of the molecule will overcome the solvation possibilities of water and the organic molecules may aggregate to form a separate phase.

4. Kinetics of Zeolite Crystallization

The kinetic nature of the crystallization of silicate frameworks is clearly observed if a structure, less stable than quartz, is formed without the occlusion of organic molecules. This is a common experimental observation when the product of a zeolite synthesis is tridymite or cristobalite, as often occurs during the preparation of high-silica zeolites. Thermodynamic data show that these dense phases are less stable than quartz at synthesis temperature and pressure [45]. However, tridymite and cristobalite are often the product of a zeolite synthesis, indicating that the kinetics of crystal nucleation and growth are the controlling factors determining the structure of the crystallization product.

In the following section we discuss issues involving the effects of alkali-metal cations on the rate of nucleation and crystal growth of high-silica zeolites. We also briefly illustrate how one structure-directing agent can direct the crystallization of different zeolites through seeding of the synthesis gel.

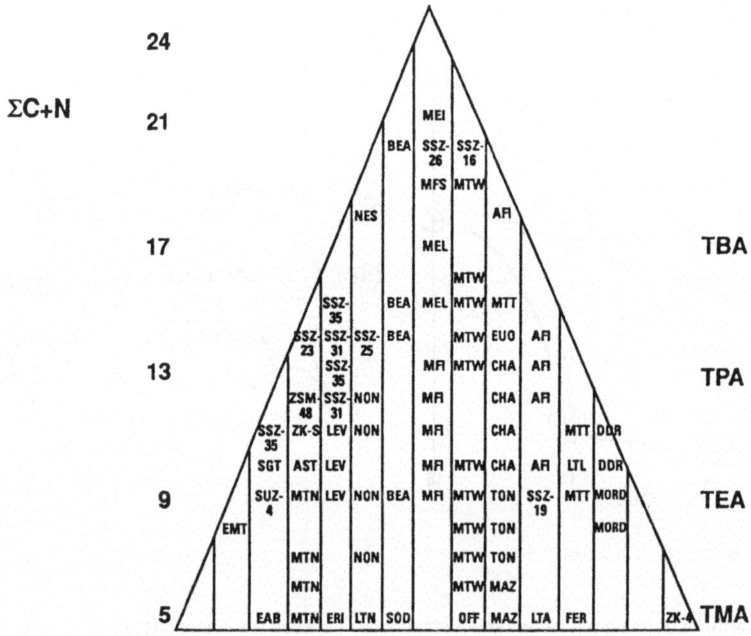

Figure 3. Illustration of the increased selectivity of different structure-directing agents as the size of the molecule (C+N) is increased.

4.1 EFFECT OF ALKALI-METAL CATIONS

The synthesis of most zeolites is carried out under basic conditions using sodium or potassium cations as one of the main components of the reaction mixture. Other mineralizing agents, fluoride in particular, are also often used (this method is not covered in this review) [see ref. 12]. The effect of alkali-metal ions on the synthesis of pure-silica zeolites was not systematically studied until it was found that the incorporation of titanium into zeolite frameworks was negatively affected by the presence of alkali-metal cations. Titanium-containing zeolites (TS-1 and TS-2) are very active catalysts for the selective oxidation of alkanes and alkenes using hydrogen peroxide as the oxidant [46]. The search for titanium-containing zeolites with new structures prompted the study of zeolite syntheses in the absence of alkali-metal cations.

It was found that most zeolite syntheses are sensitive to the presence of alkali-metal cations in the synthesis gel. Except for the MFI and the MEL structures using TPA or tetrabutylammonium (TBA) as structure-directing agents, other zeolites are very difficult to synthesize in the absence of alkali-metal cations. For example, Goepper et al. [47]

studied the synthesis of pure-silica MTW in the presence and absence of alkali-metal cations at constant hydroxide concentration. They found that the addition of sodium and potassium decreased the crystallization time from 84 days to 7 days and suggested that alkali-metal ions participate in both the nucleation and crystal growth processes.

The study of the dissolution of quartz into aqueous solutions of alkali-metal cations aids the understanding of the role of sodium and potassium in the synthesis of zeolites. Brady and Walters [48] and Dove and Crerar [49] investigated the rate of dissolution of quartz in aqueous solutions of NaCl, KCl, LiCl, and $MgCl_2$. They found that small concentrations of electrolytes increase the rate of dissolution by as much as 15 times, compared to the rates measured for deionized water. The effect is greatest for NaCl and KCl and increases for concentrations up to 0.05 M. The faster dissolution rates observed for sodium and potassium relative to magnesium and lithium are probably related to their potential for adsorption onto the quartz surface. Mg^{2+} and Li^{+} have a low potential for adsorption because they are strongly coordinated to water molecules in the solution as the result of their large charge to size ratio. Extrapolation of these results to zeolite synthesis suggests that sodium and potassium are involved in the polymerization and depolymerization of the silicate species in the gel and that this is at least one of the processes through which they increase the nucleation and crystal growth rates of high-silica zeolites.

The effect of high concentrations of sodium on the crystallization of zeolites has recently been reported by Zones [50,51]. Zones studied the effects of the concentration of the inorganic species in the synthesis of large-pore borosilicates using the tricyclo[5.2.1.0]decanes as structure-directing agents:

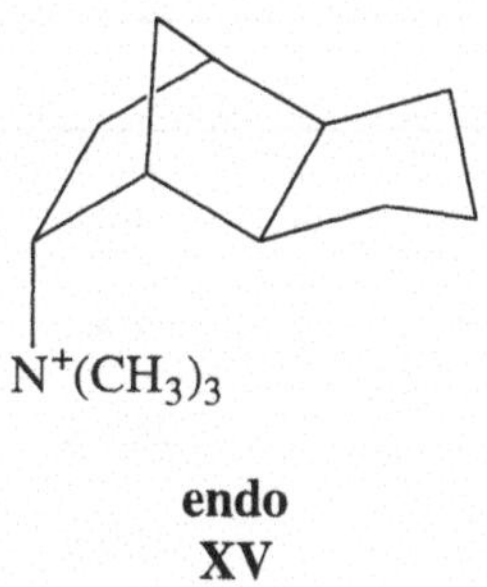

endo
XV

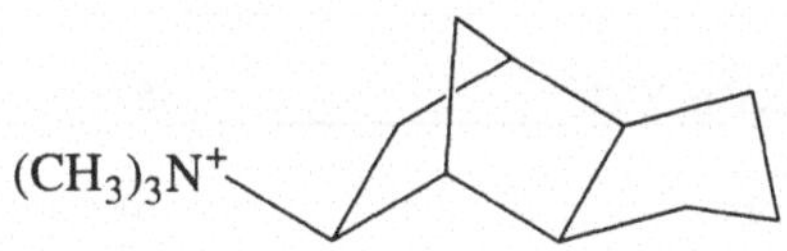

exo
XVI

The endo derivative produced zeolites SSZ-33 and SSZ-31 at high and low boron substitution, respectively. Similarly, the exo isomer yields boron zeolite beta or MTW in going from high to low boron substitution. Zones investigated the effect of increasing the sodium concentration in the synthesis gel and found that at Na/Si ratios of 0.1, microporous molecular sieves were formed. However, as the sodium concentration was increased to Na/Si ratios of 0.18-0.26, Kenyaite (a layered silicate) is obtained as the final product. It seems that at high concentrations, alkali-metal cations begin to compete with the organic molecules in a structure-directing role.

These examples illustrate the compromise that is needed in zeolite synthesis to obtain an optimum balance among the different effects that the components of the synthesis gel may have on the final product. Alkali-metal cations are needed to increase the rate of crystallization of high-silica zeolites, but excessive amounts of these components in the synthesis mixture will override the structure-directing effects of the organic molecules, and favor the formation of layered compounds.

4.2 SELECTIVITY OF A STRUCTURE-DIRECTING AGENT TOWARDS ZEOLITE FORMATION

As was illustrated in the section on clathrasils, one structure-directing agent can be used to synthesize several clathrate structures depending on the synthesis conditions. Zones et al. [52] studied how perturbations on the synthesis procedure of SSZ-33 (using molecule **XV**) dramatically affect the product formed. Table 3 shows the effect of mixing molecules **XV** and **XVI** in the synthesis gel. The synthesis of SSZ-33 is sensitive to the presence of molecule **XVI**. When the ratio of **XV/XVI**~5, zeolite beta starts to be detected in the product. The yield of crystalline product and ^{13}C MAS-NMR of these samples indicate that both molecules are occluded inside the pores of zeolite beta for

Table 3. Products from syntheses using mixtures of structure-directing agents **XV** and **XVI**.

Ratio XV to XVI	SSZ-33	B-Zeolite Beta
19	100%	0%
9	100%	0%
5.7	Major	Minor
4	Minor	Major
3	0%	100%

XV/XVI<3. These data indicate that molecule **XV** can stabilize SSZ-33 and zeolite beta. The data also indicate that the nucleation of crystals of SSZ-33 is very sensitive to the geometry of the structure-directing agent, but that the crystal growth process is more flexible and can be accomplished simultaneously by several structure-directing agents. If seeds of SSZ-33 are added to the same set of experiments, zeolite beta is not observed any more in the final product. The effect of seeding is to bypass the need for the spontaneous formation of SSZ-33 nuclei and effectively stops the formation of zeolite beta [52]. Harris and Zones [53] who investigated the synthesis of pure-silica nonasil, have reported a correlation between the tightness of the fit of the organic molecule in the cage and the rates of nucleation. The faster nucleation rates were observed for syntheses in which the agreement between the size of the molecule and the size of the cage was especially good, indicating that a tighter fit of the organic molecule in the clathrasil cage promotes faster nucleation rates.

5. Pure-Silica Large-Pore Materials

There are three known pure-silica, large-pore zeolites: ZSM-12 (MTW), SSZ-24 (AFI) and SSZ-31. All these zeolites can be synthesized under the same synthesis conditions except for the use of different structure-directing agents. MTW has relatively small elongated pores (5.5 x 5.9 Å) [3] and can be synthesized using several structure-directing agents (Table 4). The MTW structure is typical of high-silica zeolites and contains 4-, 5- and 6-rings. In contrast, the pure-silica polymorph of $AlPO_4$-5, SSZ-24, contains only 4- and 6-rings and a large 12-ring pore of 7.3 Å in diameter. In spite of these differences in shape and size, the structures of MTW and AFI appear very similar when they are viewed as an assemblage of tubular building units. From this perspective, the differences in the structures of MTW and AFI arise from assembling tubes, in two different forms, with walls formed only of 6-rings. According to the description of Kirchner and McGuire [54], both zeolites have the type 1 pores in which the pore axis is parallel to the edge of a 6-ring.

Comparison of the structure-directing agents for MTW and the structure-directing agents for AFI indicates two main differences: firstly, in general the structure-directing agents for MTW are formed from smaller organic molecules than the structure-directing agents for AFI (cyclohexane and benzene rings vs. adamantane) and secondly, the structure-directing agents for MTW are very flexible compared to the ones for AFI. The combination of these two factors suggests that the structure-directing agents used for the synthesis of AFI are large enough to prevent (by steric forces) the small elongated 12-ring pores of MTW from forming and seem to shift the kinetics of crystal nucleation from MTW to AFI.

SSZ-31 is a pure-silica zeolite synthesized using the structure-directing agents shown in Table 4. Sorption experiments suggest that SSZ-31 is a one-dimensional 12-ring pore system [55]. NCL-1 is also a pure-silica large-pore zeolite [56] synthesized with bis-(triethylammonium)butane. The X-ray diffraction patterns of NCL-1 are very similar to the pattern of SSZ-31. The structure-directing agents for SSZ-31 are also rigid and can be loosely described as containing a large polycyclic body with a protuberance formed by the N-methyl groups. These 'bumps' suggest that the structure-directing agents

Table 4. Structure-directing agents used for the synthesis of MTW, AFI (SSZ-24), and SSZ-31.

MTW[a]	AFI[b]	SSZ-31[c]

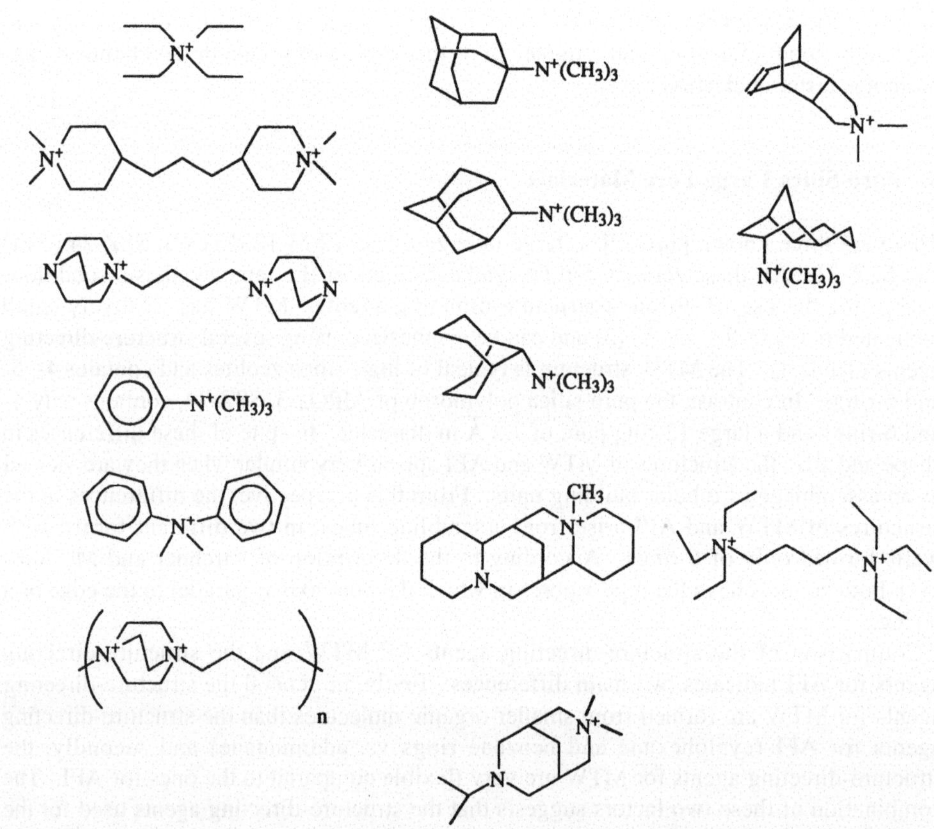

a. See ref. 13, 47.

b. See ref. 32, 56, and refs therein and Y. Nakagawa, *U. S. Patent 5,271,921* (1993).

c. See ref. 37, 57.

destabilize the formation of both MTW and AFI. In short, based on the size and shape of the structure-directing agent, the pores of SSZ-31 should have 'grooves' or 'pockets' to accommodate the shape of the organic guest molecules along the 1-dimensional pores.

It is noteworthy that, except for only one report (vide infra), large structure-directing agents have not given pure-silica materials with new, large, multidimensional pore systems. The reasons behind this observation are not understood, especially if contrasted with the easily synthesized multidimensional medium-pore silicalite-1 and silicalite-2. It can be speculated that T—O—T angles needed for the formation of a multidimensional pore system are not favored in pure-silica materials. Histograms of experimentally measured T—O—T angles (adapted from ref. 8) in Si—O—Si, Si—O—Al and the Si—O—B are shown in Figure 4. It can be observed that the variation in the Si—O—Si angle is larger than the variation in the Si—O—Al and Si—O—B. These histograms indicate that the substitution of Al or B for Si in the framework stiffens the T—O—T angle and shifts the optimum T—O—T angle to a value lower than expected in silicates.

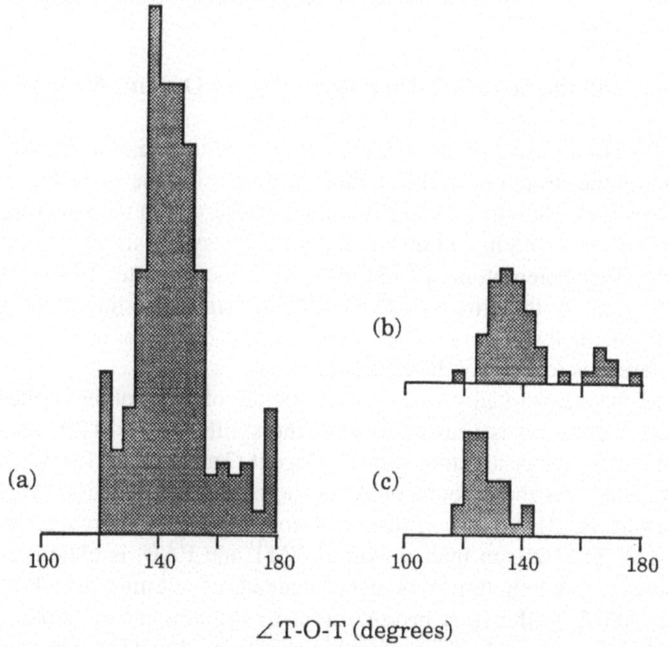

Figure 4. Histogram of experimentally observed T—O—T angles in a) Si—O—Si, b) Si—O—Al, c) Si—O—B, (adapted from ref. 9).

A second reason for the absence of pure-silica materials with intersecting large pores may be the effect of defects on the stability of the silicate framework. As suggested before, in pure-silica materials, the charge of the structure-directing agent is balanced by a deprotonated silanol group in the zeolite framework. For TPA in the MFI structure, this is equivalent to a defect for each 24 Si atoms. As the void volume of the zeolite framework is increased, the ratio Si/N^+ is drastically reduced, for example, for SSZ-33 synthesized using **XV**, Si/N^+~15 and for zeolite beta synthesized using TEA the Si/N^+~7. The charge of the structure-directing agent is largely balanced by B or Al in these two examples, but these numbers suggest that for certain Si/N^+, the density of defects necessary to balance the organic charge in the pure-silica material creates a large destabilization of the structure, thus inhibiting its formation. The exception to this rule was recently reported by van der Waal et al. [72], who synthesized the pure-silica polymorph of zeolite beta using dibenzyldimethylammonium hydroxide as structure-directing agent. The organic molecule contains a C/N=16, instead of C/N=8 in TEA, which therefore gives a Si/N^+~14. The number of defects in this preparation of zeolite beta would be about one half of the number of defects expected in a pure-silica sample of TEA. The smaller number of defects increases the stability of the final product and in this way allows the formation of the pure-silica polymorph of zeolite beta.

6. Heteroatoms and the Structure-Directing Effect of Organic Molecules

The addition of small amounts of tetrahedral atoms other than silicon (B, Al, Zn, etc.) has a striking effect on the structures of the zeolites formed using the *same* structure-directing agents. In this section, we will look at the changes observed in the structures of zeolites when Al, B, or Zn are added to an otherwise pure-silica synthesis gel. In addition to Al, B, or Zn, many other heteroatoms [2,13] have been incorporated in high-silica zeolite frameworks. In general, the effects of Ga and Fe are similar to the effect of Al. Ti has been found to have small structural effects, and Ge can substitute for Si. The reader may examine the references for more details [2,13].

An example of the effect of added Al on the structure of the zeolites formed when other variables are kept constant is illustrated with the syntheses of MTW and BEA using tetraethylammonium ions as structure-directing agent (Table 5). When the $SiO_2/Al_2O_3 >$ 50, MTW is formed. As the amount of Al is increased beyond $SiO_2/Al_2O_3 = 50$, the product observed is BEA, and if the fraction of Al is increased even further (SiO_2/Al_2O_3~15), ZSM-20, an intergrowth of EMT and FAU, is obtained. Analogous results are observed when boron is used instead of aluminum. For low boron concentrations, MTW is the final product and, as the amount of boron is increased, zeolite beta is then crystallized. This effect is not restricted to TEA, but is also observed for other structure-directing agents (see Table 5).

The structure-directing effect of other organic guest molecules that form large-pore pure-silica molecular sieves is also strongly coupled to the presence of heteroatoms in the synthesis gel. Trimethyl-2-adamantylammonium forms the one-dimensional AFI structure with pure-silica synthesis gels but forms SSZ-33 [57] when boron is added.

Table 5. Effect of aluminum, boron, and zinc on the structure of zeolites obtained using organic structure-directing agents.[a]

	SiO_2	SiO_2/Al_2O_3 <50	SiO_2/B_2O_3 <30	SiO_2/ZnO 20
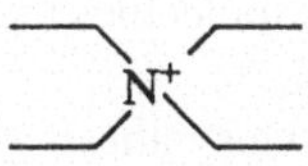	MTW	BEA	BEA	VPI-8
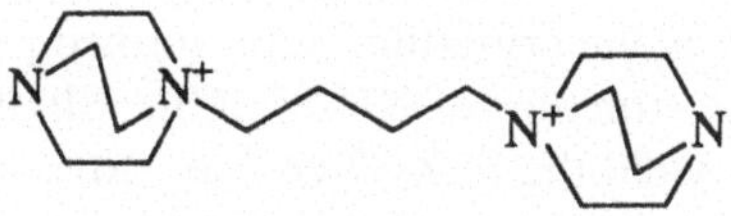	MTW	BEA	BEA	VPI-8
	MTW	MOR	BEA	Layered
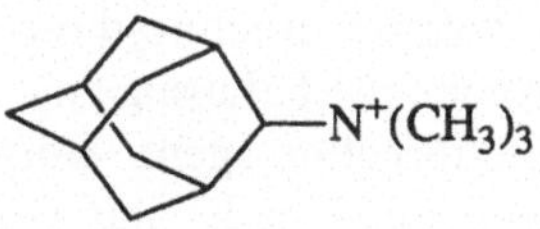	AFI	SSZ-25	SSZ-33	VPI-8
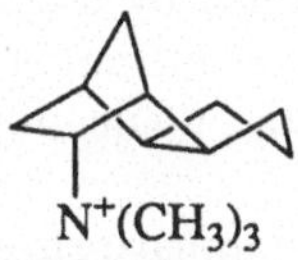	SSZ-31	MOR	SSZ-33	—[b]
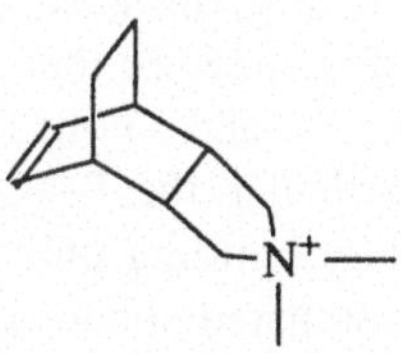	SSZ-31	SSZ-37	SSZ-33	—[b]

a. From references 37, 51, 56, 61, 62, 63. Some of these results are from experiments carried out in our laboratories and not yet published.

b. No product formed after several weeks.

SSZ-33 is a new zeolite with intersecting 10- and 12-ring pores that can be synthesized using several structure-directing agents [58] when boron is in the synthesis gel. Other organic molecules that form SSZ-31 with a pure-silica gel, also form SSZ-33 when boron is added to the reaction mixture (Table 5). For these cases, molecules that form different structures with pure-silica, form a material with the same framework structure when boron is included in the synthesis gel. On the other hand, the molecules that form MTW, only crystallize BEA when boron is added to the synthesis gel.

A different result is observed when aluminum is added to the synthesis gel if molecule **XIV** is used as the structure-directing agent. Neither BEA nor SSZ-33 are observed as the final product but rather a new material, SSZ-37, is the crystalline solid formed [59]. The structure of SSZ-37 seems to be similar to the structure of NU-87 [60]. Slightly bigger or slightly smaller organic molecules do not yield the SSZ-37 zeolite. Note that the effect of Al and B are not the same. Although the addition of Al or B when using TEA as the structure-directing agent gives BEA in both cases, the structures of the products formed when using the heterocyclic molecule **XIV** are very different. Also, note that the molecules that form SSZ-31 and SSZ-33 have not been reported to form any crystalline material when Al is substituted for B.

As is observed with aluminum and boron, the addition of zinc to a synthesis gel also radically changes the structure of the zeolite product. In all cases, however, the molecular sieve VPI-8 is observed, regardless of the organic structure-directing agent used for its synthesis. VPI-8 was first synthesized by Annen [61,62] using zinc acetate as the source of framework zinc and TEA as the structure-directing agent. Sorption data indicate that VPI-8 is a 12-ring, large-pore, 1-dimensional molecular sieve. The formation of VPI-8 in the presence of a variety of organic guest molecules suggests that in these syntheses, the structure of the final product is controlled by the divalent zinc cation and that the principal role of the organic molecule is the stabilization of the voids forming the pores [63]. The Zn atoms appear to inhibit the nucleation of zeolite structures, other than VPI-8.

Trivalent and divalent T atoms in the synthesis gel are bound to have strong influence on the syntheses of high-silica zeolites for several reasons. Firstly, the substitution of Al, B, or Zn for Si in the zeolite produce a negatively charged framework that will tend to coordinate much more strongly to the inorganic cations (Na, K, and so on) found in the synthesis gel. Secondly, the T—O bond lengths and, in particular, the T—O—T angles are expected to influence the formation of certain secondary building units such as 4-rings. Finally, instead of the formation of a charged, deprotonated silanol group in the zeolite structure to balance the charge of the organic structure-directing agent, the charge can be balanced by the anionic framework of the zeolite.

It is enlightening to compare the structures of MTW and BEA. The projection along the 12-ring pores of MTW is the same as the projection along the *a* and *b* axes of zeolite beta. Both structures are the result of the disorder stacking of silicate layers [35,64] and there are more 4-rings in the structure of BEA than in the structure of MTW. The presence of additional 4-rings in the structure of BEA may be regarded as the consequence of smaller T—O—Si angles promoted by the presence of Al or B. As was found in MTW and BEA, the projection along the 12-ring pores of AFI is the same as the projection along the 12-ring pores of SSZ-33. Also, the structures of SSZ-33 and BEA are related by the presence of similar secondary building units and both are highly faulted structures with open 3-dimensional pore systems.

The trends that have been outlined here have to be regarded as tentative. They are the result of a limited number of experiments using a small number of organic molecules and relatively narrow ranges of all the synthesis parameters. Further work, in particular, the structure solution of SSZ-31 and VPI-8 are needed in order to be able to understand further the observed experimental results.

7. Examples of Structure-Direction

The concepts of structure-direction enumerated above are now discussed in terms of recent examples. These examples show a complementary relationship between the geometry of the structure-directing agent and the geometry of the void space in the zeolite.

Zeolite SSZ-26 is an aluminosilicate molecular sieve with intersecting 10- and 12-ring pores [65]. It is synthesized using the propellane molecule **XVII** [66].

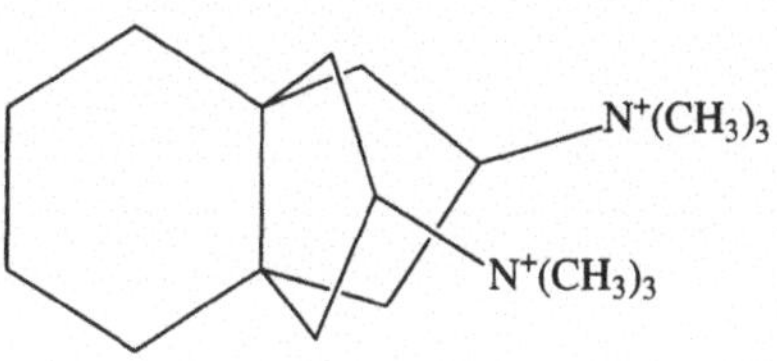

XVII

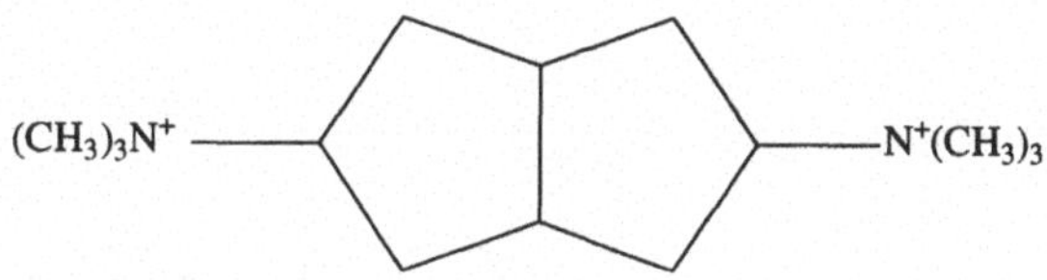

XVIII

Zones and Santilli [66] found that molecule **XVIII** formed MTW and argued that the inclusion of an additional ring as in **XVII** would force the formation of an intersecting pore system as illustrated in Figure 5. Thermogravimetric and chemical analyses indicate (based on the solution of the structure [59]) that there is one guest molecule per channel intersection. Lobo et al. [67] carried out molecular force-field calculations on the propellane molecule **XVII** inside the pores of SSZ-26 and found a very good agreement between the geometry of the pore intersection and the geometry of the organic molecule. The minimum-energy conformation of the molecule inside the pore intersections (Fig. 6) closely resembles the conformation suggested by Zones and Santilli [66] (Fig. 5). Therefore, the synthesis of SSZ-26 appears to be the first case known to the authors where the geometry of a pore-architecture has been designed *a priori* and exemplifies the usefulness of the concepts of structure-direction as an aid in the synthesis of new molecular sieves. This example, however, is not the only one where there is a very good fit between geometry of the guest molecule and the geometry of the pore.

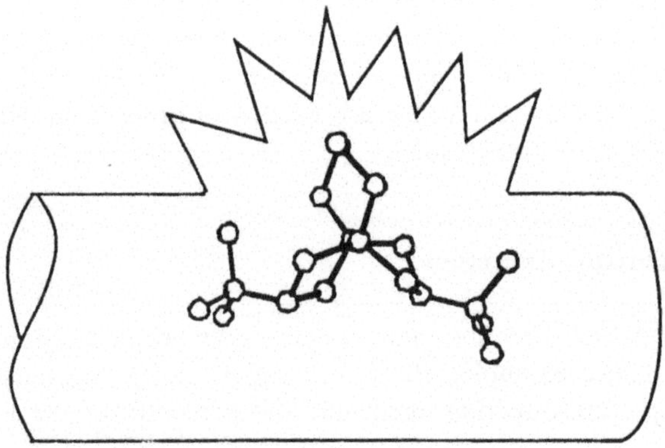

Figure 5. Proposed model for the formation of pore intersections using the propellane molecule **XVII**.

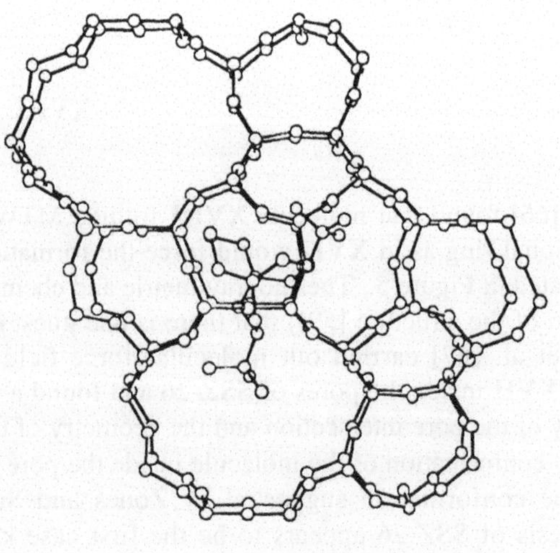

Figure 6. Minimum-energy conformation for the propellane molecule (**XVII**) in the pores of SSZ-26 as obtained from force-field calculations.

The structure of ZSM-18 (MEI) [68] is unique since it is the only aluminosilicate to contain 3-rings in the framework. With ZSM-18, a very good fit between the organic guest molecule and the zeolite cage is observed. Figure 7 shows an ORTEP drawing of the triquaternary ammonium molecule used as structure-directing agent inside the large cage of MEI. The rigidity, the size, and in particular, the three-fold axis of rotation of the organic molecule are the characteristics that favor the formation of the cage of ZSM-18. Hong et al. [19] have carried out CP/MAS ^{13}C and static NMR on this material and found that the triquaternary ammonium ion is tightly held inside the MEI cage. Recently, Schmith and Kennedy [73] reported the synthesis of ZSM-18 using molecules **XIX** and **XX.**

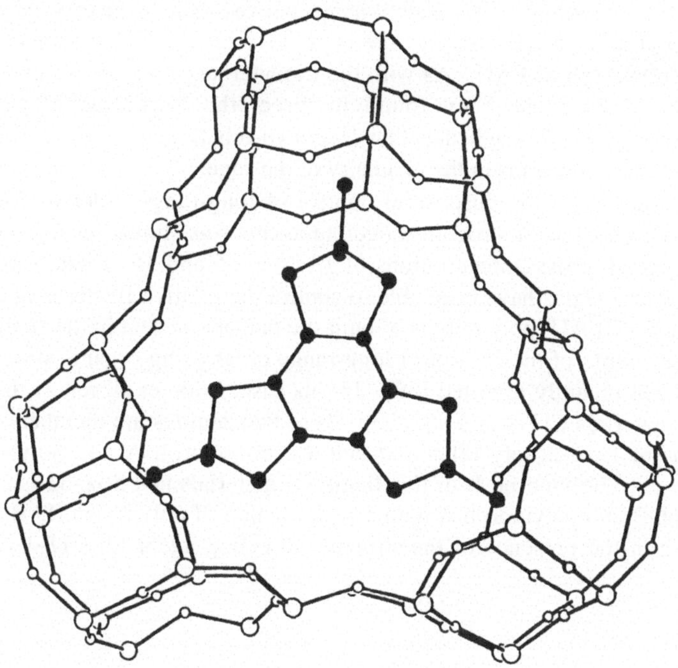

Figure 7. Position of the triquaternary ammonium molecule used as structure-directing agent in ZSM-18.

Their rationale was based on keeping the same total charge and symmetry, ease of synthesis and retaining only acyclics. The various possibilities were screened using molecular dynamics, which reduced the list to cations **XIX** and **XX**. This paper is the first example of the use of molecular dynamics to 'design' a structure-directing molecule.

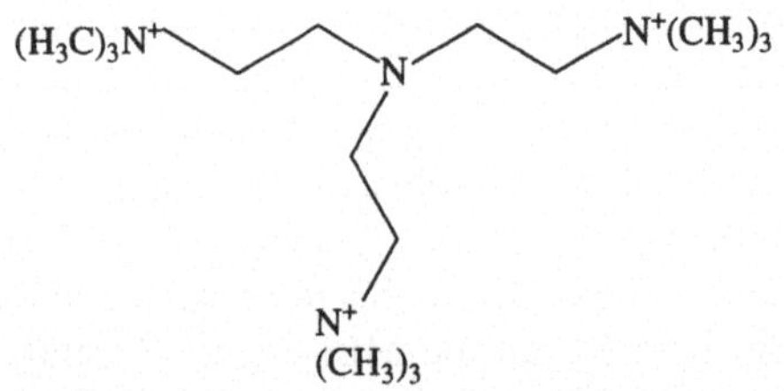

XIX

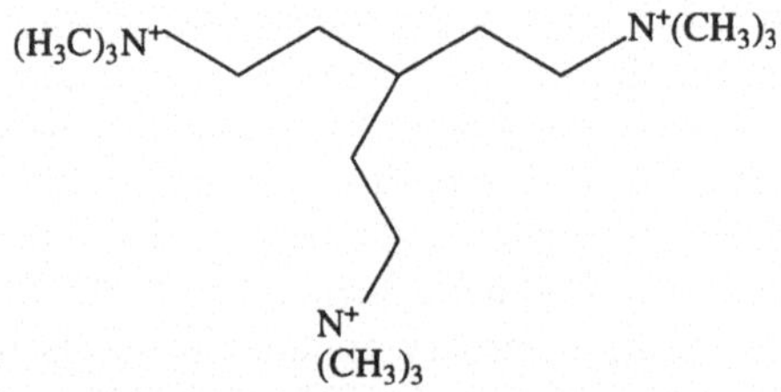

XX

Another good example of the relationship between the geometry of the structure-directing agent and the geometry of the cage is 18-crown-6 in the cages of EMT, the hexagonal polymorph of FAU. As with the triquaternary ammonium molecule in MEI, only the 18-crown-6 has been found to direct the formation of EMT. Recent crystallographic work [71] indicates that 18-crown-6 is found in the small cages of EMT and approximately conforms to the geometry of the cage.

The first example of the purposeful control of long-range order in the synthesis of zeolites through the use of structure-directing agents was shown by Arhancet and Davis [70]. By specifying the concentrations of 18-crown-6 and 15-crown-5 in the synthesis gel, Arhancet and Davis have been able to control the relative fractions of the hexagonal (EMT) and cubic (FAU) polymorphs of faujasite that are present in the zeolite product.

A second example of the control of long-range order using organic structure-directing agents has been recently reported [58]. In this case, Lobo et al. report the synthesis of two related materials CIT-1 and SSZ-33. These two zeolites are members of a family of intergrowths of two closely related structures, polymorph A and polymorph B [57]. CIT-1 is the pure polymorph B of this family of materials and SSZ-33 is an intergrowth of polymorph B and polymorph A with a 70% fraction of polymorph B. By synthesizing the materials in the presence of the structure-directing agent for CIT-1 (**XXI**) and the

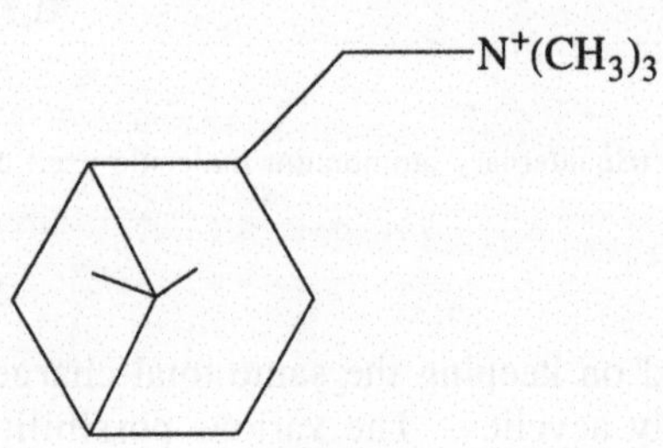

XXI

structure-directing agent for SSZ-33 (**XV**), the fraction of polymorph B in the final material can be carefully controlled from materials which are 70% of polymorph B and 30% polymorph A (SSZ-33) to materials which are 100% polymorph B (CIT-1) [57]. Figure 8 shows the experimental XRD patterns of materials with different fractions of polymorph B. These two examples also illustrate how the long-range order in zeolites can be controlled using organic guest molecules.

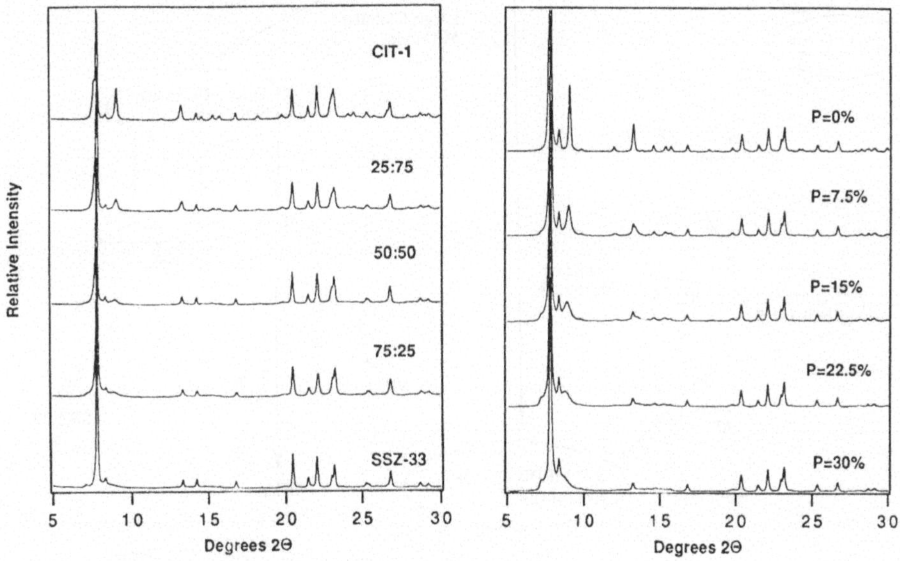

Figure 8. Experimental (A) and calculated (B) XRD patterns for intergrowths of polymorph A and polymorph B of SSZ-33.

8. Summary

We have discussed the myriad of factors that influence, in one way or another, the synthesis of zeolites. In Figure 9, we attempt to illustrate the effects that these factors have on the structures formed.

In the absence of structure-directing agents, silicate syntheses form dense (quartz, tridymite, cristobalite) or layered materials. Using only silica as the source of tetrahedral atoms, small molecules - amines in particular - direct the formation of clathrasils. The structure of the clathrasil obtained depends on the size of the molecule, on the synthesis temperature and to a minor extent on the specific geometry of the molecule. Linear molecules form in general 1-D, 10-ring pore molecular sieves. The most important factor that defines one structure from the several that can be synthesized with pure-silica, appears to be the temperature.

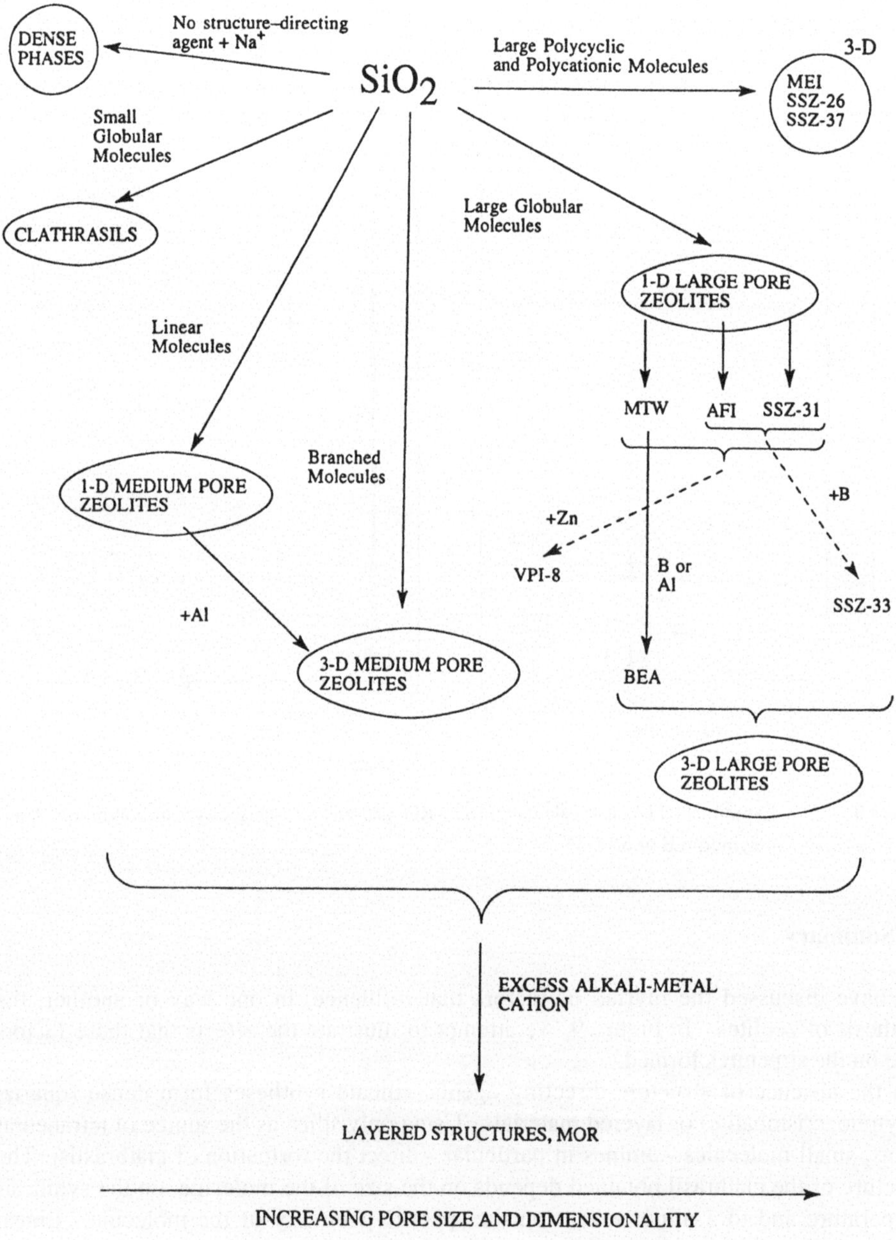

Figure 9. Summary of the trends between structure-directing agents and zeolite structures.

Branched molecules (TPA and TBA) form multidimensional medium-pore zeolites while large polycyclic molecules produce MTW, AFI, or SSZ-31 depending on their size and shape. No multidimensional, pure-silica, large-pore zeolites have been synthesized. Nakagawa, however, has indicated that, based on adsorption experiments [71], the pure-silica zeolite SSZ-35 is a multidimensional large-pore material. The structure of SSZ-35 must be solved to answer this question.

The selectivity of the structure-directing agents towards a specific zeolite structure is increased as the size of the molecule is increased. The most selective molecules are large, i.e., are formed of more than 16 carbon and nitrogen atoms, and contain two or three charges per molecule.

Small amounts of trivalent (Al, B, Ga, Fe) or divalent (Zn, Be) tetrahedral atoms have striking effects on the structures of zeolites. In general, the structures formed have multidimensional pore systems, except for Zn that (thus far) direct the crystallization of VPI-8, regardless of the structure-directing agent. Often, the projection along the 12-ring pore of 1-D pure-silica zeolites is the same as the projection along the 12-ring pores of multidimensional materials obtained after the addition of Al or B.

The thermodynamic stability of all pure-silica molecular sieves is relatively constant. This is an encouraging fact that suggests that the synthesis of many other zeolites with novel pore architectures is feasible.

The concepts of structure-direction have shown to be of tremendous power as guiding principles for the synthesis of new molecular sieves. The results presented in this review indicate that no limits are apparent to the number of structures that can be synthesized. The results also suggest that new zeolite structures will require the synthesis of larger structure-directing agents than the ones that have been prepared so far.

Acknowledgments

We are grateful to Chevron Research and Technology Company for financial support. We also acknowledge Dr. Y. Nakagawa from Chevron Research and Technology Company for helpful discussions and for supplying one of the amines used in the synthesis of clathrasils.

References

1. M. E. Davis, *Ind. Eng. Chem. Res.*, **30**, 1675 (1991).
2. J. M. Newsam, in *Solid State Chemistry: Compounds*, A. K. Cheetham and P. Day (Eds), Oxford University Press, New York (1992).
3. W. M. Meier and D. H. Olson, *Atlas of Zeolite Structure Types*, Butterworths-Heineman, Stoneham, MA (1992).
4. J. V. Smith, *Chem. Rev.*, **88,** 149 (1988).
5. J. Higgins, *Catal. Today*, in press.

6. L. B. McCusker, C. Baerlocher, E. Jahn, and M. Bulow, *Zeolites*, **11**, 308 (1991)
7. M. E. Davis, C. Montes, P. E. Hathaway, and J. P. Arhancet, *J. Am. Chem. Soc.*, **111**, 3919 (1989).
8. G. V. Gibbs and E. P. Meagher, in *Structure and Bonding in Crystals*, H. O'Keeffe and A. Navrotsky (Eds), Academic Press, New York, Vol 1, pp 195-223 (1980).
9. M. D. Newton and G. V. Gibbs, *Phys. Chem. Miner.*, **6**, 221 (1980).
10. M. E. Davis and R. F. Lobo, *Chem. Mater.*, **4**, 756 (1992).
11. B. M. Lok, T. R. Cannon, and C. A. Messina, *Zeolites*, **3**, 282 (1983).
12. J. C. Jansen, in *Introduction to Zeolite Science and Practice*, H. van Bekkum, E. M. Flanigen, and J. C. Jansen (Eds), Elsevier, Amsterdam, pp 77-135 (1991).
13. R. Szostak, *Handbook of Molecular Sieves*, Van Nostrand-Reinhold, New York, (1992).
14. H. Gies, in *Inclusion compounds,* Academic Press, London, Vol. 5, pp 1-35 (1991).
15. H. Gies and B. Marler, *Zeolites*, **12**, 42 (1992).
16. F. Liebau, in *Silicon Chemistry*, E. R. Corey, J. Y. Corey, and P. P. Gaspar (Eds) Ellis Horwood Limited, England, pp 308-323 (1988).
17. D. M. Bibby and M. P. Dale, *Nature*, **317**, 157 (1985).
18. J. Keisper, C. J. J. den Ouden, and M. F. M.Post, *Stud. Surf. Sci. Catal.,* **49**, 237-248 (1989)
19. C. Y. Chen, personal communication (1994).
20. J. J. Pluth and J. V.Smith, *Stud. Surf. Sci. Catal.,* **49**, 835-844 (1989).
21. F. Liebau, *Structural Chemistry of Silicates*, Springer-Verlag, Berlin (1985).
22. S. B. Hong, H. M. Cho and M. E. Davis, *J. Phys. Chem.*, **97**, 1622 (1993).
23. S. B. Hong, H. M. Cho and M. E. Davis, *J. Phys. Chem.*, **97**, 1629 (1993).
24. S. Burkett and M. E. Davis, *Microporous Mater.*, **1,** 265 (1993).
25. H. Gies, *Z. Kristal.*, **164**, 247 (1983).
26. H. Gerke and H. Gies, *Z. Kristal.*, **166**, 11 (1984).
27. G. Miehe, T. Vogt, H. Fuess, and U. Muller, *Acta Cryst.*, *Sect. B*, **49**, 745 (1993).
28. H. Gies, *Z. Kristal.*, **167**, 73 (1984).
29. H. Gies, *Z. Kristal.*, **175**, 93 (1986).
30. L. B. McCusker, *Acta. Cryst.*, *Sect. A*, **47**, 297 (1991).
31. L. B. McCusker, *J. Appl. Crystallography*, **21**, 305 (1988).
32. R. F. Lobo and M. E. Davis, *Microporous Mater.*, accepted.
33. R. M. Dessau, K. D. Smith, G. T. Kerr, G. L. Woolery and L. B. Alemany, *J. Catal.*, **104**, 484 (1987).
34. C. A. Fyfe, H. Gies, G. T. Kokotailo, B. Marler, and D. E. Cox, *J. Phys. Chem.*, **94**, 3718 (1990).
35. C. A. Fyfe, H. Gies, G. T. Kokotailo, C. Pasztor, H. Strobl, and D. E. Cox, *J. Am. Chem. Soc.*, **111**, 2470 (1990).
36. F. Liebau, *Structural Chemistry of Silicates,* Springer-Verlag, Berlin, p 242 (1985).
37. Y. Nakagawa and S. I. Zones, in *Synthesis of Microporous Materials*, M. L. Occelli and H. Robson (Eds), Vol 1, Van Nostrand Reinhold, New York, pp 222-239 (1992)

38. C. D. Chang and A. T. Bell, *Catal. Lett.*, **8**, 305 (1991).
39. R. M. Barrer, *Hydrothermal Synthesis of Zeolites*, Academic Press, London, p 359 (1982).
40. I. Petrovic, A. Natvrotsky, S. I. Zones, and M. E. Davis, *Chem. Mater.*, **5**, 1805 (1993).
41. N. J. Henson, A. K. Cheetham, and J. D. Gale, *Chem. Mater.*, submitted (1994).
42. X. Hu and W. Depmeier, *Z. Kristal.*, **201**, 99 (1992).
43. A. Shimizu and Y. Taniguchi, *Bull. Chem. Soc. Jpn.*, **63**, 1572 (1990).
44. A. Shimizu and Y. Taniguchi, *Bull. Chem. Soc. Jpn.*, **63**, 3295 (1990).
45. D. T. Griffen, *Silicate Crystal Chemistry*, Oxford University Press, Oxford, p 15 (1992).
46. C. B. Khouw, H. X. Li, C. B. Dartt, and M. E. Davis, *ACS Symp. Ser.*, **523**, 293 (1993).
47. M. Goepper, H. X. Li, and M. E. Davis, *J. Chem. Soc., Chem. Commun.*, 1665 (1992).
48. P. Brady and J. V. Walther, *Chem. Geol.*, **8**, 253 (1990).
49. P. M. Dove and D. A. Crerar, *Geochim. Cosmochim. Acta*, **54**, 955 (1990).
50. S. I. Zones, *Microporous Mater.*, in press.
51. S. I. Zones and Y. Nakagawa, *Microporous Mater.*, in press (1994).
52. S. I. Zones Y. Nakagawa, G. Yuen and T. Harris, *J. Am. Chem. Soc.*, submitted.
53. T. V. Harris and S. I. Zones, in *Proceedings of the 10th International Zeolites Conference., Garmisch-Partenkirchen, Germany*, Elsevier, Amsterdam, Vol A, pp 29-36 (1994).
54. R. M. Kerchner and N. K. McGuire, in *Extended Abstracts and Program, 9th International Zeolite Conference*, J. B. Higgins, R. von Ballmoos and M. M. J. Treacy (Eds), Butterworth-Heinemann, Stoneham, MA (1992).
55. S. I. Zones, L. T. Yuen, Y. Nakagawa, R. A. Van Nordstrand and S. D. Toto, in *Proceedings of the 9th International Zeolites Conference*, R. von Ballmoos, J. B. Higgins and M. M. Treacy (Eds) Butterworth-Heinemann, Stoneham, MA, Vol.1, pp 163-170 (1993).
56. R. Kumar, K. Ramesh Reddy, A. Raj and P. Ratnasamy, in *Proceedings of the 9th International Zeolites Conference*, R. von Ballmoos, J. B. Higgins and M. M. Treacy (Eds), Butterworth-Heinemann, Stoneham, MA, Vol.1, pp 189-196 (1993).
57. R. F. Lobo, M. Pan, I. Chan, H. X. Li, R. C. Medrud, S. I. Zones, P. A. Crozier, and M. E. Davis, *Science*, **262**, 1543 (1993).
58. R. F. Lobo, S. I. Zones, and M. E. Davis, in *Proceedings of the 10th International Zeolites Conference., Garmisch-Partenkirchen, Germany*, Elsevier, Amsterdam, Vol A, pp 461-468 (1994).
59. Y. Nakagawa, *U.S. Patent 5,254,514* (1993).
60. Y. Nakagawa, in *Proceedings of the 10th International Zeolites Conference., Garmisch-Partenkirchen, Germany*, Elsevier, Amsterdam, Vol A, pp 323-330 (1994).
61. M. J. Annen, Ph.D. Thesis, Virgina Polytechnic Institute, Blacksburg, VA (1992).

62. M. J. Annen and M.E. Davis, *Microporous Mater.*, **1**, 57 (1993)
63. M. A. Camblor, S. I. Zones and M. E. Davis, in preparation.
64. J. M. Newsam, M. M. J. Treacy, W. T. Koetsier and C. B. de Gruyter, *Proc. R. Soc. Lond., Ser. A,* **420** 375 (1988).
65. S. I. Zones, M. M. Olmstead, and D. S. Santilli, *J. Am. Chem. Soc.,* **114**, 4195 (1992).
66. S. I. Zones and D. S. Santilli, in *Proceedings of the 9th International Zeolites Conference*, R. von Ballmoos, J. B. Higgins, and M. M. Treacy (Eds), Butterworth-Heinemann, Stoneham, MA, Vol.1, pp 171-179 (1993).
67. R. F. Lobo, M. Pan, I. Chan, S. I. Zones, P. A. Crozier and M. E. Davis, *J. Phys. Chem.*, in press.
68. S. L. Lawton, and W. J. Rohrbaugh, *Science*, **247**, 1319 (1990).
69. L. B. McCusker, *Microporous Mater.,* in press.
70. J. P. Arhancet and M. E. Davis, *Chem. Mater.*, **3**, 567 (1991).
71. Y. Nakagawa, *U.S. Patent 5,271,921* (1993).
72. J. C. van der Waal, M. S. Rigutto, and H. van Bekkum, *J. Chem. Soc., Chem. Commun.*, 1241 (1994).
73. K. D. Schmith and G. J. Kennedy, *Zeolites*, in press.

Abstract. The concepts of structure-direction in the synthesis of clathrasils and high-silica molecular sieves are reviewed. The effects of size, geometry, and chemical nature of the organic structure-directing agent on the crystalline structures that are formed are discussed beginning with clathrasils (0-dimensional pore systems) and ending with 12-ring zeolites with 3-dimensional pore systems. Emphasis is focused on the energetic interactions between the organic guest and the inorganic framework. The energetic stability of porous frameworks is compared to the stability of dense pure-silica phases and the effects of trivalent (Al, B) and divalent (Zn) tetrahedral heteroatoms on the structure of zeolites formed is reviewed. The application of structure-directing concepts are described using the syntheses of ZSM-18 and SSZ-26 as examples, and the control over long-range order in zeolites by structure-directing effects is illustrated by the purposeful variation of the stacking probability of SSZ-33–CIT-1 and FAU–EMT intergrowths.

Journal of Inclusion Phenomena and Molecular Recognition in Chemistry **21**: 79–112, 1995.

Structural Case Studies of Inclusion Phenomena in Zeolites: Xe in RHO and Stilbene in ZSM-5

JOHN B. PARISE

1. Introduction

Zeolite molecular sieves, firmly established in industry as selective heterogeneous catalysts for cracking, reforming, and separations, have more recently been used as platforms for the inclusion chemistry described in this volume. Their unique structures [1] result in the shape selectivity [2] desirable in both classes of applications. The characterization of the position and orientation of absorbed species, their preferred sorption sites, and their interactions with the host framework are fundamental to a proper understanding of the chemistry of zeolites.

X-Ray, neutron, and electron [3] scattering are recognized as the most appropriate and potentially unambiguous tools available for the study of atomic structure, in both crystalline and non-crystalline materials. These techniques are not expedient, but they are often the most efficacious. Zeolitic materials present a number of special problems in crystallographic investigations [4,5] which become magnified in attempting an accurate description of the structure of zeolite/inclusion complexes. These problems include poor crystallinity, pseudo-symmetry, positional and orientational disorder, and the uncertainty in composition. Fortunately, recent developments in scattering [6-9], spectroscopic [10,11], and molecular modeling techniques [12,13] have mitigated against some of these problems. The latter two techniques are discussed in detail elsewhere in this volume.

1.1 RECENT INNOVATIONS IN X-RAY AND NEUTRON TECHNIQUES

Laboratory-based structural studies will continue to play important roles in developing our understanding of zeolite-inclusion phenomena. However, the advent of more powerful X-ray and neutron sources has provided opportunities to perform experiments revealing a level of detail not available in the laboratory. For example, synchrotron X-ray sources provide bright, tunable beams. This allows the study of smaller single crystals and smaller quantities of powder. The ability to tune energy has made routine X-ray absorption and anomalous scattering measurements [14], which were either laborious or impossible to perform using laboratory-based sources [15]. The high angular resolution available at these sources has aided the ab initio solution of complex structures from powdered samples [16]. Investigations of structural changes in real time are feasible and one example is discussed in this chapter.

1.1.1 *Single Crystal Studies*

When crystals are suitably sized, homogeneous, and ordered, single crystal crystallography is the most powerful, reproducible, and testable technique for the determination of the structure of zeolite/inclusion complexes. Many single crystal studies have concentrated on the interaction between the organic template species, added in hydrothermal crystallization of molecular sieves, and the framework; a recent review of this work in the aluminophosphate systems is given by Pluth and Smith [17]. Another excellent recent example is the work of Koningsveld and co-workers [18-22]. Carefully prepared single crystals of ZSM-5 [19] were examined and the positions and orientations of absorbed species, along with subtle framework distortions, were determined. One important result from these studies was the comparative ease of detection of pseudo-symmetry in the ZSM-5 system using single crystal techniques; subtle changes in framework symmetry can be extremely difficult to detect using spectroscopies and routine powder diffraction. Studies using single crystal diffraction can provide a confident point of departure for subsequent investigations with powder diffraction [23].

The size range of samples suitable for single crystal diffraction has been decreasing steadily with the availability of brighter laboratory and synchrotron sources. Rotating anode sources operating at 18 kW have been available for some time and have allowed routine data collection on zeolite crystals in the 30 μm size range. Recently, these high power conventional sources have been coupled with area detectors, such as image plates and charge coupled detectors (CCD's). The problems associated with X-ray fluorescence not withstanding, the dynamic range and sensitivity of the former devices make them particularly well-suited to the study of small single crystals of inclusion complexes at low temperatures. For example, we have absorbed a number of gases into 35 μm crystals of zeolite X and have been able to locate density within the channels related to the absorbed species. The data for these studies were collected at 120 K using a 6 kW source and a commercially available imaging-plate system. A full data set was collected in 2 hours raising the possibility of real-time studies involving sorption and desorption as a function of temperature and pressure. With the advent of more powerful conventional and synchrotron sources, and more sensitive detection systems, the line between single crystals and powders continues to blur.

1.1.2 *Powder Diffraction Studies*

Because suitable single crystals of all systems of interest are not available, powder diffraction [24] has gained a new prominence in the study of the structure of zeolite/inclusion complexes. This allows the study of many molecular sieves which have not been synthesized as single crystals. Further, the processes of dehydration and loading of the materials to be included can cause a decrease in crystal quality, making single crystal studies difficult to perform. Powder diffraction also has certain inherent advantages, including the comparative ease with which the molecular sieve can be loaded and handled.

Recent investigations of the structures of zeolite/inclusion complexes have benefited from the availability of dedicated synchrotron radiation sources, coupled with the design of flexible powder diffractometers at these facilities [6,16]. The advantages for the study of inclusion complexes are: high brightness, high peak to background discrimination and routine cooling capabilities to 20 K using Displex® refrigerators. Often molecules included into zeolitic materials disorder above a "freezing point"; data collected at low temperatures give the best chance of locating these species. Other improvements include the design and construction of readily interchangeable beamline-optics to optimize resolution to the problem at hand and more sensitive energy discriminating detectors, such as the position sensitive detector (PSD) recently described by Smith [25]. Provided some resolution can be sacrificed, the PSD is a powerful tool for the study of the zeolite/inclusion complex as a function of temperature or loading.

Most inclusion complexes involve some degree of disorder, either static or dynamic; in this case the powder diffraction experiment suffers an inevitable loss of information. The supplementing of diffraction data with information from NMR, molecular modeling, and other spectroscopic techniques is important in this work.

1.1.3 *EXAFS*

Diffraction techniques are most suited to structural studies of ordered materials. If the occluded species bind to a variety of sites on the walls of the zeolitic cavity, a more local probe is often appropriate. Extended X-ray absorption spectroscopy (EXAFS) is a popular technique for exploring the structure of occluded species in zeolites and has been used to postulate the geometry of these species and their relationship to the host framework. The ability of the EXAFS to yield local structural information has been reviewed elsewhere [26]. Essentially, the amplitude and frequency of oscillations, observed on the high energy side of the X-ray absorption edge for a particular element, are a function of the local environment for that element. By taking the Fourier transform of these oscillations, one obtains a radial distribution function. Refinement of parameters to best fit the observed oscillations for each shell of neighbors surrounding the absorber atom yields distances, coordination numbers, and displacement parameters. Prior knowledge of composition and possible geometries can be useful in translating this information into a credible model. For example, selenium exists in a number of allotropes, a helical chain form, and a number of ring structures. The allotrope stabilized in a variety of zeolitic hosts was shown to depend upon the geometry of the pore structure [27]. The EXAFS results, when combined with observations using optical absorption spectroscopy and NMR, showed the local environments for Se are consistent with the helical form of this element when it is constrained in zeolites containing 1-D channels [27] (see Chapter 10 in this volume).

More recent examples of the use of EXAFS, in combination with a battery of complementary techniques, are reported by Ozin and co-workers [28 and references therein]. The systems of interest in these studies are stabilized semiconductor nanoclusters which have been induced to grow topotactically within the pores of selected zeolites. Disorder and partial occupancy of the available pore spaces required the interpretation of data, from a variety of sources, to deduce the nature of the semiconductor nanoclusters and their relationship to the host framework. Some of the techniques employed included reflectance spectroscopy, luminescence, elemental analysis, Rietveld refinement, as well as, EXAFS.

1.1.4 *Neutron Scattering*

Neutron scattering is the tool of choice when the positions of hydrogen atoms are of importance [17]. Recently the self-assembly of species in aluminosilicate gels impregnated with Pr_4N^+ ions was studied using small angle neutron scattering [29]. The scattering length densities of the gel particles and their texture were determined by using contrast variation methods [7]. Gels formulated with Pr_4N^+ template molecules promptly transform into an amorphous "embryonic" structure, and crystallization to zeolite ZSM-5 ensues via a solid hydrogel transformation mechanism. Gels formulated from colloidal silica without template show different scattering behavior from which a liquid-phase transport mechanism was inferred. Along with neutron spectroscopic studies of molecules trapped in zeolites using energy-loss spectroscopy, such investigations offer much promise for characterizing the interactions of absorbed species in microporous materials [7,29,30].

In this chapter, two systems studied using synchrotron X-ray powder diffraction are described. In both cases, the strengths of diffraction for determining site occupation, position and orientation are emphasized. In both studies, the importance of complementary analytical and spectroscopic techniques, along with insights provided by theoretical calculations and molecular modeling, are emphasized.

2. Xenon in Zeolite Rho

Distortions of the framework along with the type and number of adsorption sites, can affect the performance of microporous catalysts. Because the effects of physical and chemical changes on the framework of aluminosilicate RHO (Figure 1) are large [31-51] and the number of adsorption sites is apparently limited [52-56], it is an excellent model system for studying the interactions between zeolite frameworks and absorbed species.

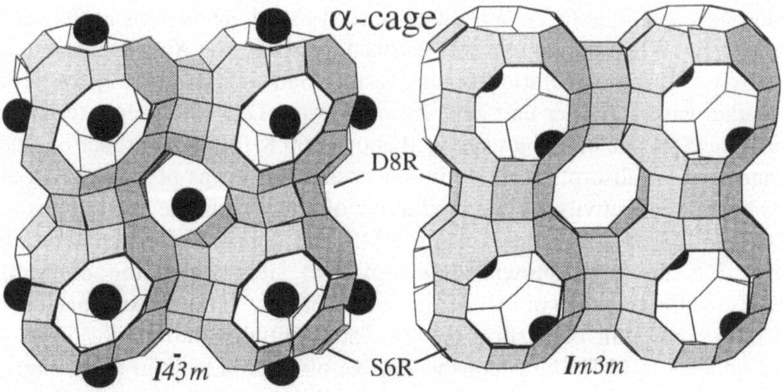

Figure 1. Representation of the acentric (left) and centric forms of zeolite rho. The acentric form is favored at low temperature, with exchange of small non-polarizable cations, and upon dehydration [39]. Upon heating, many divalent ion-exchanged forms undergo cation relocation. In the case represented here, Cd^{2+} moves from the center of the double 8-ring (D8R) to the wall of the α-cage near the single 6-ring (S6R) site.

Some of the most intriguing changes in RHO occur upon heating; one of these involves cation relocation [49]. Movement of exchangeable cations upon dehydration is a response to a loss of the H_2O ligand and a consequent need to satisfy coordination requirements [57-61]. In RHO, this movement can be dramatic [49,50] and affords an opportunity to study the interaction of absorbed species with mobile, exchangeable cations.

2.1 XENON IN H-RHO

As a prelude to studying the absorption and controlled release of Xe in the "trap-door" system Cd-rho described below, we decided to test the sensitivity with which we could locate Xe in H-rho. Diffraction studies of RHO [44-48] indicate it can adopt either a high or lower symmetry structure (Figure 1) depending on the conditions of treatment [36], hydration state, and extra-framework cations exchanged into the structure [37]. This structural change is accompanied by an elliptical distortion of the D8R pore opening (Figure 1), the degree of which is dependent on the unit cell parameters [48], and the nature of the framework cations [51]. A number of studies have examined the relationship between the location of Xe in zeolite rho and such parameters as the number of Xe per unit cell, the temperature, and the degree of ion-exchange.

Working with a sample of RHO containing about 6 Xe per unit cell, or one for each D8R in the unit cell (Figure 1), Wright et al. [55] found Xe adsorbs at the center of the D8R-site at 5 K. When the sample was warmed to 245 K, the Xe was located at the S8R site (Figure 1). In an elaboration of this result, Gameson et al. [62] found that for a sample loaded with 2 Xe per unit cell, Xe prefers the D8R-site at low temperature and migrates to the S8R site on the α-cage wall above 160 K [62]. A number of studies have concentrated on the absorption of Xe in ion-exchanged forms of RHO in an attempt to clarify its unusual selectivity for the production of dimethylamine in comparison to other 8-ring zeolites [63,64]. In studies using NMR and temperature dependent absorption [54,56,65], Xe has been shown to occupy two sites depending on temperature, Xe-loading, and the presence of extra-framework cations. For ion-exchange of more than 2.67 Cs atoms per unit cell, sorption was "substantially inhibited" [54]. In recent theoretical studies [52,53], the occupancy by Xe of a second site in the α-cage, near the single six-ring (S6R, Figure 1), has been predicted to occur after the 8R-sites have been fully occupied. One intriguing possibility raised by this work is the existence of ordered Xe-clusters at very high loadings [53].

To test the capabilities of the diffraction experiment to locate Xe in zeolite H-rho, we chose a series of samples with an accurate chemical analysis and with well-determined absorption characteristics [54,56]. For this purpose, it was imperative that any residual Cs^+, which is extremely difficult to remove completely by ion-exchange [54], be accounted for; Cs^+ and Xe, since they are isoelectronic, are indistinguishable in the X-ray experiment. Xe was loaded into approximately 0.3 g batches of H-rho. The chemical analysis obtained for this sample, based on inductively coupled plasma spectroscopy [54], showed it to have the unit cell formula $H_{10.77}Na_{0.48}Cs_{0.75}$-$Al_{12}Si_{36}O_{96}$; the H^+ are added to the formula to charge balance the analyzed value for Al, which is presumed to occupy only framework positions. Samples were loaded at 1, 50, 100, 200 and 400 torr in glass capillaries 1 mm in diameter. The isotherm for Xe absorption in this system has been determined at 300 K allowing the calculation of the number of Xe expected to sorb per unit cell (Figure 2). Three sites can host Xe: the D8R-site at the center of the octagonal prism, the S8R near the surface of the α-cage and about 1.5 Å from the D8R-site, and the S6R on the surface of the α-cage (Figure 1).

X-Ray powder diffraction data for Xe-loaded H-rho were collected at 300 K at the National Synchrotron Light Source (NSLS) beamline X-7A. This facility can be configured to operate in a variety of diffraction geometries by replacing optic elements and detectors [6,16]. For this experiment, a positron sensitive detector designed and built at the Brookhaven National Laboratory (BNL) [25] was used. The unique features of this detector design have been described by Smith [25] and include exceptional angular resolution and peak-to-background discrimination; the ability to discriminate against fluorescent background is particularly effective in obtaining data suitable for structural refinement [66].

The data were analyzed using the Rietveld Technique [66] implemented in the GSAS (Generalized Structure Analysis System) suite of programs [67]. The diffraction data could be indexed on the basis of a body centered cubic cell approximately 15 Å on edge, suggestive of the ***Im*$\bar{3}$*m*** symmetry structure (Figure 1) of RHO [37,48,51].

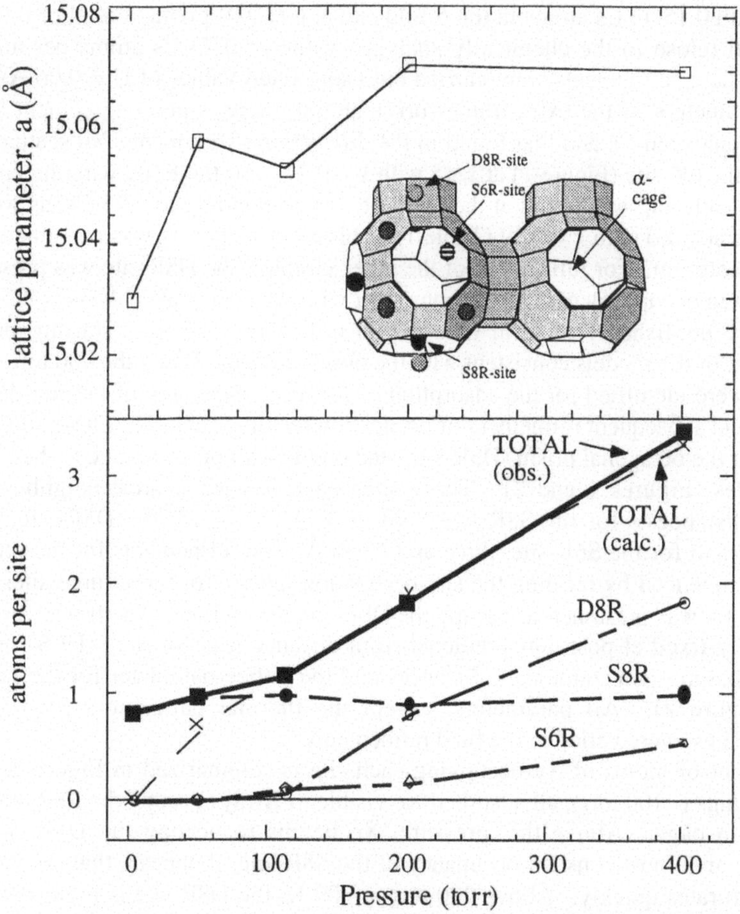

Figure 2. Variation in the lattice parameter (upper) and a summary of the results of site occupancy refinements for H-rho (0.75 Cs) loaded with Xe (lower) at the pressures indicated (see Table 1). Up to 200 torr, the 8-ring site on the wall of the α-cage is preferred over the site in the center of the octagonal prism (Figure 1). Above this loading, while the occupancy of the former site remains constant, the Xe loads preferentially into the D8R site in the octagonal prism. A third site, the S6R on the wall of the α-cage, is occupied after a loading pressure of 100 torr and increases in occupancy as the pressure increases.

The powder diffraction pattern for the H-rho loaded at 1 torr clearly contained contributions from two cubic unit cells with different cell parameters (Figure 3); these were 15.033(1) and 14.947(1) Å. This sample afforded an unexpected opportunity to test the accuracy of the chemical analysis, and to locate the residual Cs found in the chemical analysis of the H-rho. The model best-fitting the RHO with the smaller unit cell was found to contain scattering from only the D8R-site (Figure 1). This scattering was best modeled with 0.15(1) Cs atoms in the 6-fold site at ($^1/_2$,0,0) giving a total of 0.90(6) Cs per unit cell (close to the chemically analyzed value of 0.75 Cs atoms per unit cell). Given that these refinements were carried out using fixed values of U = 0.06 Å^2 for the thermal parameters of the extra-framework material, there is good agreement between chemically analyzed Cs and that found in the diffraction experiment. No scattering was found near the 6R-site (Figure 1) at x,x,x with $x = 0.37$. For the RHO with the larger cell parameter, scattering was found at the D8R-site, corresponding to 0.75 Cs, as well as at the S8R-site at x,0,0 with $x = 0.39$ (Table 1). This extra scattering was presumed to arise from Xe absorption. For refinement of the other samples, the D8R-site was presumed to contain a fixed occupancy of 0.75 Cs, consistent with the chemical analysis.

Previously published structural data [31,47,68] were used as a starting point for development of the models consistent with the observed data. Using the 400 torr data set, three sites were identified for the adsorption of Xe from inspection of Fourier difference maps [67] and subsequent refinement of occupation factors (Table 1). These sites were at the center of the octagonal prism (D8R-site) and on the wall of the α-cage at the S6R- and the S8R-sites (Figures 1 and 2). These sites have oxygen nearest neighbors at the following distances: for the S8R-site, four at 3.281(6) Å; for the D8R-site, eight at 3.754(7) Å; and for the S6R-site, three at 3.85(8) Å The refinements for the remaining samples commenced by refining the site occupation factors for these three sites (Figure 2); the 0.75 Cs was presumed to occupy the D8R-site throughout. The framework atoms were initially fixed at positions predicted from distance least squares (DLS) modeling [48,67,69] based upon framework distances and the lattice parameter for the particular sample (Figure 2). All parameters, except the thermal parameters for the extra-framework sites, were varied in the final refinements.

The number of atoms of Xe occupying each site is summarized in Figure 2. At the lowest loadings (<100 torr), all adsorbed Xe visible to X-rays occupy the S8R-site on the wall of the α-cage. Above this pressure, Xe begins to occupy the D8R-site while maintaining an almost constant occupancy of the S8R-site; it appears that, *at 300 K*, the S8R-site saturates quickly. Above 200 torr at 300 K, the D8R site is preferred site for adsorption by a ratio of 2:1, compared with the S8R at 400 torr. The occupancy of the second site on the α-cage, the S6R (Figure 1), also increases above 100 torr. Essentially all the Xe, predicted to be absorbed from previously published isotherms [54], was found in the diffraction experiment. The exception is the non-equilibrated sample loaded at 1 torr. This sample contains two phases, one with and one without Xe; equilibration would distribute the Xe into a larger quantity of H-rho (Figure 2) and bring the observed loading closer to the curve calculated from the measured isotherms [54].

Table 1. Refined* positional and occupation parameters for Xe in H-rho.

Pressure	1 torr¶	50 torr	110 torr	200 torr	400 torr
x_T†	0.1016(1)	0.1011(1)	0.1008(1)	0.1013(1)	0.1019(2)
$x_{O(1)}$†	0.1673(1)	0.1682(2)	0.1685(2)	0.1680(2)	0.16736(2)
$z_{O(1)}$†	0.3741(4)	0.3762(4)	0.3770(4)	0.3758(4)	0.3739(5)
$y_{O(2)}$†	0.2187(4)	0.2169(4)	0.2159(4)	0.2177(4)	0.2198(5)
$z_{O(2)}$†	0.3772(2)	0.3788(2)	0.3791(3)	0.3789(2)	0.3780(2)
occ Xe(1)	-	-	-	0.135(4)	0.310(7)
$z_{Xe(2)}$†	0.390(3)	0.375(3)	0.400(2)	0.380(3)	0.389(4)
$occ_{Xe(2)}$	0.080(4)	0.072(3)	0.085(3)	0.072(3)	0.085(4)
$x_{Xe(3)}$†	-	-	0.375(9)	0.373(4)	0.376(3)
$occ_{Xe(3)}$	-	-	0.007(3)	0.020(4)	0.036(6)

* in all refinement chemically analyzed $Cs_{0.75}$ fixed at position 6*b*, (0,0,$^1/_2$) with occupancy 0.125 .

† space group $\boldsymbol{Im\bar{3}m}$, T at site 48*i* ($^1/_4$, *y*, $^1/_2$-*y*), randomly occupied by $Al_{12}Si_{36}$; O(1) at site 48*k*, (*x*,*x*,*z*); O(2) at site 48*j*, (0,*y*,*z*); Xe(1) at site 6*b*, (0,0,$^1/_2$); Xe(2) at site 12e, (0,0,*x*); Xe(3) at site 16f, (*x*,*x*,*x*).

¶ nominal pressure, since Xe-containing phase occurs in the same sample as unloaded H-rho ($Cs_{0.75}$; see Figure 3); the refined phase fractions [67] for the unloaded and loaded phases are 0.70(1) and 0.30(1), respectively; positional and occupancy parameters for the Xe-loaded phase reported here.

The behavior of Xe above 200 torr (Figure 2) is in agreement with previous studies at fixed loading but at lower temperatures [55,62]. It is also consistent with recent theoretical work [52]. At 300 K and at loadings above 200 torr, Xe prefers the D8R site (Figures 1 and 2). However, at 300 K, the S8R-site is occupied by a constant amount of Xe, regardless of loading pressure, up to the maximum studied here, 400 torr. At a loading below 200 torr, the preference for the S8R-site is at odds with predictions based upon theory [52].

This preliminary study showed we could locate small quantities of Xe and discern small differences in occupancy between the D8R-sites and those on the walls of the α-cage. It also showed that at 300 K essentially all the Xe is localized, on the time scale of the X-ray experiment. This confirms the high interaction energy with the aluminosilicate framework proposed by Tsaio et al. [54].

2.2 XENON IN Cd-RHO

The powder X-ray diffraction patterns of some divalent cation-exchanged RHO display abrupt changes at high temperature [37]. Structural studies [49] reveal that these "transitions" are not associated with water loss, but rather accompany relocation of the cation from the D8R-site. For example, the unit cells for the Ca^{2+}-, Sr^{2+}-, and Ba^{2+}-exchanged materials underwent a rapid decrease in volume associated with movement

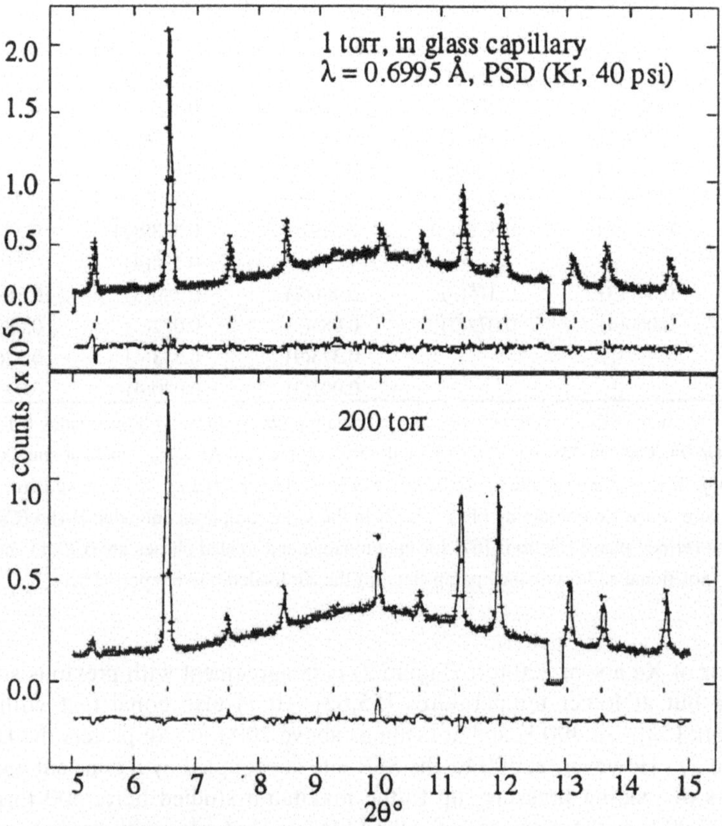

Figure 3. Comparison of the fit between observed (crosses) and calculated (continuous line) profiles [67] for H-rho loaded with Xe at 1 (top) and 200 torr. The difference between the observed and calculated profiles is plotted beneath each on the same scale. Two distinct phases are present in the sample loaded at 1 torr; the vertical dashes indicating the 2θ–positions of the Bragg reflections for each phase. The broad hump in the diffraction pattern centered at 10°, arises from scattering from the glass capillary containing the sample.

(Figure 1) of the cations within the 8R-sites [34,35,49]. However, in the case of the Cd^{2+}-exchanged RHO the unit cell parameter *increased* abruptly from 14.5 Å to 15.0 Å. A complete structure analysis using data collected in situ at temperatures above and below the transition [49,50] revealed the consequences of this change in cell volume; the Cd^{2+} ion undergoes relocation from the 8R-site in the octagonal prism to the S6R-site on the wall of the α-cage (Figure 1). This relocation is the largest in a zeolite (some 5Å) and involves the opening of access to the α-cage above the transition temperature of 400° C. Upon cooling to 300° C, Cd^{2+} moves back to the 8R, thereby restricting access to the α-cage. More recently, we have also observed this behavior in Zn-exchanged RHO and are pursuing structural studies on this material as well.

This property suggested Cd^{2+} could behave as an ionic trap door. While the trap door is open, above the transition to the $Im\bar{3}m$ form of RHO (Figure 1), sorption to the interior of the zeolite by molecules small enough to negotiate the D8R is permitted. Following sorption above 400° C, cooling below the transition temperature traps the sorbed guest; the Cd^{2+} moves back into the 8R as the framework distorts to the lower volume $I\bar{4}3m$ form of RHO (Figure 1). Provided the Cd^{2+} trap door remains in place, the sorbed guests, reactants, or products should be retained in the α-cage (Figure 1). Previously determined isotherms for Cs-RHO have shown that, for more than 2.67 Cs per unit cell, sorption is "substantially inhibited" [54], suggesting less than complete Cd^{2+}-exchange was required to test the philosophy outlined above.

2.2.1 *Static Studies*

The ability to trap and release Xe was demonstrated by thermogravimetric/mass spectral analysis (TG/MS) and ^{129}Xe-NMR [70]. A sample of Cd,NH_4-rho, $[Cd_{3.3}(NH_4)_{5.42}Na_{0.03}Cs_{0.25}]Al_{12}Si_{36}O_{96}$ (NH_4^+ content is assumed for charge balance), was prepared by conventional NH_4^+-exchange followed by Cd^{2+}-exchange of Na,Cs-rho [71,72]. The sample of Cd,NH_4-rho was out-gassed for 10 days at 400° C under 10^{-5} torr. It was then placed in a bomb, charged with 200 psi of Xe, and heated to 350° C for 2 hours. After cooling to room temperature, the pressure was decreased to 25 psi and the sample was transferred to a dry-box. Here, samples for NMR, TG/MS and powder diffraction were prepared for analysis.

In the TG/MS results, given in Figure 4, the event at 340° C was attributed solely to desorption of xenon (m/z=129). Ion peaks corresponding to all of the isotopes of xenon as well as the +2 ions were observed. The appearance of xenon at this temperature correlates well with the in situ X-ray powder diffraction results [49,50]. These show the Cd is relocated away from the 8-ring opening allowing the entrapped Xe to egress from the α-cage above this temperature (Figure 4). The weight loss of 5.7% associated with this escape corresponds to a loading of approximately 0.85 Xe/α-cage or 1.7 Xe/unit cell. This corresponds to approximately half of the Xe found in the diffraction experiment described below; considering the amount of water sorbed, significant amounts of Xe may have been lost during the transfer to the TG apparatus. Diffraction experiments which follow the desorption of Xe in real time are described below. They confirm the loss of about half of the Xe within 1 hr of the sample being exposed to the atmosphere.

Additional events are observed at approximately 100° C and 520° C, with weight losses of 9.8 and 0.93%, respectively. The ion chromatograms associated with water (m/z=18, 17) show the low temperature event, along with small events at about 290° C and 390° C, are associated with the desorption of water, most likely sorbed during transfer from the dry box to the TG apparatus. In addition, the event at 520° C appears to be associated with loss of ammonia (m/z=17). Apparently, the out-gassing at 400° C was not sufficient to completely deammoniate the Cd,NH_4-rho.

To compare the siting of Xe in this case with that found in the case of Xe loaded into H-rho described above, we carried out diffraction experiments at room temperature (300 K) and at 13 K. A sample from the batch of Cd/Xe-rho described above (Figure 4) was loaded into thin-walled glass capillaries in a dry box. Constraints imposed by the geometry of the Displex® low temperature unit at X-7A made the task of sealing these 0.7 mm OD x 20 mm capillaries a challenge. Several were surveyed at room temperature before one was deemed suitable for data collection at 13 K. The rapid survey was accomplished using a linear position sensitive detector designed and built at the BNL [25]. This gas proportional detector, when operated with X-rays near 40 keV and 40 psi of Kr gas, gives good resolution and high peak-to-background discrimination by gating on the Kr escape peak [25]. Samples, which had been improperly sealed, were found to give patterns which could be indexed on the basis of a 15 Å unit cell; this was indicative of the Cd-rho having hydrated and lost Xe. Some capillaries were found to have the dehydrated form intact. One of these was placed in the Displex® refrigeration unit and a full data set was collected at 300 K. The sample was then cooled to 13 K.

The 13 K data were collected in high resolution mode [6,16]. The positional parameters for the TO_4 framework (T = Al/Si) were fixed, in accordance with the lattice parameters and the results of DLS modeling [48,67], as were the isotropic thermal parameters.

Unconstrained refinement of the occupation factor for a dummy oxygen atom placed in the D8R, S8R, and two S6R-sites (Figure 1, Tables 1 and 2) suggested these sites are occupied. After assigning the scattering factor for Cd to those sites with occupancies greater than one, the occupation factors were once more refined. From this and from the results of calculated Fourier Difference maps, six atoms were located in the 8R-octagonal prism. The scattering is equally distributed between two sites, close to the center of the D8R and offset from the center of the S8R (Figure 5, Table 2). This implies that 2.7 Xe, along with the 3.3 Cd found in the chemical analysis, occupy this structural unit. In order to determine possible site distributions, a series of models was tested. In one such trial, the occupancy of the S8R- and D8R-sites were refined without constraints; the sites were assigned the scattering factors of Cd and Xe, respectively. This resulted in a calculated formula of $Cd_{3.44(2)}Xe_{2.78(4)}$ for the octagonal prism when all the Cd is located in the S8R and all the Xe is in the D8R-site. This is close to the expected total of six, for the six octagonal prisms per unit cell, and close to the total Cd found in the chemical analysis. Although the large contributions to the scattering from both Xe and Cd aid in their location in the X-ray experiment, their closely matched scattering factors makes distinguishing between them difficult. Indeed, refinements of the fully hydrated/ completely Xe-desorbed sample described in the next section, suggests that Cd is partially disordered over the two sites, but with a preference for the S8R-position. On the basis of a model in which all octagonal prisms were occupied with either Cd at the S8R-site or Xe at the D8R-site, a complete refinement was obtained based upon the 13 K data set (Table 2). The assignment of the Xe and Cd to the sites near $(x,0,0)$ with $x \approx 0.5$ and $x \approx 0.35$ respectively, was on the basis of an ordered arrangement of Xe and Cd within the octagonal prism. When the positions of the Cd and Xe were reversed from those given in Table 1, the refinement led to a composition $Cd_{2.5}Xe_{3.6}$, a worse match to the observed chemistry ($Cd_{3.3}$) than that found in the final refinement (Table 1) of $Cd_{3.8}$ for the 13 K data set.

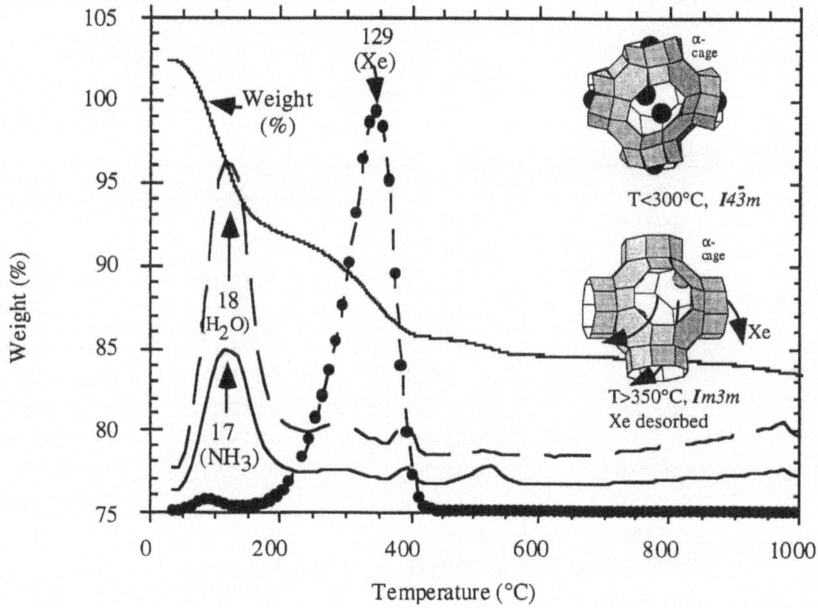

Figure 4. TG/MS plots for Xe-loaded Cd,NH_4-rho as a function of temperature. The weight loss refers to the left hand scale while the relative intensity of the desorption peaks is referred to the scale to the right.

The possibility that Cd and Xe are partially disordered over these two sites can not be ruled out based on these results. A neutron diffraction experiment would also probably fail to resolve the two site chemistries. Not only is Cd highly absorbing to neutrons, but the Cd and Xe scattering lengths differ by only 2%; they differ by 12% at $\sin\theta/\lambda = 0$ for X-radiation. This problem could be resolved using anomalous scattering techniques [16]. Because of the energies of the edges involved (≈30 keV), this experiment is probably best carried out on the third generation synchrotron sources currently being prepared for use.

In addition to the two sites in the 8-rings, two other sites were found to be occupied. One, which was at *(x,x,x)* with $x \approx 0.15$, amounted to the equivalent of 1.4 Xe per unit cell, while the other site, at *(x,x,x)* with $x \approx 0.35$, was found to be occupied by 0.1 equivalent atoms of Xe per unit cell. This later site, clearly less preferred by Xe, may be residual H_2O not removed during the dehydration process. For the final refinements, utilizing the 13 K and 300 K data sets, Cd was located at site 24g *(x,y,y;* $x \approx 0.35$, $y \approx -0.06$) Xe(1) at site 12e *(x,0,0;* $x \approx 0.54$), Xe(2) at 8d *(x,x,x;* $x \approx 0.14$) and O(H_2O) at 8d *(x,x,x;* $x \approx 0.31$). The total content of the octagonal prisms was constrained to equal six. The results for the final refinements of both the 13 K and 300 K data are given in Tables 1 and 2.

Table 2. Positional, thermal and occupation factors for Cd/Xe-RHO* at 13 (top) and 300 K.

Atom	site/occ.	*x*	*y*	*z*
T(1)†	48*h*/1.0	0.2673(2)	0.1183(2)	0.4170(3)
		0.2676(2)	0.1180(2)	0.4174(2)
O(1)	24*g*/1.0	0.2121(5)	0.2121(5)	0.3951(7)
		0.2077(3)	0.2077(3)	0.3926(5)
O(2)	24*g*/1.0	0.1341(3)	0.1341(3)	0.6194(5)
		0.1349(3)	0.1349(3)	0.6228(4)
O(3)	48*h*/1.0	0.0263(3)	0.2128(3)	0.3864(4)
		0.0266(2)	0.2136(3)	0.3857(3)
Cd(1)¶	24*g*/0.158(1)	0.3487(8)	-0.0569(5)	-0.0569(5)
	24*g*/0.163(1)	0.3365(9)	-0.0568(5)	-0.0568(5)
Xe(1)¶	12*e*/0.184(3)	0.541(1)	0	0
	12*e*/0.174(3)	0.540(2)	0	0
Xe(2)	8*c*/0.173(4)	0.1460(7)	0.1460(7)	0.1460(7)
	8*c*/0.158(4)	0.144(1)	0.144(1)	0.144(1)
O(H_2O)	8*c*/0.18(4)	0.318(5)	0.318(5)	0.318(5)
	8*c*/0.21(4)	0.304(5)	0.304(5)	0.304(5)

* RHO framework, $I\bar{4}3m$, a=14.6886(6) and 14.712(1) Å, for 13 K and 300 K data, respectively. Refinement statistics (%) [67] with values for 300 K data in parentheses: R_{wp} = 8.00(6.72), R_p = 6.65(5.23), R_{Bragg} = 5.61(5.23), χ^2 = 0.75(0.70). Isotropic thermal parameters for all extra-framework atoms fixed at U = 0.06 Å^2; those for the framework (T and O-sites) constrained to be equal and refined to U = 0.007(1) at 13 K and U = 0.002(1) Å^2 at 300 K.

† Scattering adjusted to match framework cation composition [$Al_{12}Si_{36}$] from chemical analysis.

¶ total occupancy constrained to 6 per unit cell.

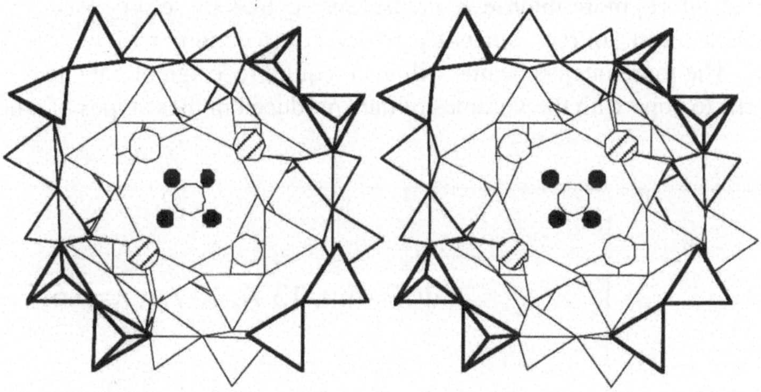

Figure 5. Stereo representation of the interior of the α-cage showing the positions of Xe (blank circles), Cd (black), and either water, ammonia or low occupancy Xe site (striped) marked $O(H_2O)$ in Tables 2 and 3. The Xe at the center of the 8-ring and at the S6R-site are marked Xe(1) and Xe(2), respectively, in Tables 1 and 2. Interatomic distances are given in Table 3.

The preference of Xe for the 8R site is consistent with the previous theoretical work on H-rho. However, in that work, account was not taken of cases in which the 8R-site was partially occupied by a tightly-bound cation. In the present study, this situation leads to occupancy of one the S6R-site on the wall of the α-cage (Figure 5, Table 2). The occupancy of only one of these sites may be indicative of a Cd-Xe or Xe-Xe interaction since this is possible only from the Xe(2)-site (Figure 5).

2.2.2 *Real-Time Studies of Xe Desorption*

During the dismounting of one of the samples used to study Xe in Cd-exchanged zeolite rho, the capillary in which it was contained broke. Changes were observed to occur as data in the PSD were being collected. This afforded an unexpected opportunity to study the desorption of Xe from the sample in real time.

The PSD detector being used for this work [25] is capable of collecting data in less than 10 minutes; these data would be to a resolution of 1 Å and suitable for Rietveld [66] refinement. Unfortunately, a certain lack of urgency on our part limited the time resolution to about 40 minutes. During this time period, the diffraction patterns were evolving with the lattice parameters of phases changing and new phases appearing (Figure 7). The implications of this type of study are obvious, however. Real-time monitoring of sorption/desorption and of ion-exchange processes is feasible. All that is required is a means of placing materials under controlled atmosphere, temperatures, and pressure. In the future, more intense X-ray beams, such as those proposed for the third generation synchrotron sources currently under construction, will greatly aid these investigations. The more intense beams will also require redesign of data acquisition and analysis systems to cope with the volumes of data produced in these types of studies.

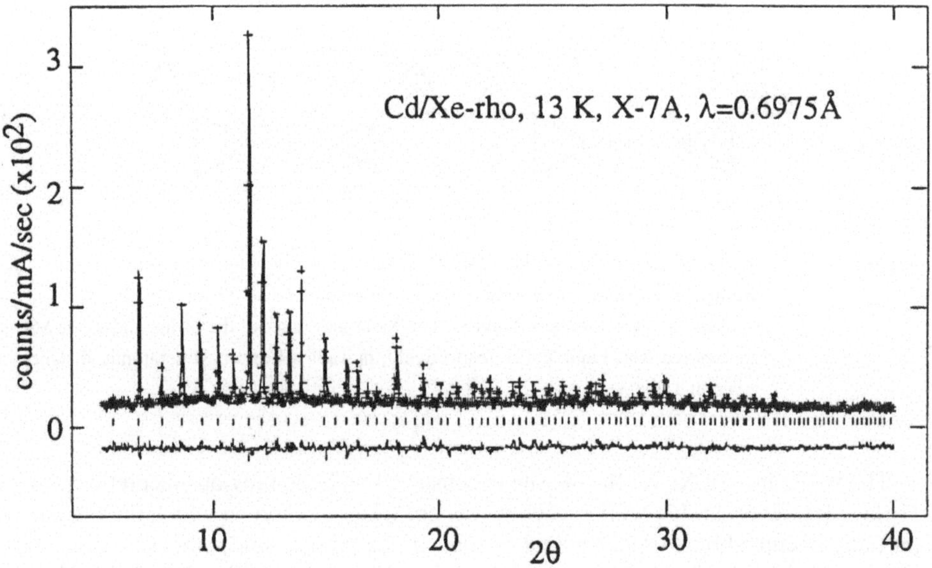

Figure 6. The X-ray diffraction data for Cd,NH_4-rho loaded with Xe (referred to as Cd,Xe-rho) at 13 K. Points shown by (+) represent observed data. The continuous lines through the sets of points are the data calculated from the model given in Table 2. The set of tic marks below the data indicate the positions of the allowed reflections. The lower curve in each panel represents the difference between observed and calculated profiles.

Since there are distinct phases contributing to the powder diffraction patterns, and not just one or two with continuously varying lattice parameters, this implies a competition between sites as Xe is desorbed and H_2O is sorbed. This is similar to the situation observed in Cd,NH_4-rho at high temperature where the competition is between occupancy of the 8R- and 6R-sites as the framework transforms from $I\bar{4}3m$ to $Im\bar{3}m$ symmetry at 350° C [49,50,70].

For the structural refinements, the framework atomic positions were initially fixed at values obtained from DLS modeling. Following refinement of site occupation factors and in the final stages of model refinement, soft constraints on the interatomic distances between atoms in the aluminosilicate framework were introduced to supplement the diffraction data and allow adjustment of all positional parameters.

Table 3. Selected interatomic distances (Å) and angles (°) for Xe in Cd,NH_4-rho at 13 and 300 K.

atoms	13 K	300 K	atoms	13 K	300 K
T(1)-O(1)	1.632(1)	1.628(1)	O(1)-Si(1)-O(2)	110.0(3)	109.6(1)
T(1)-O(2)	1.630(1)	1.629(1)	O(1)-Si(1)-O(3)	113.6(6)	109.9(2)
T(1)-O(3)	1.634(1)	1.630(1)	O(1)-Si(1)-O(3)	108.6(6)	109.9(2)
T(1)-O(3)	1.634(1)	1.631(1)	O(2)-Si(1)-O(3)	108.5(3)	109.2(2)
			O(2)-Si(1)-O(3)	107.1(3)	108.5(2)
Cd(1)-O(2) x2	3.063(8)	3.104(7)	O(3)-Si(1)-O(3)	108.9(3)	109.7(2)
Cd(1)-O(3) x2	2.398(6)	2.458(7)	Si(1)-O(1)-Si(1)	143.3(5)	145.8(3)
Cd(1)-Cd(1)	2.36(2)	2.36(2)	Si(1)-O(2)-Si(1)	144.8(5)	146.1(4)
Cd(1)-Xe(1)	3.07(2)	3.21(3)	Si(1)-O(3)-Si(1)	137.4(3)	137.7(3)
Cd(1)-Xe(1)	2.00(1)	2.17(2)			
Cd(1)-Xe(2)	3.50(1)	3.36(1)	Xe(1)-O(2) x2	3.013(8)	3.06(1)
			Xe(1)-O(2) x2	3.65(1)	3.68(1)
Xe(2)-O(1) x3	3.91(2)	3.90(2)	Xe(1)-O(3) x4	3.88(1)	3.89(1)
Xe(2)-Cd(1) x3	3.50(1)	3.36(1)	Xe(1)-O(3) x4	3.323(8)	3.35(1)
O(H_2O)-O(1) x3	2.48(6)	2.39(5)	Xe(1)-Xe(1)	1.21(3)	1.16(5)

The occupation factors derived for four phases identified in this study are summarized in Table 4. Occupancies are given as numbers of atoms per unit cell. The scattering factors were arbitrarily assigned to conform to the distribution found in the study of the fully loaded materials (Tables 2 and 3). The fully hydrated centric form of Cd,NH_4-rho, designated D (Table 4, Figures 7 and 8) can be used as a baseline in proposing a scenario for the Xe-desorption process. It was presumed this phase represented the fully desorbed form and that any scattering found arose from either Cd or H_2O; scattering from N and O in NH_4^+ or H_2O were not distinguished. This assumption was reinforced by the closeness of the match between the amount of Cd found in the refinement of the phase D structure model, 2.8 Cd (Table 4), and that determined from the chemical analysis, 3.3 Cd. Another aspect of the refinement for phase D concerns the position of Cd in the hydrated material; Cd is found to prefer the S8R-site, but also to partially occupy the D8R-site. Indeed, the changes in the occupancy of these two sites from phase A' through D (Table 4, Figure 7) suggests that upon Xe loss and hydration there is considerable mobility of Cd.

Most of the Xe (40%) is lost within the first hour following exposure to the atmosphere and explains why only half the Xe expected is found in the TG/MS experiment (Figure 4). This causes a dilation in the unit cell as the sorbed gas is lost (Table 4). Following the initial loss of all Xe from the 8R site in phase A' (Table 4, Figure 7), a comparatively slow rearrangement and loss of Xe from the S6R occurs, possibly as a response to competition with sorbed H_2O. Phase B, for example, has about the same total Xe in the S6R as the sealed sample but equally distributed over the two available sites rather than

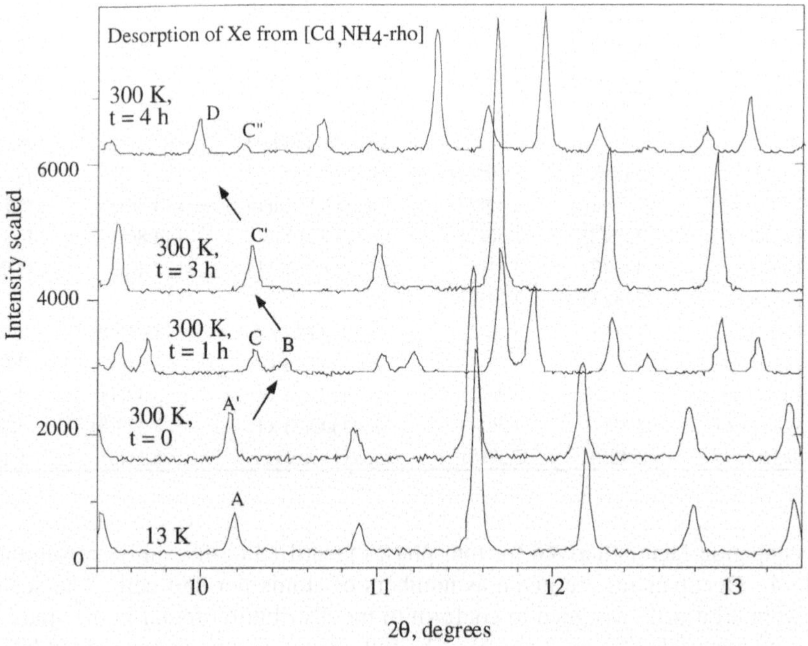

Figure 7. Stack-plot of representative diffraction patterns collected from Xe-loaded Cd,NH_4-rho. The times (t, in hours) refer to the start time for the particular pattern after the contents of the capillary had been exposed to the atmosphere.

concentrated on the site at (x,x,x); $x = 0.15$. The unit cell edge (Table 4) gradually increases as the sample hydrates and loses Xe.

Clearly, this type of transformation is worthy of further investigation using other cation/gas combinations in RHO, and possibly other zeolite systems. Structure analysis using the data collected here was complicated by the lack of time resolution obtained during data collection. The contributions from particular phases (Figures 7 and 8) are changing with time and the lattice parameters of those phases are also changing. This situation would be greatly improved by shorter data collection times and such studies are planned. The broader implications for this study are clear. Given a gas pressure cell, the technology now exists to collect and analyze X-ray data taken in real time on desorbing zeolites, and to obtain meaningful structural information from these data. This philosophy can easily be extended to studies of ion-exchange in real time as well.

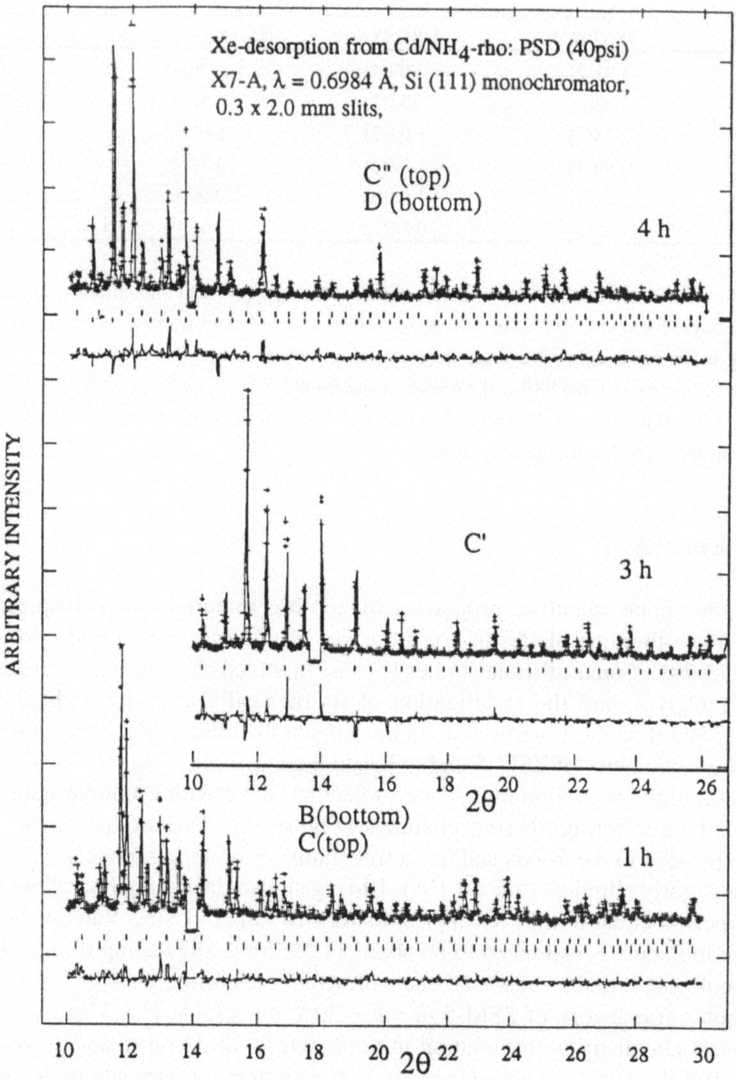

Figure 8. Observed (crosses) and calculated (continuous line) profiles for three samples of Xe desorbing from Cd,NH_4-rho. The difference curves are plotted below each pattern and tick marks refer to the Bragg positions of the diffraction maxima for the phases indicated here, in Figure 7 and Table 4.

Table 4. Results of site occupancy refinements* for the phases shown in Figures 7 and 8.

	sites			
phase/*a* (Å)	S8R† (Cd)	D8R (Xe)	S6R (Xe @*x*=0.15)	S6R (Xe@*x*=0.35)
A'/14.712(1)	3.91(2)	2.09(4)	1.26(4)	¶
B/14.332(1)	1.53(3)	0.57(2)	0.88(4)	0.72(4)
C/14.557(1)	1.44(3)	1.04(2)	1.03(6)	0.24(6)
C'/14.631(1)	0.48(3)	1.08(3)	0.72(4)	0.56(4)
C"/14.657(1)	1.2(2)	1.63(3)	0.90(6)	0.80(4)
D/15.033(2)	1.92(2)	0.84(2)	<·········$8H_2O$§·········>	

* During these refinements soft constraints were used to supplement the diffraction data; thermal parameters were fixed at U=0.02 and 0.06 Å^2 for framework and extra-framework sites, respectively.

† Scattering factor set to Cd or Xe based upon results of 300 K refinement of dehydrated and hydrated material which suggested the S8R is the preferred site for Cd.

¶ occupied by 1.7 H_2O or equivalent scattering material (NH_4^+ for example).

§ space group $Im\bar{3}m$, S6R-sites are equivalent.

3. Stilbene in ZSM-5

Recently, the shape selective properties of zeolitic materials have inspired organic chemists to use them as platforms for a variety of polymerization and photochemical reactions [73-78]. Some of these studies [77,78] involved the sorption of *trans*-stilbene ($C_6H_5C_2H_2C_6H_5$), and the stabilization of its thermally-activated radical cation, in Na-ZSM-5. We thought it worthwhile to investigate the effects of stilbene absorption on the framework geometry of ZSM-5 and to locate the molecule within the structure.

ZSM-5 is a high silica aluminosilicate molecular sieve with a unique pore geometry consisting of intersecting 10-ring channels (Figure 9). One of these, the "straight" channel, is parallel to the ***b***-axis and has a free diameter of approximately 5.5 Å [1]. The second is a slightly elliptical (5.5 x 5.1 Å), 10-ring sinusoidal channel parallel to ***a***, with a half-wave period equal to ***a***/2; or approximately 10 Å [1]. Using Van der Waals radii, the stilbene molecule is estimated to be about 11.5Å long, suggesting it may be difficult to fit this molecule into the sinusoidal channel without distorting it.

The complex framework of ZSM-5 ($\boldsymbol{a} \approx \boldsymbol{b} \approx 20$Å, $\boldsymbol{c} \approx 13.4$Å, $\boldsymbol{U} \approx 5360$ Å^3) displays a rich structural chemistry with variation in temperature, Si/Al-ratio, and sorbate loading [18,20,21,79-84]. The as-synthesized material, containing a template molecule such as tetrapropylammonium, is orthorhombic, ***Pnma*** [1]. It distorts to monoclinic ($\boldsymbol{P2_1/m}$) upon removal of the template and conversion to the H-form [18], which in turn exhibits a reversible phase transition at about 340 K. The orthorhombic form is stable above and the monoclinic form stable below this temperature. Further, the crystal symmetry is affected by the amount and type of sorbate [81] (Figure 10).

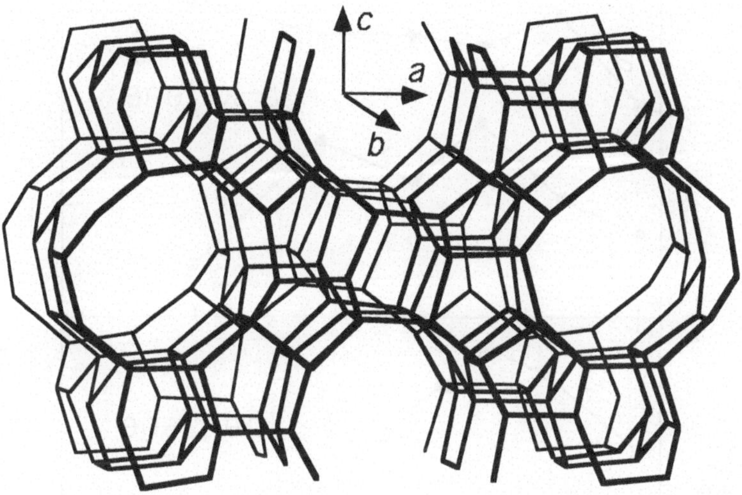

Figure 9. Structure of the ZSM-5 framework (***Pnma***) viewed along [*010*]. The straight 10-ring channels are parallel to the ***b***-axis while the direction of the sinusoidal 10-ring channel is parallel to the ***a***-axis. For the sake of clarity only the tetrahedrally coordinated Si/Al (T) atoms in the framework are shown; oxygen lies approximately at the midpoint of the lines joining the T-sites.

Single crystal investigations are well-suited to problems in which subtle changes in crystal symmetry produce weak reflections or in which complex structures are involved. If sufficiently large single crystals are not available, crystal quality is poor, or diffusion of the sorbate of interest into large single crystals is limited, powders offer a favorable alternative. When only the metric of the unit cell changes, high resolution powder diffraction is particularly useful, as outlined below.

The structures of several sorbate/ZSM-5 complexes have been investigated by both single crystal [20,21,85] and powder diffraction [11,23,86-88]. In those diffraction studies in which small aromatic molecules were absorbed into ZSM-5 [11,20], the location of the sorbate was found to depend on loading. In the most comprehensive of these investigations, Koningsveld et al. [20] found that *p*-xylene was absorbed into both the straight and the sinusoidal channels (Figure 9) at loadings of eight molecules per unit cell. At this loading, the framework undergoes a distortion to $\boldsymbol{P2_12_12_1}$ symmetry. The relationship between these structures is described in detail elsewhere [20].

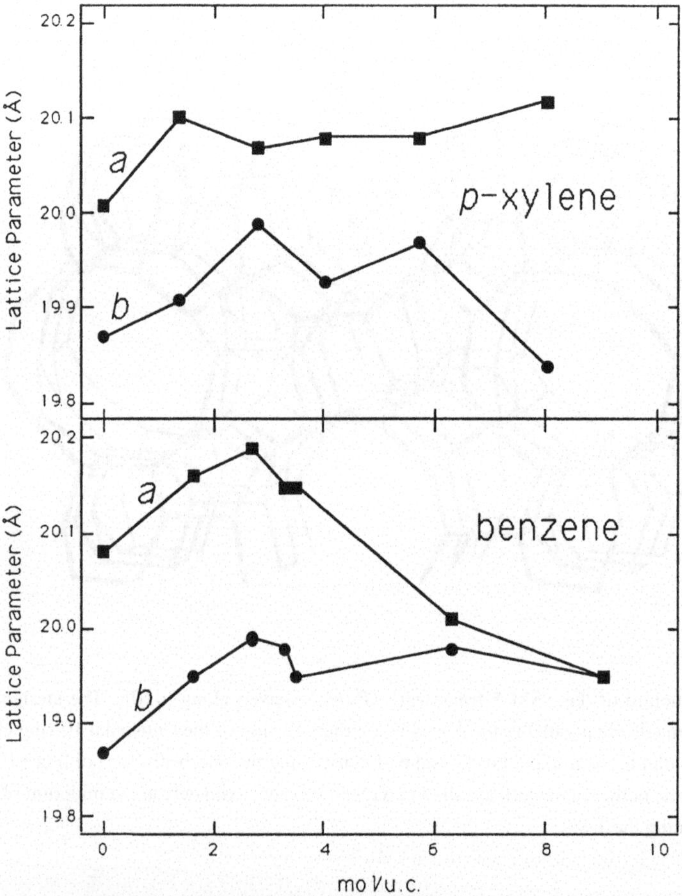

Figure 10. Variation in the ***a*** and ***b*** lattice parameters for ZSM-5 loaded with *p*-xylene (top) and benzene (bottom). Data taken from Kokotailo et al. [81].

3.1 PRELIMINARY INVESTIGATIONS

Initially, attempts were made to load stilbene into single crystals of ZSM-5; the intention being to collect data on a conventional single crystal diffractometer. Unfortunately, the prohibitively slow rates of absorption for stilbene into crystals larger than 50 μm in length necessitated the use of powdered samples. The lattice parameters of crystals exposed to stilbene were found not to have changed; as described below and elsewhere [81], the unit cell parameters are sensitive indicators of the degree of sorption into the ZSM-5 channels.

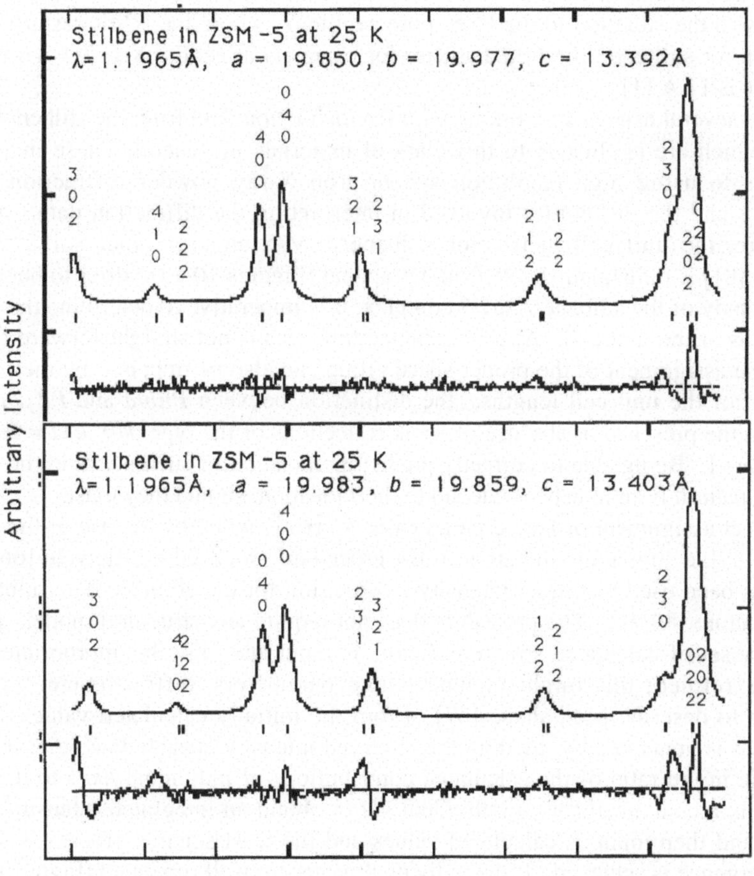

Figure 11. Results of the Le Bail [89] extraction of intensities for two relative settings of the lattice parameters for stilbene loaded in ZSM-5 at 25 K. Refinement from a starting model with $a>b$ converges quickly, but to a clearly inferior fit compared to the model in which $a<b$.

Trans-stilbene was obtained commercially and was recrystallized several times from ethanol. Na,TPA-ZSM-5 was prepared according to the procedure of Rollman and Valyocsik [90]. The stilbene was introduced into a sample with a Si/Al ratio of 550/1 according to the procedure described previously [77,78].

The inevitable loss of information involved in using powder diffraction can be mitigated against by using spectroscopic information in order to constrain composition and degree of order. A thermogravimetric analysis suggested the samples contained 2 molecules of stilbene and one molecule of water per unit cell of ZSM-5. ^{2}H-NMR results [77] suggested little dynamic disorder at room temperature for the sorbed stilbene. This contrasts with the situation in the large-pore zeolite Y, where the NMR results indicate free motion for stilbene; the free diameter for channels in zeolite Y is 7.4 Å while that for the pore is 11 Å [1].

There are several aspects of working with the diffraction data from the stilbene/ZSM-5 complex which are applicable to this class of materials in general. These include the advantages to using high resolution synchrotron X-ray powder diffraction at low temperature and the difficulties involved in interpreting the diffraction data. At room temperature, the unit cell metric for stilbene/ZSM-5 and for other sorbate/ZSM-5 complexes [81] is indistinguishable from tetragonal (Figures 10-13). Prior to beginning a structural study of the stilbene/ZSM-5 complex, it is imperative to determine the relative axial lengths of the unit cell. At room temperature, this is not straight-forward (Figures 10,13). The assignment of the proper space group can also be impaired by the incorrect assignment of the unit cell lengths; the distinction between ***Pnma*** and $\boldsymbol{P2_12_12_1}$ [20] depends on the presence or absence of weak reflections of the type $0kl;\ k+l = 2n+1$ and $h0l;\ h = 2n+1$. Being able to correctly predict the position of such peaks in the powder diffraction pattern is an indispensable aid to their location, should they exist.

The correct assignment of lattice parameters is greatly aided by the use of the Le Bail technique [89], a simple modification of the procedure proposed by Rietveld [66]. This method has been used to extract intensity information for the ab initio determination of crystal structure [89,91]. The procedure does not require any structural model. Instead, an arbitrary set of calculated structure factors is input and only the appropriate profile parameters refined; this might include lattice parameters, diffractometer zero, and parameters to describe peak shape [67]. From the initially calculated values, a set of I(*hkl*) values is generated by dividing the observed intensity at each data point in a step-scan profile in the ratio of the calculated contributions of individual *hkl*'s to that point. These points are summed for an individual *hkl* to obtain an integrated intensity for that reflection and then input as calculated values and the least squares process is repeated until convergence is achieved. If the individual peaks are well separated (Figure 11), this is a rapid process since the correlations between parameters is unimportant. On the other hand, when, as in the case of ZSM-5 at room temperature (Figures 12,13), peaks are not well-resolved, they will tend to be assigned similar intensities; the relative lengths of the lattice parameters must then be decided on the basis of weak intensities which unambiguously distinguish between models in which the $\boldsymbol{a}$-axis is longer than $\boldsymbol{b}$, and one in which $\boldsymbol{a} < \boldsymbol{b}$.

3.2 VARIATION IN UNIT CELL OF STILBENE/ZSM-5 WITH TEMPERATURE

The diffraction experiments were performed in high resolution mode [6,16] at the NSLS on beamlines X-7A and X-3B. In order to follow changes in the unit cell with temperature, the sample was loaded into a quartz capillary of 1 mm OD and placed first in a Displex® closed-cycle refrigerator fitted with a Be window for measurement from 20 to 280 K. A cylindrical furnace [92,93] was used for measurements from 300 K to 450 K. Step scans were carried out over selected angular regions covering the (*110*), (*020/200*), (*321/231*), (*040/400*), and (*410*) reflections at several temperatures (Figures 12 and 13). At about 440 K, stilbene was lost from the sample. The wavelengths used for the experiments were obtained by calibration with a reference sample of CeO_2 (***a*** = 5.4113Å).

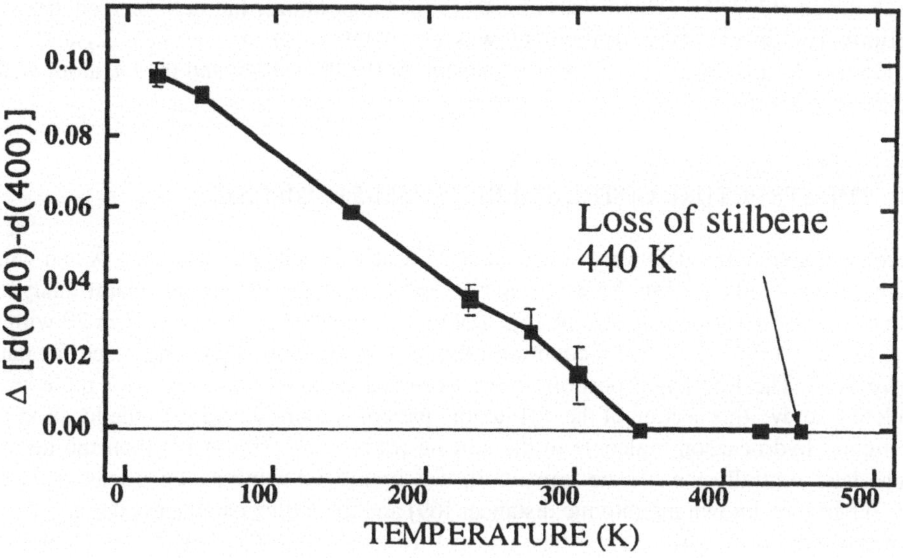

Figure 12. Variation in the splitting of the (*040*)-(*400*) reflections, shown in Figure 13, as a function of temperature. The wavelength, determined from the CeO_2 standard, is 1.1965Å.

Diffraction data for the (*040/400*) pair of reflections along with the (*410*) are reproduced in Figure 13 to show the relative changes in the ***a***- and ***b***-axes in the temperature range 25 K to 300 K. The assignment of the (*040*) to the lower angle peak in the (*040/400*) pair is made on the basis of the final Rietveld refinement described below. The assignment at the higher temperatures was made on the basis of the Le Bail extraction [89]. The relative lengths of the ***a***- and ***b***-axes is unusual in that other orthorhombic cells reported for ZSM-5 [4,18,20,21,23,80,81,83,84,86-88,94,95] have $a > b$. It should be noted that, because of the pseudo-tetragonal metric for this material, the position of a few weak reflections of the type *hk0; h* = 2*n* help to define the unit cell (Figure 11). With laboratory-based diffractometers, these reflections can be easily lost in the background or appear as poorly resolved shoulders (Figure 11). This situation is exacerbated at room temperature where, even with the superior resolution afforded by the

X-7A diffractometer, the ***a*** unit cell parameter is indistinguishable from ***b*** (Figures 12 and 13). In this case, Rietveld refinement [66,67,89] is required to resolve the ambiguity.

At 25 K, the orthorhombic nature of the cell is clearly evident (Figure 13). Further, there is no broadening of the *(111)*, which would indicate monoclinic symmetry; the diffraction pattern was fully indexed on the basis of an orthorhombic cell (***a*** = 19.8526(1), ***b*** = 19.9801(2) and ***c*** = 13.3921(2) Å) with absences corresponding to diffraction symmetry ***Pn-a***. The mirror plane perpendicular to the ***b***-axis was assumed throughout the subsequent investigations; there was no diffraction evidence for the space group $P2_12_12_1$. Upon warming to room temperature, the cell derived from a preliminary Rietveld-type refinement was ***a*** = 19.9601(2), ***b*** = 19.9907(2) and ***c*** = 13.4155(2) Å; this represents a linear expansion of 0.54%, 0.05% and 0.17% along ***a***, ***b***, and ***c***, respectively.

3.3 THE STRUCTURE OF THE STILBENE/ZSM-5 COMPLEX

More extensive X-ray data were collected at 25 K at a wavelength of 1.1965 Å and with beam defining slits 1.1 mm by 8.0 mm. A Ge(111)/Ge(220) monochromator/analyzer with a Kevex® detector was used [6,16]. Data were collected from 5° to 70° in 2θ with a step size of 0.008° in 2θ and a counting time of 4 sec for low angle and 5 sec for data above 30°. The low angle portion of the extended data set was used in a molecular modeling study; this section of the diffraction pattern is more sensitive to the location of the lighter hydrocarbon molecule in the silicate framework (Figure 14) than the higher angle data. Initially the silicate framework was optimized using distance least squares (DLS) based on known interatomic distances [69] and the lattice parameters derived from a Le Bail-type [89] refinement. The unit cell contents were fixed at two molecules of stilbene per unit cell (an occupancy of 0.25 per general position in ***Pnma***) consistent with results of thermogravimetric measurements.

Final adjustments were carried out to give a good visual fit between observed and calculated patterns and the positional parameters for the framework atoms and sorbate molecule (Figure 14) were exported for further refinement using GSAS [67]. The positional parameters for the framework were adjusted using the Rietveld method [66]; the diffraction data were supplemented with 107 inter-atomic T-O, O-O and T-T distances taken from the recent accurate single crystal study of orthorhombic ZSM-5 loaded with *para*-xylene [20]. For the final refinement the positions of all framework atoms were varied while their thermal parameters and all parameters for the stilbene were fixed. The diffractometer zero, lattice parameters, and 5 profile coefficients [67] were also varied. Figure 15 gives a graphical representation of the fit in the low-angle region of the 25 K data and the model is given in Table 5. The usual discrepancy indices were R_{wp} = 0.185 and χ^2 = 3.2. The background was estimated at points between the diffraction maxima (Figure 15).

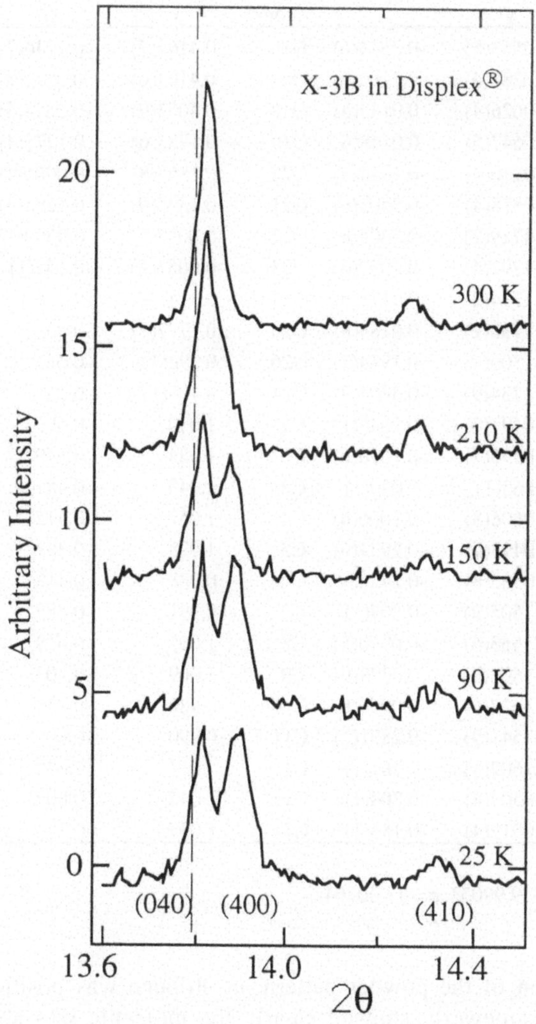

Figure 13. Synchrotron X-ray scans through the (*040/400*) and (*410*) reflections from stilbene-loaded ZSM-5 as a function of temperature at a wavelength 1.1956Å on the X-3B beamline. The (*140*) reflection, absent for diffraction symbol ***Pn-a***, is not observed. It would appear just to the low angle side of the (*410*) reflection. Note that the ***a***-axis (Figure 9) is almost invariant with temperature.

Table 5. Refined framework* parameters and positions of atoms in stilbene in the ZSM-5/stilbene complex.

atom	*x*	*y*	*z*	atom	*x*	*y*	*z*
T1	0.4211(4)	0.0553(5)	-0.3514(6)	O16	0.4163(7)	0.1246(7)	-0.410(1)
T2	0.2993(4)	0.0288(4)	-0.2102(6)	O17	0.4103(6)	-0.0042(8)	-0.4281(9)
T3	0.2818(4)	0.0626(4)	0.0143(5)	O18	0.4053(9)	-0.1324(5)	-0.4407(8)
T4	0.1241(5)	0.0647(5)	0.0148(6)	O19	0.1783(6)	0.1275(4)	-0.4076(7)
T5	0.0707(5)	0.0270(4)	-0.1934(5)	O20	0.1863(8)	-0.0006(4)	-0.429(1)
T6	0.1775(4)	0.0578(4)	-0.3503(6)	O21	0.1895(9)	-0.1295(3)	-0.4383(8)
T7	0.4209(6)	-0.1719(2)	-0.3404(8)	O22	-0.0057(5)	0.0500(7)	-0.1964(7)
T8	0.3001(5)	-0.1292(4)	-0.2052(6)	O23	-0.0053(6)	-0.155(1)	-0.1969(9)
T9	0.2801(6)	-0.1734(2)	0.0174(5)	O24	0.4147(9)	-0.250	-0.360(1)
T10	0.1238(6)	-0.1738(2)	0.0183(8)	O25	0.1777(8)	-0.25	-0.3711(9)
T11	0.0707(5)	-0.1309(4)	-0.1914(7)	O26	0.296(1)	-0.25	0.0397(9)
T12	0.1790(6)	-0.1734(9)	-0.3407(7)	O27	0.109(1)	-0.25	0.0462(9)
O1	0.3670(6)	0.0497(8)	-0.264(1)	C1	1.030	-0.593	-0.522
O2	0.2986(8)	0.0596(7)	-0.1010(6)	C2	1.024	-0.519	-0.518
O3	0.2029(4)	0.063(1)	0.030(1)	C3	1.045	-0.481	-0.592
O4	0.1071(6)	0.0606(8)	-0.1006(6)	C4	1.052	-0.412	-0.589
O5	0.1079(5)	0.0493(7)	-0.2931(6)	C5	1.028	-0.381	-0.511
O6	0.2370(6)	0.0565(8)	-0.272(1)	C6	0.988	-0.418	-0.438
O7	0.3685(7)	-0.1505(9)	-0.257(1)	C7	0.991	-0.487	-0.441
O8	0.3002(8)	-0.1563(6)	-0.0940(5)	C8	1.005	-0.629	-0.444
O9	0.2021(6)	-0.1602(8)	0.0329(9)	C9	1.009	-0.703	-0.447
O10	0.1055(7)	-0.1627(6)	-0.0960(8)	C10	0.988	-0.740	-0.366
O12	0.1097(8)	-0.1541(9)	-0.2891(8)	C11	0.990	-0.809	-0.370
O13	0.2388(8)	-0.1602(8)	-0.265(1)	C12	1.014	-0.840	-0.454
O14	0.2944(9)	-0.0502(4)	-0.204(1)	C13	1.045	-0.803	-0.535
O15	0.0737(9)	-0.0519(4)	-0.183(1)	C14	1.033	-0.734	-0.532

* ***Pnma***, ***a*** = 19.86537, ***b*** = 19.99053, ***c*** = 13.40364.

Real-time calculation of the powder pattern, as stilbene was positioned within the channels of a rigid framework, showed clearly the molecule was absorbed into the straight-channel (Figure 14) rather than the sinusoidal channels. Translation along the straight-channel located the molecule with one phenyl ring close to the intersection of the straight- and sinusoidal-channels (Figures 9 and 14) and the other at $y = 1/2$. This is close to the position found for *para*-dichlorobenzene [11].

The refinement shows that stilbene, at a loading of two molecules per unit cell, is located with one phenyl ring close to the mirror planes at $y = 1/4$ and $3/4$ at the

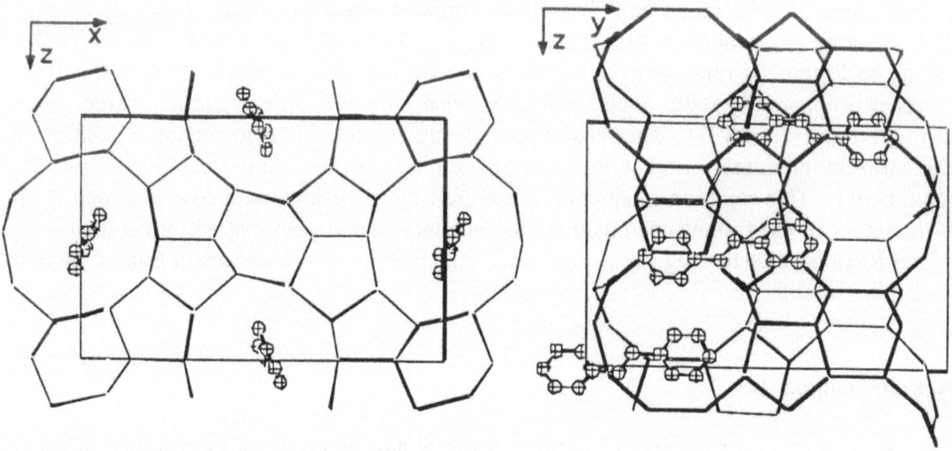

Figure 14. Projection of the structure of the ZSM-5/stilbene complex at 25 K along the ***b***-axis (left) and the ***a***-axis (right) The position of the sorbate molecule, *trans*-stilbene, is that obtained from the structure modeling procedure (see text). For clarity, only selected orientations of the molecule, which is disordered about the mirror planes at $y = 1/4$ and 3/4, are shown.

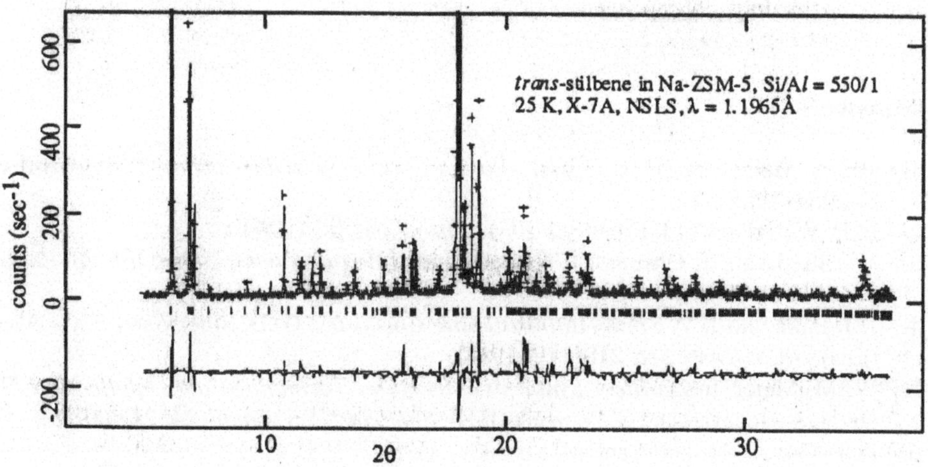

Figure 15. Calculated (line) and observed X-ray powder diffraction profiles for *trans*-stilbene (occupancy = 0.25) in Na-ZSM-5. The difference pattern is given below. Note that the intensity is truncated at 600 counts normalized to 100 mA. The strongest peak in the pattern, the *(011)*, has an intensity of 1500 counts.

intersections of the straight and sinusoidal channels; the center of the molecule (Figure 14) is located at a general position at $x = 1.02$, $y = -0.611$, and $z = -0.483$. The molecules shown in Figure 14 represent one of two possible orientations about the mirrors. This disorder is apparently static since NMR experiments suggest little dynamic disorder, even at room temperature [77]. The second phenyl ring (Figure 14) is located at a position in the straight channel where contacts between the sorbate and the framework are maximized. This position is apparently favored by *para*-xylene at low loadings [11]. Refinement of the framework positional parameters also suggests an elliptical distortion of the 10-rings, constituting the straight-channels, in response to the absorption of the two molecules of stilbene.

Acknowledgments

This work has benefited greatly from the financial and moral support of collaborators at The DuPont Company. These include Drs. L. Abrams, J. Calabrese, D. Corbin, R. Harlow, G. Jones and M. Smith (University of Delaware). Additional financial support from NSF grant DMR-90-24249 is much appreciated. The X-7A and X-3 beamlines at the NSLS, Brookhaven National Laboratory, are supported by the DOE. The diffraction work reported here owes much to my fellow X-7 Participating Research Team (PRT) members at X-7A: L. Finger, J. Hriljac, B. Toby, and most particularly David Cox who, through his singular skill in building and developing X-7A, has demonstrated how productive the PRT concept can be.

References

1. W. M. Meier and D. H. Olson, *Atlas of Zeolite Structure Types*, Butterworths, London (1987).
2. P. B. Weisz and V. J. Frilette, *J. Phys. Chem.*, **64**, 382 (1960).
3. V. Alfredsson, T. Ohsuna, O. Terasaki, and J. Bovin, *Angew. Chem. Int. Ed. Engl.*, **32**, 1210 (1993).
4. J. J. Pluth and J. V. Smith, in *Intrazeolite Chemistry*, G. D. Stucky and F. G. Dyer (Eds), *ACS. Symp. Ser.* **218**, 119 (1983).
5. W. M. Meier, in *Zeolites: Synthesis, Structure, Technology and Application*, B. Drzaj, S. Hocevar, and S. Pejovnik (Eds), *Stud. Surf. Sci. Catal.*, **24**, 217 (1985).
6. D. E. Cox, B. H. Toby, and M. M. Eddy, *Aust. J. Phys.*, **41**, 117 (1988).
7. L. E. Iton, F. Trouw, T. O. Brun, J. E. Epperson, J. W. White, and S. J. Henderson, *Langmuir*, **8**, 1045 (1992).
8. J. Newsam and D. E. W. Vaughan, in *Zeolites: Synthesis, Structure, Technology and Application*, B. Drzaj, S. Hocevar, and S. Pejovnik (Eds), *Stud. Surf. Sci. Catal.*, **24** (1985).
9. J. M. Newsam, *Mat. Sci. Forum*, **27-28**, 385 (1987).

10. C. A. Fyfe, H. Stobl, H. Gies, G. T. Kokatailo, and G. Barlow, *J. Am. Chem. Soc*, **66**, 1942 (1988).
11. H. Gies, B. Marler, C. Fyfe, G. Kokotailo, Y. Feng, and D. E. Cox, *J. Phys. Chem. Solids*, **52**, 1235 (1991).
12. C. R. A. Catlow and C. M. Freeman, *Royal Institution Proceedings*, **63**, 51 (1992).
13. J. M. Newsam, in *Proceedings of the Ninth International Zeolite Conference*, R. B. v. Ballmoos, J. B. Higgins, and M. M. J. Treacy (Eds), Butterworth-Heineman, London, p 127 (1992).
14. A. P. Wilkinson, A. K. Cheetham, S. C. Tang, and W. J. Reppart, *J. Chem. Soc., Chem. Comm.*, 1485 (1992).
15. P. Coppens, *Synchrotron Radiation Crystallography*, Academic Press, London, (1992).
16. D. E. Cox, in *Synchrotron Radiation Crystallography*, P. Coppens (Eds), Academic Press, London, Vol. 186 (1992).
17. J. J. Pluth and J. V. Smith, in *8th International Zeolite Conference*, P. A. Jacobs and R. A. van Santen (Eds), Vol. 49, p 835 (1989).
18. H. v. Koningsveld, J. C. Jensen, and H. v. Bekkum, *Zeolites*, **10**, 235 (1990).
19. H. v. Koningsveld, J. C. Jensen, and H. v. Bekkum, *Zeolites*, **9**, 253 (1989).
20. H. v. Koningsveld, F. Tuinstra, H. v. Bekkum, and J. C. Jensen, *Acta Crystallog.*, **B45**, 423 (1989).
21. H. v. Koningsveld, F. Tuinstra, H. v. Bekkum, and J. C. Jensen, *Acta Crystallog.*, **B45**, 127 (1989).
22. H. v. Koningsveld, J. C. Jensen, and H. v. Bekkum, *Zeolites*, **7**, 564 (1987).
23. B. F. Mentzen and M. Sacerdote-Peronnet, *Mat. Res. Bull.*, **28**, 1161 (1993).
24. D. L. Bish and J. E. Post, *Modern Powder Diffraction*, Mineralogical Society of America, Washington, DC (1989).
25. G. C. Smith, *Synch. Rad. News*, **4**, 24 (1991).
26. P. A. Lee, P. H. Citrin, P. Eisenberger, and B. M. Kincaid, *Rev. Mod. Phys*, **53**, 769 (1981).
27. J. B. Parise, J. E. MacDougall, N. Herron, R. Farlee, A. W. Sleight, Y. Wang, T. Bein, K. Moller, and L. M. Moroney, *Inorg. Chem.*, **27**, 221 (1988).
28. G. A. Ozin, M. R. Steele, and A. J. Holmes, *Chem. Mater.*, **6**, 999 (1994).
29. C. J. Drummond and B. W. Ninham, *Chem. Aust.*, **59**, 529 (1992).
30. J. M. Newsam, T. O. Brun, F. Trouw, L. E. Iton, and Curtiss, in *New Catalytic Materials and Techniques*, R. T. K. Baker and L. L. Murrell (Eds), *ACS. Symp. Series* **437**, 25 (1990).
31. W. H. Baur, R. X. Fischer, R. D. Shannon, R. H. Staley, A. J. Vega, L. Abrams, D. R. Corbin, and J. D. Jorgensen, *Z. Kristallogr.*, **179**, 281 (1987).
32. W. H. Baur, R. X. Fischer, and R. D. Shannon, in *Innovation in Zeolite Materials Science*, P. J. Grobet (Ed), Elsevier, Amsterdam p 281 (1988).
33. W. H. Baur, A. Bieniok, R. D. Shannon, and E. Prince, *Z. Kristallogr.*, **187**, 253 (1989).
34. A. Bieniok and W. H. Baur, *J. Solid State Chem.*, **90**, 173 (1991).

35. A. Bieniok and W. H. Baur, *Mat. Sci. Forum*, **79-82**, 721 (1991).
36. A. Bieniok and W. H. Baur, *Acta Cryst.*, **B49**, 812 (1993).
37. D. R. Corbin, L. Abrams, G. A. Jones, M. M. Eddy, W. T. A. Harrison, G. D. Stucky, and D. E. Cox, *J. Amer. Chem. Soc.*, **112**, 4821 (1990).
38. D. R. Corbin, L. Abrams, G. A. Jones, T. E. Gier, R. L. Harlow, J. B. Parise, and P. J. Dunn, in *Synthesis/Characterization and Novel Applications of Molecular Sieve Materials*, R. L. Bedard, T. Bein, M. E. Davis, J. Garces, V. A. Maroni, and G. D. Stucky (Eds), Materials Research Society, Pittsburgh, PA, p 23 (1990).
39. D. R. Corbin, L. Abrams, G. A. Jones, M. M. Eddy, W. T. A. Harrison, G. D. Stucky, and D. E. Cox, *J. Amer. Chem. Soc.*, **112**, 4821 (1990).
40. R. X. Fischer, W. H. Baur, R. D. Shannon, R. H. Staley, A. J. Vega, L. Abrams, and E. Prince, *J. Phys. Chem.*, **90**, 4414 (1986).
41. R. X. Fischer, W. H. Baur, R. D. Shannon, R. H. Staley, A. J. Vega, L. Abrams, D. R. Corbin, and J. D. Jorgensen, *Z. Kristallogr.*, **179**, 281 (1987).
42. R. X. Fischer, W. H. Baur, R. D. Shannon, R. H. Staley, L. Abrams, A. J. Vega, and J. D. Jorgensen, *Acta Cryst.*, **B44**, 321 (1988).
43. R. X. Fischer, W. H. Baur, and R. D. Shannon, *Acta Cryst.*, **C45**, 983 (1989).
44. L. B. McCusker, and C. Baerlocher, in *Sixth International Zeolite Conference*, D. Olson and A. Bisio (Eds), Butterworths, London, p 812 (1984).
45. L. B. McCusker, *Zeolites*, **4**, 51 (1984).
46. J. B. Parise, and E. Prince, *Mat. Res. Bull.*, **18**, 841 (1983).
47. J. B. Parise, L. Abrams, T. E. Gier, D. R. Corbin, J. D. Jorgensen, and E. Prince, *J. Phys. Chem.*, **88**, 2303 (1984).
48. J. B. Parise, T. E. Gier, D. R. Corbin, and D. E. Cox, *J. Phys. Chem.*, **88**, 1635 (1984).
49. J. B. Parise, X. Liu, D. R. Corbin, and G. A. Jones, in *Synthesis/Characterization and Novel Applications of Molecular Sieve Materials*, R. L. Bedard, T. Bein, M. E. Davis, J. Garces, V. A. Maroni, and G. D. Stucky (Eds), Materials Research Society, Pittsburgh, PA, p 267 (1991).
50. J. B. Parise, X. Liu, and D. R. Corbin, *J. Chem. Soc., Chem. Commun.*, **3**, 162 (1991).
51. J. B. Parise, D. R. Corbin , L. Abrams, P. Northrup, J. Rackovan, T. M. Nenoff, and G. Stucky, *Zeolites*, **14**, 24 (1993).
52. A. V. Vernov, W. A. Steele, and L. Abrams, *J. Phys. Chem.*, **97**, 7660 (1993).
53. A. Loriso, M. J. Bojan, A. Vernov, and W. A. Steele, *J. Phys. Chem.*, **97**, 7665 (1993).
54. C. -J. Tsiao, J. S. Kauffman, D. R. Corbin, L. Abrams, E. Carroll, and C. Dybowski, *J. Phys. Chem*, **95**, 5586 (1991).
55. P. A. Wright, J. M. Thomas, S. Ramdas, and A. K. Cheetham, *J. Chem. Soc., Chem. Comm.*, 1338 (1984).
56. M. L. Smith, D. R. Corbin, and C. Dybowski, *J. Phys. Chem.*, **97**, 9045 (1993).
57. R. Rinaldi, J. J. Pluth, and J. V. Smith, *Acta Cryst.*, **B31**, 1603 (1975).
58. J. L. Schlenker, J. J. Pluth, and J. V. Smith, *Mat. Res. Bull.*, **14**, 751 (1979).

59. J. L. Schlenker, J. J. Pluth, and J. V. Smith, *Mat. Res. Bull.*, **14**, 849 (1979).
60. W. J. Mortier, J. J. Pluth, and J. V. Smith, *Nature*, **256**, 718 (1975).
61. W. J. Mortier, *Comparison of Extraframework Sites in Zeolites*, Butterworths, London (1983).
62. I. Gameson, P. A. Wright, T. Rayment, and J. M. Thomas, *Chem Phys. Lett.*, **123**, 145 (1986).
63. M. Keane Jr., G. C. Sonnichsen, L. Abrams, D. R. Corbin, T. E. Gier, and R. D. Shannon, *Appl. Catal.*, **32**, 361 (1987).
64. T. E. Gier, R. D. Shannon, G. C. Sonnichsen, D. R. Corbin, and M. Keane, Jr., *US Patent 4,806,689* (1989).
65. T. Ito and J. Fraissard, *Zeolites*, **7**, 554 (1987).
66. H. M. Rietveld, *J. Appl. Crystallog.*, **2**, 65 (1969).
67. A. C. Larson and R. B. Von Dreele, *GSAS: General Structure Analysis System,* Los Alamos Nat. Lab., Rpt. No.:LAUR86-748 (1986)
68. R. X. Fisher, W. H. Baur, R. D. Shannon, R. H. Staley, L. Abrams, A. Vega, and J. Jorgensen, *Acta Cryst.*, **B44**, 321 (1988).
69. C. Baerlocher, A. Hepp, and W. M. Meier, *DLS Manual*, Institut für Kristallographie und Petrographie: ETH, Zurich (1977).
70. D. R. Corbin, L. Abrams, M. L. Smith, C. R. Dybowski, J. B. Parise, and J. Hriljac, *J. Chem. Soc., Chem. Comm*, 1027 (1993).
71. H. E. Robson, D. P. Shoemaker, R. A. Ogilvie, and P. C. Manor, *ACS. Advan. Chem. Series*, **121**, 106 (1973).
72. H. E. Robson, *U.S. Patent 3,904,738* (1975).
73. N. J. Turro and P. Wan, *J. Am. Chem. Soc.*, **107**, 678 (1985).
74. N. J. Turro and Z. Zhang, *Tetrahedron Letters*, **28**, 5637 (1987).
75. N. J. Turro, X. Lei, and C. -C. Cheng, *J. Am. Chem. Soc.*, **107**, 5824 (1985).
76. J. V. Caspar, V. Ramamurthy, and D. R. Corbin, *J. Am. Chem. Soc.*, **113**, 600 (1991).
77. V. Ramamurthy, J. V. Caspar, D. R. Corbin, D. F. Eaton, J. S. Kauffman, and C. Dybowski, *J. Photochem. Photobiol*, **31**, 47 (1990).
78. V. Ramamurthy, J. V. Caspar, and D. R. Corbin, *J. Am. Chem. Soc.*, **113**, 594 (1991).
79. H. Nakamoto and H. Takahashi, *Chem. Letters*, 1013 (1981).
80. D. G. Hay and H. Jaeger, *J. Chem. Soc., Chem. Comm.*, 1433 (1984).
81. G. Kokotailo, L. Riekert, and A. Tissler, in *Zeolites as Catalysts, Sorbents and Detergent Builders*, H. G. Karge and J. Weitkamp (Eds), Elsevier Scientific, Amsterdam, p 821 (1989).
82. J. B. Parise, J. A. Hriljac, D. E. Cox, D. R. Corbin, and V. Ramamurthy, *J. Chem. Soc., Chem. Comm.*, 226 (1993).
83. J. Taylor, *Zeolites*, **7**, 311 (1987).
84. E. L. Wu, S. L. Lawton, D. H. Olson, A. C. Rohman, and G. T. Kokotailo, *J. Phys. Chem.*, **83**, 2777 (1979).
85. G. D. Price, J. J. Pluth, J. V. Smith, and T. Araki, *Nature*, **292**, 818 (1981).

86. C. Baerlocher, in *Proceedings of the 6th International Zeolite Conference*, D. Olson and A. Bisio (Eds), Butterworths, Guildford, UK, p 823 (1984).
87. B. F. Mentzen, *Mat. Res. Bull.*, **22**, 489 (1987).
88. B. F. Mentzen, *J. Appl. Crystallog.*, **22**, 100 (1989).
89. A. Le Bail, H. Duroy, and J. L. Fourquet, *Mat. Res. Bull.*, **23**, 447 (1988).
90. L. S. Rollman and E. K. Volyocsik, *Inorganic Syntheses*, Wiley, New York, **22**, 61-68 (1983).
91. J. A. Hriljac, J. B. Parise, G. H. Kwei, and K. B. Schwartz, J. *Phys. Chem. Solids*, **52**, 1273 (1991).
92. Y. Zhao, D. J. Weidner, J. B. Parise, and D. E. Cox, *Phys. Earth Planet. Inter.*, **76**, 1 (1993).
93. Y. Zhao, D. J. Weidner, J. B. Parise, and D. E. Cox, *Phys. Earth Planet. Inter.*, **76**, 17, (1993).
94. E. R. Geus, M. J. Den Exter, and H. v. Bekkum, *J. Chem. Soc., Faraday Trans*, **88**, 3101 (1992).
95. D. H. Olson, W. O. Hagg, and R. M. Lago, *J. Catal.*, **61**, 390 (1980).

Abstract. The development and increased availability of brighter synchrotron sources has stimulated investigations of zeolite-inclusion complexes using powder X-ray diffraction data. This is illustrated with two examples of studies facilitated by these sources. Both involve the use of Rietveld refinement to obtain structure from data collected at a versatile synchrotron beamline. The first example describes the structural changes commensurate with Xe desorption from a fully Xe-loaded powder sample of Cd-exchanged zeolite rho. After exposure to the atmosphere, data were collected in real-time as Xe desorbed from the sample over a four hour period. The results of Rietveld refinement indicate that Xe is not lost continuously; instead within the first hour, Xe desorbs completely from the 8-ring and is replaced by H_2O. Following rearrangement of the remaining Xe between the available 6-ring sites, it is lost as the sample hydrates fully. This behavior is probably the result of competition with H_2O, which prefers the 8-ring site. In the second example, the high resolution and signal-to-background discrimination afforded the synchrotron powder X-ray diffraction experiment, was used to advantage to study the stilbene-ZSM-5 complex. Stilbene was located, using modeling techniques, in the straight channels with one phenyl ring at the intersection of the two channel system. The structure is pseudo-tetragonal (*a a b*) at room temperature but distorts at lower temperatures to give $b > a$. Distinguishing these subtle structural changes is facilitated by the superior resolution available at a beamline configured with both incident and diffracted beam monochromators.

Journal of Inclusion Phenomena and Molecular Recognition in Chemistry **21**: 113–136, 1995.

Zeolite-Included Molecules Studied by NMR

CECIL DYBOWSKI

1. Introduction

Zeolites, like silicas and aluminas, are used widely in a variety of catalytic and noncatalytic applications in the chemical industry. To define the nature of processes occurring while molecules are absorbed, it is necessary to measure the properties of these absorbed molecules. Very frequently, the absorbed phases are neither liquid, gas, nor solid, being strongly influenced by the proximity of the zeolite structure. Aside, then, from the practical importance of these materials, the study of an unusual state of matter is yet another impetus for examining absorbed molecules. One may investigate these intermediate states in many ways, ranging from classical thermodynamic studies to sophisticated instrumental techniques, to define the structural and dynamic properties of absorbed molecules.

NMR spectroscopy has a particular strength in defining structural and dynamic properties of molecules, since NMR parameters depend strongly on the molecule's state. For example, the resonance position is sensitive to the local electronic environment, which changes with molecular conformation and with intermolecular arrangement. Dynamic parameters such as relaxation times are reporters of molecular internal and overall mobility. In addition, line widths of resonances can often be used to infer such useful information as the average distance from the location of one nuclear spin to another. Thus, NMR spectroscopy provides a sensitive spectroscopic tool with which to probe the absorbate's environment.

The information content of an NMR spectrum is quite specific [1]; however, because the technique is relatively insensitive, requiring substantial numbers of spins in the resonance region for a signal to be detected, its use is limited to the study of relatively high-surface-area materials loaded to reasonably high loadings [2]. In addition, since it is not inherently a surface tool, NMR spectroscopy may be subject to interferences, either from the material whose surface is being probed or from the spectrometer system. Therefore, most NMR spectroscopy has been performed on high-surface-area materials such as zeolites.

The extreme sensitivity of NMR parameters to slight differences in the local environment means that NMR spectra of molecules in contact with amorphous solids are often broadened by the distribution of structures and/or environments, with a concommitant loss of detail. Crystalline zeolites are generally rather homogeneous, and molecules absorbed in them can be expected to have NMR spectra of higher resolution than molecules in an amorphous material such as a silica-alumina. Hence, zeolites are ideal systems to be investigated with NMR spectroscopy.

2. Effect of Exchange in NMR Spectroscopy

The principal effect of a surface is to provide sites for localization of an absorbed molecule. From the NMR standpoint, each site may be considered a unique state of the molecule with unique NMR parameters. If the molecule were localized at a site sufficiently long, one would detect only the NMR parameters of the molecule in that state. Because surfaces, and catalytic surfaces in particular, may not bind molecules strongly enough that residence times at a site are long on the NMR time scale (milliseconds to seconds), the detected signal may result from a molecule exchanging among various sites.

Exchange imparts interesting NMR properties to a system, which have been discussed in many elementary texts [3]. Here, I review simple concepts which, nevertheless, provide a useful framework for understanding the NMR spectroscopy of molecules absorbed in a catalyst such as a zeolite when exchange is present.

2.1 RAPID EXCHANGE AND CHEMICAL SHIFTS

In general, exchange can result in rather involved NMR results. In the limit that residence times are long and exchange rates are low, one detects individual resonances for molecules in all the various sites. In the range of intermediate exchange, the lineshapes can be rather complex, and will depend on the nature of the exchange. However, if exchange is rapid on the time scale of NMR interactions, the results are simple averages.

Consider the observed chemical shift of some nucleus on a molecule rapidly exchanging among several environments. Let the shift of the nuclear species be δ_i when it is localized for a long time on site i. Because of exchange between sites, one detects only an average chemical shift:

$$\delta = p_1\delta_1 + p_2\delta_2 + p_3\delta_3 + ... \quad (1)$$

where p_i is the probability of the molecule residing at site i. In cases in which only a few sites participate in exchange on the NMR time scale, one need only consider those sites in performing the average of Equation 1.

The problem of the effect of exchange between two sites on chemical shift is illustrative of the effects of adsorption on the detected NMR spectroscopy. In this case, the exchange may be represented by the following process:

$$A + S \overset{K}{\rightleftarrows} S - A \quad (2)$$

where S is a vacant site, A is the unbound adsorbate, S-A is the complex formed by adsorption, and K is the equilibrium constant for this exchange. As examples of systems where this model might be appropriate, one may cite the interaction of a molecule in the voids of the material with a specific adsorption site on the surface of the void space, and the interaction of a molecule from the bulk gas or liquid interacting with a site at a surface. Considering the total number of adsorption sites to be N*, the total number of molecules to be N_0, and the equilibrium constant for the exchange process to be K, one can calculate the probability that the molecule is in the adsorbed state by Equation 3:

$$p_a = \frac{1}{2}\left(1 + \frac{N^*}{N_0} + \frac{1}{KN_0}\right) \pm \frac{1}{2}\left(\left(1 + \frac{N^*}{N_0} + \frac{1}{KN_0}\right)^2 - 4\,\frac{N^*}{N_0}\right)^{1/2} \tag{3}$$

The probability that the molecule is not absorbed in this two-site case is $1 - p_a$. Using this result and Equation 1, one can predict the average chemical shift of the nuclear species as a function of N_0, the total concentration:

$$\delta = \delta_0 + (\delta_{ads} - \delta_0)p_a. \tag{4}$$

In the limit that adsorption is strong or there are many adsorption sites [$KN^* >> 1$], Equation 4 simplifies to

$$p_a = \frac{N^*}{N_0} \tag{5}$$

and so

$$\delta = \delta_0 + (\delta_{ads} - \delta_0)\left(\frac{N^*}{N_0}\right). \tag{6}$$

The shift approaches that of the unabsorbed material as N_0 becomes very large. For the case that there are fewer molecules than there are adsorption sites, i.e., $N_0 << N^*$, the value of p_a approaches one. In this condition, one measures a seemingly coverage-independent chemical shift, which is the shift of the molecule while bound at the site.

It is important to note that some or all of the bulk gas phase may NOT be in rapid exchange with the sorbed material, in which case one detects a resonance for that material not participating in exchange at the unshifted position. N_0 is then the sum of material absorbed and that part of the bulk phase participating in exchange with the sorbed state, not the total amount including the material not in exchange. In many cases, the exchange may be between sites in the zeolite, perhaps one being a physisorption site and the other a more strongly bound site.

2.2 SPIN-LATTICE RELAXATION IN AN EXCHANGING SYSTEM

The spin-lattice relaxation rate is affected by exchange between sites. Considering two sites A and B, we let $x(t) = M_A(t) - M_A^0$ and $y(t) = M_B(t) - M_B^0$, where $M_i(t)$ represents the magnetization at site i and M_i^0 is the equilibrium magnetization at that site, one obtains the following two differential equations:

$$\left(\frac{dx}{dt}\right) = -K_A x + k_{AB} y \tag{7}$$

$$\left(\frac{dy}{dt}\right) = -K_B y + k_{BA} x \tag{8}$$

$$K_i = \frac{1}{T_{1i}} + k_{ij} \tag{9}$$

where k_{AB} is the rate constant for exchange from A to B, k_{BA} is the rate constant for the exchange from B to A and where i and j refer to the sites A and B. $1/T_{1i}$ is the intrinsic relaxation rate if the molecule remained at site i indefinitely. Equations 7 and 8 may be solved simultaneously to give solutions for x(t) and y(t):

$$x(t) = \exp(-\bar{\lambda}t)[X_1 e^{-\Delta t} + X_2 e^{\Delta t}] \tag{10}$$

$$y(t) = \exp(-\bar{\lambda}t)[Y_1 e^{-\Delta t} + Y_2 e^{\Delta t}] \tag{11}$$

$$\bar{\lambda} = \frac{K_A + K_B}{2} \tag{12}$$

$$\Delta = \left(\frac{K_A + K_B}{2}\right)\sqrt{1 + \frac{4k_{AB}k_{BA}}{(K_A - K_B)^2}} \tag{13}$$

$$Y_1 = \frac{k_+ y(0) - x(0)}{k_+ - k_-} \quad \text{and} \quad Y_2 = \frac{x(0) - k_- y(0)}{k_+ - k_-} \tag{14}$$

$$X_1 = k_- Y_1 \quad \text{and} \quad X_2 = k_+ Y_2 \tag{15}$$

and

$$k_\pm = \frac{K_B - \bar{\lambda} \pm \Delta}{k_{BA}} . \tag{16}$$

Understanding the NMR spectroscopy of absorbed species often hinges on understanding the interchange of molecules among several adsorption sites. Thus, the theory of the effects of chemical exchange on line positions, line shapes, and relaxation parameters will be important in the analyses discussed below, because zeolites have various sites in which absorbed molecules may reside.

3. Xenon in Zeolites

3.1 GENERAL QUALITIES OF XENON NMR

Xenon gas has two stable isotopes that can be investigated with NMR spectroscopy: ^{129}Xe (26.4%) and ^{131}Xe (21.2%), having spins of 1/2 and 3/2, respectively. NMR spectroscopy of ^{129}Xe is relatively easy to perform. Owing to the adequate natural abundance (26.8%) and its magnetic moment, its receptivity relative to ^{13}C is 31.8. Thus, detection of resonant absorption in a sample of natural abundance is straightforward, neglecting differences in relaxation time and volume density. The resonance occurs at a frequency about 10% higher than ^{13}C at the same magnetic field strength; thus, experiments can be performed in a spectrometer designed for detection of the ^{13}C resonance by a slight change of the circuit tuning. ^{129}Xe in a variety of environments has been studied with NMR spectroscopy, and the range of shifts is over 5000 ppm [4], although the majority of shifts lie in a range of about 500 ppm. The dominant effect in ^{129}Xe-gas NMR is the strong dependence of parameters on gas density - the result of collisions experienced by xenon atoms. Owing to xenon's high polarizability, collisions affect the average chemical shielding due to the electrons. The resonance frequency of the gas at constant magnetic field, for example, changes dramatically with the density. It is this sensitivity to collisions and interactions with the local environment that has recently made the study of the ^{129}Xe NMR spectrum of the absorbed gas a topic of intense study in a number of laboratories.

The shift of the gas at infinitely low density is used as a reference against which shifts of xenon in other environments are reported; practical standard reference materials include xenon gas at various pressures and $XeOF_4$. The resonance of an isolated xenon atom has been estimated to be 5642 ppm more shielded (i.e., to lower frequency at constant magnetic field) than the bare xenon nucleus [5]. Similar calculations [6] for the isolated ^{1}H atom compared to the proton and ^{13}C relative to the bare ^{13}C nucleus give shifts of 18 ppm and 260 ppm, respectively.

The resonance shift of pure liquid xenon ranges between ~240 ppm at 160 K and ~160 ppm at 244 K, and the resonance of the pure solid occurs between ~270 ppm to ~305 ppm, depending on the temperature [7].

Gas-phase NMR spectra contain a great deal of information. In macroscopic spaces xenon-xenon collisions greatly exceed xenon-"wall" collisions. The ^{129}Xe resonance condition strongly depends on the gas pressure (or density). It becomes less shielded (shifts to higher frequency at constant magnetic field) the higher the gas pressure [8,9]. Over a reasonably wide range of density, the shift can be represented by a virial expansion:

$$\delta(T,\rho) = \delta_0 + \delta_1(T)\rho + \delta_2(T)\rho^2 + \delta_3(T)\rho^3 + ... \tag{17}$$

where $\delta(T,\rho)$ is the resonance shift at a temperature T and a density ρ, δ_0 is the resonance position at an infinitely low density, and $\{\delta_i(T)\}$ are the virial coefficients of the resonance shift in density [10]. The linear term results from binary collisions and the higher-order terms arise from collisions involving three- and four-body collisions. For densities below 100 amagats, terms with powers higher than 1 are essentially negligible. Over the wider range from 0 to 250 amagats [11], higher-order terms contribute and the virial coefficients have the following values for pure xenon gas:

$$\begin{aligned}
\delta_1(298.15\text{ K}) &= 0.548\ (\pm 0.004)\text{ ppm/amagat} \\
\delta_2(298.15\text{ K}) &= 1.69\ (\pm 0.2) \times 10^{-4}\text{ ppm/amagat}^2 \\
\delta_3(298.15\text{ K}) &= -1.63\ (\pm 0.1) \times 10^{-6}\text{ ppm/amagat}^3
\end{aligned}$$

One can predict the magnitude of δ_1 by calculating the interaction of two xenon atoms during a binary collision [12]. The perturbation due to repulsive interactions between two xenon atoms giving a contribution to δ_1 of approximately 0.28 ppm/amagat at 298.15 K. The attractive part of the potential also contributes [11], but not to the extent it does for hydrogen or fluorine gas, for which the dominant effects on shift result from attractive interactions. The coefficients $\{\delta_i\}$ are temperature dependent [11]. The extrapolated shift, δ_0, of an isolated xenon atom is, however, rigorously temperature-independent [13]. This fact allows one to calibrate the temperature-dependent shifts of standard reference materials such as tetramethylsilane (TMS) [14]. The shift of the xenon resonance depends on the collision partner. For mixtures of xenon with other gases at sufficiently low densities, one may generalize to a multi-dimensional virial expansion of the shift, keeping only linear terms due to binary collisions:

$$\delta((T),\rho_{Xe},\rho_A) = \delta_0 + \delta_{1,Xe}(T)\rho_{Xe} + \delta_{1,A}(T)\rho_A \tag{18}$$

where $\delta_{1,i}$ is the second virial coefficient in the density of gas i [11]. Heteromolecular collisions are usually less effective in changing the xenon resonance position than are collisions with xenon atoms, the exceptions being paramagnetic species like O_2 and NO. The change in resonance position is due to bulk susceptibility and to a "contact-overlap" mechanism [15], in which Xe(5s) orbitals interact with π^* or $\pi_g{}^*$ orbitals of the collision partner and unpaired electrons contact the xenon nucleus. Just as the second virial coefficient for xenon-xenon interactions is temperature dependent, so are the $\delta_{1,A}$ for interactions with other gas molecules [16].

Xenon, like many gases, dissolves to some extent in liquids [17]. The observed shifts tend to be independent of concentration, indicating the infinitely-dilute limit. In addition to bulk susceptibility, shifts in this limit are affected by specific or nonspecific interactions with solvent molecules. Specific interactions result from definite molecular associations, e.g., hydrogen bonding. Since xenon is relatively chemically unreactive, shifts are attributable to specific interactions in only a few systems [18]. Non-specific interactions include: (1) magnetic anisotropy of the solvent molecule; (2) electric-dipole interactions; and (3) van der Waals interactions between the solute and solvent. Rummens [19] treated the contribution of the dispersive parts of the van der Waals interactions as if xenon were acted upon by a "reaction field" of a continuous medium arising from spontaneous electronic fluctuations in the solute, following the early work of Onsager [20] and Linder [21], giving a term in the medium shift proportional to $[(n^2 - 1)/(2n^2 - 1)]^2$, where n is the refractive index of the solution. The relation of the xenon shift to Rummens' function seems to hold for a rather diverse set of solvents, [17,18] as long as one considers sets of homologous materials, e.g., the linear alcohols. This is most probably because other contributions to the shift, including repulsive parts of the van der Waals interactions, are similar within an homologous series. Xenon dissolved in water shows an anomalously large shift when compared to similar materials, such as alcohols, probably because of a significant effect of local solvent structure. Similar interactions with the environment are seen in Ripmeester and Davidson's study [22] of enclathrated xenon, and in studies of polymer morphology through NMR of "dissolved" xenon [23,24].

3.2 XENON IN MICROSCOPIC SPACES

Obviously, the interaction of a xenon atom with its environment profoundly affects the observed NMR properties. In the macroscopic gas, the walls are sufficiently far apart that interactions with them are rendered statistically insignificant compared to interactions with other xenon atoms or other molecules. In solutions, the nature of the "solvent" seems to be the dominant factor, and the general trend is understood in terms of specific and nonspecific interactions with the solvent.

When one partitions the space in which xenon is confined by walls separated by dimensions of 0.5 to 10 nanometers, the interactions with the "walls" dominate, as is seen in the clathrates [25], in solution, and in intercalation compounds [26,27]. Zeolitic materials, of course, are paradigms of this effect. The regular crystalline structures formed by zeolite frameworks partition space into regions of molecular dimensions connected in regular arrays [28]. Xenon in such structures would be expected to have strong interactions with the zeolite, which should be clearly signaled in the NMR behavior. In addition to the manner in which space is partitioned microscopically, the presence of a framework bearing charge means that there also exist charge-compensating ions, with which xenon may interact. The amount and type of ion(s) associated with a zeolite have a significant effect on its catalytic and sorptive properties. Thus, it was proposed [29] that the NMR spectroscopy of absorbed xenon may detect features of the zeolite which are relevant to its catalytic properties such as where an ion resides, the kinds of coabsorbed materials, and the framework structure.

Figure 1 shows the adsorption isotherm, at 293 K, for xenon in Y zeolite containing both nickel and sodium counterions. The isotherm is approximately linear over the range. Because dehydrated Y zeolite has a typical pore volume of 0.34 cm^3/g [28], xenon absorbed at 3 atoms per supercage corresponds to a density of 0.6 g/cm^3, a reasonably high density but quite a bit lower than the 3.5 g/cm^3 of liquid xenon [30]. Loadings in the range of 3 atoms per supercage correspond to equivalent densities in the range of 100 amagats.

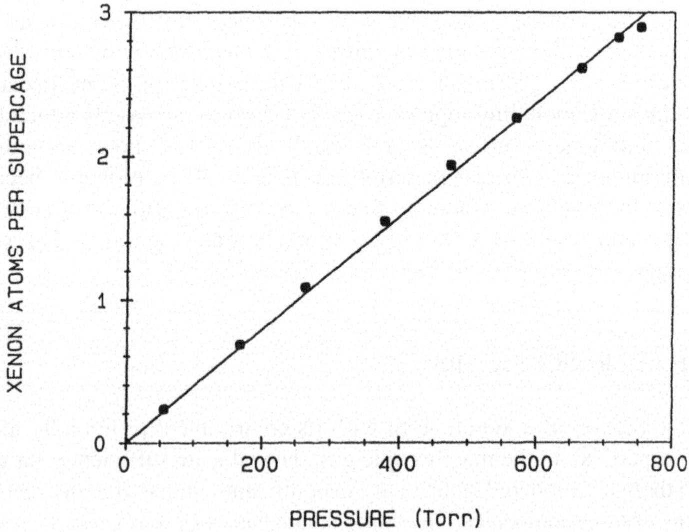

Figure 1. Isotherm for adsorption of xenon on a partially dehydrated Ni,Na-Y zeolite at 295 K. The solid line indicates the best fit to a linear isotherm for this material.

There are a number of interactions that contribute to the local environment of xenon in zeolites: (a) interaction with the "walls"; (b) interaction with other sorbed xenon atoms; (c) exchange among different adsorption sites; and (d) interaction with other absorbed material. In the absorbed state, the "walls" of the zeolite have a substantial effect on xenon. Ripmeester and Ratcliffe [31] describe a model in which two regimes are identified depending on the difference between the van der Waals radius of a xenon atom and the radius of curvature of the sphere formed by the substrate. Calculations based on the Xe-O Lennard-Jones potentials predict an optimum xenon-zeolite interaction of 20 kJ/mole in a hypothetical supercage with van der Waals radius of 12.5 Å. For zeolites with smaller cages (such as the α-cage of Na-A zeolite with a diameter of 11.5 Å) the adsorption potential for xenon has a single minimum at the center of the supercage (The energy at the minimum is 15 kJ/mole for Na-A zeolite). For zeolites with larger supercages (e.g, with a diameter of 14.5 Å), the adsorption potential for xenon has a minimum near the walls and a hump at the center of the supercage. (For Na-Y zeolite, the energy at the minimum is 15 kJ/mole at the wall with a barrier of 1.2 kJ/mole in the center.) In this model, the effect of the local environment depends on the ratio of the supercage volume to the surface area and divides into two classes: "clathrated" xenon in small supercages and "fluid" xenon in larger supercages.

Xenon diffuses through the interconnected supercages [32] and thus its NMR spectrum reflects the average of the local environments sampled during the free-induction decay of its nuclear magnetic signal. For a typical linewidth of 20 ppm in a magnetic field of 4.7 Tesla, the xenon free-induction decay persists for only a few milliseconds. The extent that the xenon diffuses during this time depends on the width and height of the potential barriers between adjacent adsorption sites and the occupancy of the sites. Consider the two classes of adsorption that emerge from the model of Ripmeester and Ratcliffe. For small supercages, adjacent minima are the centers of adjacent supercages and the potential barriers are at the windows. Assuming a barrier of 10 kJ/mole, a pre-exponential factor of 5×10^{-14} sec, and a loading of 1 xenon atom per 2 supercages, one calculates a site lifetime of 6×10^{-12} sec. A random walk in 1 msec would carry a xenon atom to many supercages. Thus, the time-averaged parameters (shifts, relaxation times, etc.) detected in an NMR experiment are averages over regions of rather large size (as compared to the supercage). Such averaging will tend to "wash out" the effects of inhomogeneities on smaller length scales, however, it should, in principle, be possible to detect large inhomogeneities with xenon NMR spectroscopy. An important result is that almost all xenon NMR information from such systems are exchanged-averaged quantities, of the sort one obtains from a simple theory as discussed in Section 2.

Such exchange-averaged quantities can then be considered to be a sum of contributions. Ito and Fraissard describe the NMR shift as arising from a series of terms, each the result of some interaction of xenon with the local environment:

$$\delta = \delta_{wall} + \delta_{xenon} + \delta_{adsorbate} + \delta_{electric} \tag{19}$$

where δ is the observed shift, δ_{wall} is a contribution to the shift from collisions with the zeolite structure, δ_{xenon} is a contribution from other xenon atoms, $\delta_{adsorbate}$ arises from interactions with other adsorbates, and $\delta_{electric}$ is a contribution due to polarization of xenon by electric fields in the zeolite pore. Each of these terms depends on physical parameters of the system such as the density of xenon in the absorbed phase, the type, number and distribution of ions, and the temperature. As a first approximation, the first term is independent of xenon density. The second is given by an expansion in powers of the density, as for the gas. The third term is analogous to the contribution of other gases in a mixture and will depend linearly on the amount of co-absorbate. The contribution from electric fields, in this model, depends on factors such as the amount and positioning of ions; for systems like Na-Y zeolite at 295 K, it is considered to be negligible. Thus, for totally dehydrated Na-Y zeolite containing no other absorbates at 295 K, this model predicts that the shift depends on xenon density in the following manner:

$$\delta = \delta_{wall} + \delta_1 \rho + \delta_2 \rho^2 + \ldots\ldots\ldots \tag{20}$$

where ρ is the density of absorbed xenon. If the isotherm is linear, then an analogous expression in the gas pressure at equilibrium can be used, as shown in Figure 2 for xenon in Na-Y zeolite [33]. The linearity of this plot and the virtual linearity of the isotherm over this range indicates that only the first two terms of Equation 20 make significant contributions to the ^{129}Xe NMR shift under these conditions. The intercept gives a value for δ_{wall} of 57.4 (±2) ppm; thus interactions with the framework do have a significant effect on the ^{129}Xe NMR properties. The parameter δ_1, when the density is expressed in amagats, has a value of 0.40 ppm/amagat, somewhat less than the value of 0.548 ppm/amagat observed for xenon in the bulk.

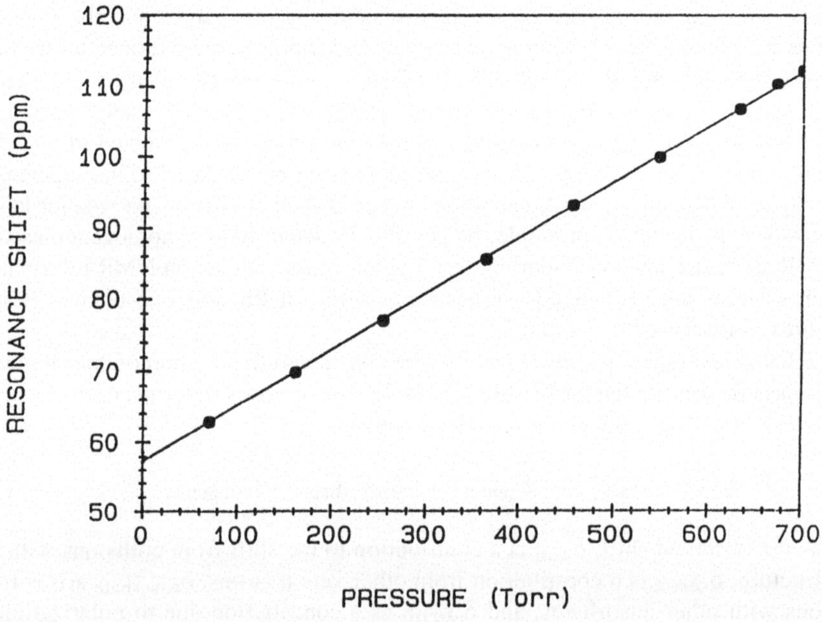

Figure 2. The dependence of ^{129}Xe NMR shift of xenon absorbed in a partially dehydrated Na-Y zeolite on pressure of the gas above the zeolite. The zero of the shift scale is the point extrapolated to P = 0 for bulk xenon by Jameson's equation (Equation 1).

Ito and Fraissard argue that this discrepancy results from restricted diffusion from one cage to another, effectively decreasing either the frequency or impact of xenon-xenon collisions, or both. If one assumes only collision frequency is altered by the presence of the zeolite (i.e., the zeolite merely acts as a steric hindrance to decrease the rate of xenon-xenon collisions but does not alter the microscopic details of the xenon-xenon collision), the ratio of these two parameters (0.73) is the relative number of collisions per unit time for absorbed xenon compared to a bulk gas at the same number density. The fact that both δ_{wall} and δ_1 are the same for Na-Y zeolites of varying silicon/aluminum ratios [34] demonstrates that neither interactions with the wall nor xenon-xenon collisions are affected by the amount of aluminum and silicon in the framework of these materials. Thus, minor changes in structural parameters accompanying changes in silicon-to-aluminum ratio of Y zeolites are not sensed by this xenon NMR technique. Indeed, Ito and Fraissard found that the Na and H forms of Y zeolite were indistinguishable by xenon NMR spectroscopy, showing that interactions with these ions are also either negligible or the same for these conditions.

The dependence of ^{129}Xe NMR shift on the xenon density in the absorbed phase can be more complex than that of xenon in Na-Y zeolite at 295 K. For example, in their early work, Ito and Fraissard observed that the dependence of shift on xenon density in a calcium Y zeolite was distinctively curved, with a minimum shift at a loading of approximately 1.75 xenon atoms per supercage. Similarly, Scharpf et al. [33,34] found that a Y zeolite containing a mixture of nickel and sodium counter-ions exhibited a similar density dependence of the shift. Two, more thorough, studies [35,36] demonstrated that this behavior may be viewed as evidence for exchange of xenon between two different adsorption sites [37,38], and could therefore be modeled as an exchange-averaged shift as discussed in Section 2. The two sites could be, for example, a site near an ion (a chemisorption site) and a site in the pore volume of the supercage (a physisorption site). They might also be a site inside the zeolite particle and one in a gas phase outside the zeolite particle. They could be a site near the wall of the pore and one farther from the wall caused by steric or energetic variation of absorption within the pore [39]. Whatever the nature of the two sites, the shift when xenon is fixed at one site must be measureably different from that at the other, and there must be rapid exchange between the two sites on the NMR time scale. The average shift can be written as follows, by application of the exchange averaging discussed in Section 2:

$$\delta = \left[\frac{N_A}{N}\right]\delta_A + \left[\frac{N-N_A}{N}\right]\delta_B \tag{21}$$

where N_A is the equilibrium population of site A, and N is the total amount of material adsorbed on the two sites. If A is a site of limited sorption capacity such that $N >> N_A$, this result approximately becomes:

$$\delta = \delta_0 + \delta_1 N + \frac{[N_A\delta_A]}{N} \tag{22}$$

where δ_0 is a density-independent term and δ_A is a shift coefficient describing the effects of "sticky" collisions with the special sites. This mechanism predicts that, under some conditions, the average shift will vary inversely with N and linearly with N under other conditions. If δ_A and δ_1 are positive numbers, the minimum value of δ is reached for $N_{min} = [\delta_A N_A/\delta_1]^{1/2}$.

There is still the question of the identity of the sites. If site B is identified as a dense gas-like material in the pores of the zeolite, δ_B will depend on uptake (like the gas depends on density), giving a complex dependence of the shift on N. If interactions at site A (presumed to be at the wall or at ions) perturb the electronic structure of xenon, this mechanism could account for the observed effects in certain zeolites. Early work by Ito and Fraissard [34] demonstrated that replacement of sodium by calcium introduced this sort of term into the density dependence of the xenon NMR shift. Our work on xenon in Ni,Na-Y zeolites [35] with varying nickel contents showed that the strength of this term in the inverse of the uptake depends sharply on the nickel content of these catalysts and on the state of hydration of the ions. Shifts in fully hydrated samples show no evidence of a term inverse in the density. This can be most easily seen in a plot of δ_{-1} [$\equiv N_A\delta_A$] as a function of the amount of sodium replaced by nickel [Figure 3].

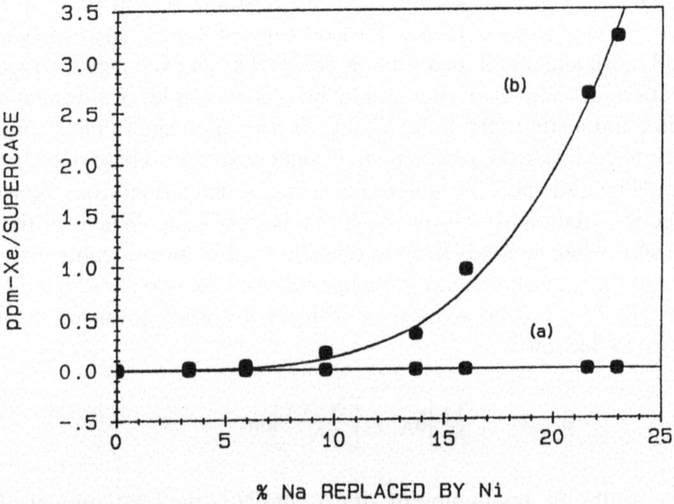

Figure 3. The coefficient, δ_{-1}, of the xenon shift term in inverse density for a series of Ni,Na-Y zeolites as a function of Na ions replaced: (a) samples dehydrated at 298 K; (b) samples dehydrated at 373 K. (After Ref. [35])

Samples dehydrated at 295 K contain a substantial amount of water which most probably sequesters the ions; samples dehydrated at 373 K do contain water, but some form of nickel ion is probably accessible to xenon in these samples. The fact that the inverse-density term appears only for these latter samples, and only at substantial nickel content, indicates that xenon *must* interact directly with some ion of nickel, implying the need for the close contact characteristic of adsorption. It has been found [40] that the xenon-density dependence of the shift even for Na-Y zeolite has a complex behavior at temperatures below room temperature, in which interactions with ions cause deviations from linearity. Xenon absorbed in a wide variety of other structures has been analyzed by the NMR method: ZSM-5, ZSM-11, L, Z, silicalite [41], mordenite [42], omega and ferrierite [43], beta [44], rho [45], and a silicoalumino-phosphate molecular sieve, SAPO-37 [46].

A Y zeolite structure with platinum ions present gives an even more complex result [47]. For xenon in a Y zeolite containing platinum as well as sodium, the NMR shift depends inversely on the xenon density, showing that the inverse term dominates in these materials. The authors attribute the large shift to electron transfer from xenon to the metal, with structures like $Pt\text{-}Xe^+$ contributing to the structure of the chemisorbed complex [47,48]. Similarly, NMR shifts of xenon gas sorbed in platinized aluminas show effects of interaction with the platinum particles [49].

From a practical point of view, xenon NMR experiments on such materials [47] provide an analytical tool to discriminate various regions in the sample. In Pt,Na-Y zeolites in which hydrogen is co-adsorbed with xenon, the xenon NMR spectrum contains several lines. By investigation of the relative intensities of these lines as a function of hydrogen content, Ito and Fraissard [47] demonstrate that each line corresponds to xenon in a region consisiting of platinum particles containing a specific number of hydrogen atoms. Thus, the xenon NMR experiment reports on the inhomogeneous distribution of hydrogen on metal particles. In further experiments using CO as a co-adsorbate, they [47] also show that xenon NMR is sensitive to the distribution of CO on platinum particles. By studying how the spectra change as a function of heat treatment, the authors show that adsorption of CO at 298 K is non-uniform, with CO adsorbing in the region where it first contacts the metal. Heating to 573 K redistributes the CO, which is again detected by a change in the spectrum of xenon co-adsorbed with the CO. Thus, xenon NMR can be used to investigate the distribution of hydrogen and CO in metal-containing zeolites, if the inhomogeneities are larger than the region over which xenon travels (and hence averages the shift).

Processes such as reduction change the ionic composition and the ion positions in a zeolite. These changes may be detected by the effects of the ions on the NMR properties of xenon absorbed in the material [33,34]. For example, as nickel in a Ni,Na-Y zeolite is reduced by treatment with hydrogen, the spectrum of co-adsorbed xenon shows evidence of an inhomogeneous distribution of xenon environments, with two positions becoming particularly prominent. One of these lines is similar in position to that of xenon absorbed in a Na- or H-Y zeolite. This line is not affected by the extent of reduction, whereas the second one is. At sufficiently high reduction temperature (643 K), only the first line is observed, although it is quite broad compared to uniform H-Y zeolites. This observation is consistent with the reduction of nickel ion by replacement with a proton and removal of the nickel from the supercage, most likely moving to the surfaces of the zeolite particles. Upon re-oxidation at high temperature, the line narrows but retains its position, suggesting that the re-oxidation process has made the magnetic environment of xenon more uniform in these particles, but it does not return the material to the state it had before reduction.

Xenon NMR of the absorbed gas is sensitive to differences in the structure of the zeolite in certain cases [50]. While the dependence on xenon density is slightly different from one zeolite structure to another, the most striking difference is the variation of the wall term, δ_{wall}, with the structure. Demarquay and Fraissard [51] assume this variation with structure results from change in the collision frequency with structure. They define a mean-free-path in the structure that describes the exchange of a xenon atom between a site on the surface of the pore and a site in the void volume of the pore. From this simple picture, they deduce that the wall term depends on this mean-free-path by the following simple relation:

$$\delta_{wall} = \frac{\delta_a}{1 + \lambda / a} \tag{23}$$

where δ_a is a shift determined by the wall adsorption, λ is the mean-free-path in the structure, and a is a coefficient characteristic of the adsorption site [37].

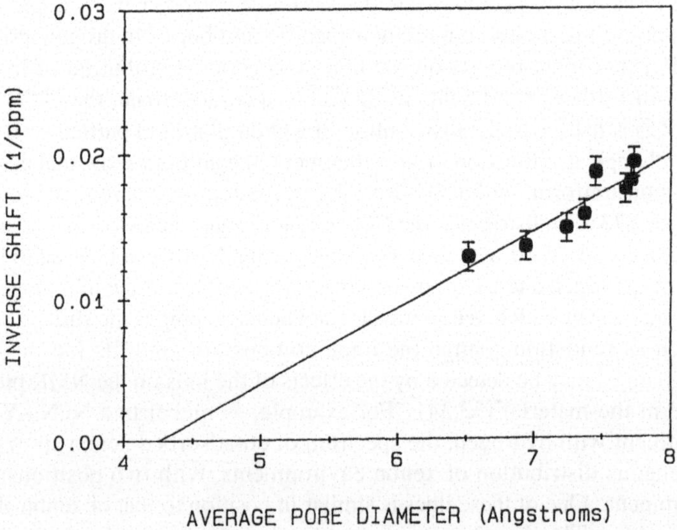

Figure 4. Dependence of $1/\delta_{wall}$ on dynamic pore diameter for a series of dehydrated zeolites, showing that this parameter obeys a relation similar to Equation (9). (After reference [52])

The data of Figure 4 compare $1/\delta_{wall}$ with average dynamic pore diameter for xenon absorbed in several partially dealuminated zeolites [52]. This has the form of the Demarquay-Fraissard equation, if we make the assumption that mean free path and dynamic pore diameter are linearly related. Interestingly, the extrapolation to infinite shift gives a dynamic pore diameter of 4.3 Å for a pore just large enough to accommodate a xenon atom, close to the accepted van der Waals diameter of the xenon atom of 4.4 Å.

Derouane and Nagy [53] propose an alternative model to predict the dependence of δ_{wall} on structure. In this static picture, the shift results from van der Waals interactions between the adsorbed xenon and oxygen anions in the curved surface of the zeolite pore that produce a paramagnetic contribution to the shift proportional to the interaction energy. The variation of the van der Waals interaction energy with distance to the wall produces a preferred adsorption position [54]. Indeed, the energy at the center of the pore is predicted to be higher than that adjacent to the pore surface, similar to the Ripmeester-Ratcliffe model. Changes in the size of the pore result in a change in position of xenon relative to the pore wall and a concommitant change in average shift detected as a dependence of δ_{wall} on structure. This picture is static, but the fact that positions near the wall are more favored than those near the center of the pore indicates that δ_{wall} is determined by adsorption effects. Since diffusion in these structures carries a xenon atom to thousands of supercages and sites within a supercage during the acquisition of an NMR spectrum [55,56], δ_{wall} surely represents a Boltzmann average over all positions available to xenon.

The density-independent term in the xenon shift, usually ascribed to the effects of the walls, may also reflect changes in the magnetic environment [35]. For example, the exchange of nickel ions for sodium ions shifts the resonance an amount proportional to the amount of nickel [35]. The variation of the shift with nickel concentration was attributed to a variation of the magnetic susceptibility of the sample.

Another application of xenon NMR as a structural probe is the dependence of the shift on calcium content in dehydrated Ca,Na-A zeolites [57]. The shifts extrapolated to the low-uptake limit are linear in the calcium concentration up to 4.5 - 5.0 calcium ions per unit cell. For more than 5.0 Ca^{2+}/unit cell, the shift is constant. This is evidence for different siting of calcium ions. The first 4 or 5 are accessible to xenon in the α-cage, but the additional ions are inaccessible to xenon and thus are unable to influence the shift. This observation is consistent with an X-ray structure of a $Na_{0.4}Ca_{5.2}$-A zeolite [58], in which four calcium ions are found to project into the α-cage and the remaining calcium ion is sequestered in the sodalite cage.

Another experimental effect is seen in the dependence of the xenon shift on the amount of water in a Na-Y zeolite [59]. δ_0 [= $\delta_{wall} + \delta_{adsorbate}$] and δ_1 are both affected by the presence of water. δ_0 ranges from 58 ppm for "completely" dehydrated Na-Y zeolite to 158 ppm for a zeolite at 85% of saturated water content. (At water contents above 85% of saturation, the xenon uptake is so low that the resonance intensity is below the detection limit of the spectrometer.) The xenon NMR spectrum depends not only on the water content, but hydration history. Absorption of water produces an inhomogeneous distribution of xenon environments (interpreted as two different regions), whereas desorption produces an apparently uniform xenon environment. δ_0 is independent of water content for concentrations less than 15% of saturation and δ_1 is independent of water content below 40% of saturation. One of the two environments obtained upon adsorption has NMR properties identical to that obtained by desorption, from which the authors conclude that adsorption results in inhomgeneous distributions of water.

Absorption of organic molecules in Na-Y zeolite has effects similar to water on the NMR parameters of co-adsorbed xenon [60]. The results are consistent with a reduction of dynamic pore diameter. As fewer organics are absorbed per cage, the dynamic pore diameter calculated from Equation 23 increases. In addition, the sorption of xenon is observed to saturate at lower pressure when co-adsorbed with organics, an observation consistent with a smaller volume of the supercage. Given that a number of mechanisms may affect the density-independent shift of xenon [35,53], such calculations of dynamic pore diameters should be viewed with caution, particularly if other factors change from sample to sample, but they provide a convenient means to parameterize trends.

One of the more interesting uses of xenon NMR spectroscopy has been the study of materials deposited in the pores of zeolites during the course of a reaction. The quintessential example of this application is the investigation of Y zeolites fouled by the deposition of coke [61]. The deposition of coke changes the pore structure into which xenon is absorbed, affecting the xenon NMR spectroscopy. For some samples, xenon NMR indicates two regions into which xenon absorbed - attributable to pores in the zeolite and to mesopores in the external coke. For the xenon in zeolite pores, the increase in δ_{wall} with coking shows that coke alters the pore structure, estimated in this case from the dependence of δ_0 on the van der Waals interaction energy [53]. The drastic change in δ_1 with coking was interpreted as a change in the void volume per gram of catalyst available to xenon after coking. The shift of the resonance line attributable to xenon in mesopores gave an estimated pore diameter greater than 25 Å. The broadness of the resonance attributable to xenon in the zeolite pores indicates a distribution of environments.

Tsiao et al. [62] used the same technique to investigate acidification of sites in ZSM-5 with various sterically hindered agents. After acidification, the catalysts were coked from 2-butene and examined by xenon NMR spectroscopy of the absorbed gas. The xenon NMR results suggested that the structure of the pores in which the gas absorbed is only slightly changed when acidification is performed with an agent too bulky to enter the channels of the ZSM-5. When a smaller acidifying agent is used, the xenon NMR spectra reveal a channel structure totally altered by the deposition of coke. Significant amounts of internal volume are blocked in this latter treatment which directly affects the NMR spectroscopy of the sorbed xenon.

In a similar system, Tsiao and Botto [63] have investigated xenon absorbed in coal; the xenon NMR spectra were interpreted as evidence for regions with pores of several different sizes in the coal. Conner et al. [64] have investigated the NMR spectroscopy of xenon in pores created by compression of nonporous silica spheres; the mesopore structure being created by agglomeration of the spheres under pressure. The xenon NMR shifts are reasonably independent of the xenon pressure and are interpreted in terms of rapid exchange between sites on the surface of the silica particles and the gas phase trapped between the spheres. It is puzzling that the observed shifts, ranging from 40 ppm to 100 ppm, are comparable to the resonances of xenon in microporous zeolites, yet the only pores are substantially larger than in zeolites.

The ability to distinguish macroscopic regions was exploited in early work in which the distribution of CO on a Pt,Na-Y zeolite was investigated with xenon NMR spectroscopy

[47]. Using a quenched-reaction technique with detection via NMR of absorbed xenon, Tsiao et al. [65] estimated the rate constant for the removal of the template molecules from a pentasil zeolite by observing the collapse of a two-peaked spectrum to a single-peak spectrum. However, the spatial resolution is exceedingly coarse-grained and the collapse of the spectrum to a single resonance only indicates that there are no inhomogeneities of an extent larger than the root-mean-square distance a xenon atom travels in the time it takes to acquire an NMR free-induction decay. If one deliberately creates a macroscopic inhomogeneity, it is possible to detect signals from the separate regions and follow the change in NMR shift as the macroscopic inhomogeneity is eliminated by processes such as diffusion of water from a wet to a dry region [66].

Most of the experiments on xenon in zeolites have been performed at temperatures near 298 K. At this temperature, the adsorbed phase in equilibrium with pressures of xenon up to 1 atmosphere corresponds to densities up to about 100 - 150 amagats, a density 1/3 to 1/2 that of liquid xenon. For Y zeolites, this upper limit corresponds to 3 - 4 xenon atoms per supercage. One can increase the absorption of xenon by performing the spectroscopy at much lower temperatures, as demonstrated by Cheung and coworkers [37,67]. Under these conditions, one can essentially achieve packing of xenon at liquid densities in the void volume of the zeolite. The resulting NMR spectra of adsorbed xenon at these low temperatures exhibit line shifts that depend on the uptake. There are changes in the shift-versus-uptake curve that correspond to the transition from a dense gas to a liquid-like phase in the pores of the zeolite. The form of the chemical-shift-versus-uptake is similar to the adsorption isotherm, suggesting that the change in xenon shift mirrors the change in density. Changes in xenon linewidth suggest that exchange is occurring in these systems, even at rather low temperatures.

4. Deuterium-Containing Aromatics in Zeolites

Understanding the structure and reactivity of organic molecules in catalysts is essential to characterizing heterogeneous catalysis [68]. Thus, a great deal of effort has been expended in analyzing them with spectroscopy, particularly NMR spectroscopy [69]. In addressing aromatic materials absorbed in zeolites, extensive studies have given considerable information about site distributions and dynamics of benzene in zeolite pores [70,71,72] . At low loadings, benzene seems to be partitioned between a window site and a cation site, where it is bound primarily through its π-cloud. Little is known of the behavior of larger aromatic molecules absorbed in zeolites. In our laboratory, we have concentrated on study of phenanthrene-d_{10} (with a long axis of ~9.2 Å), whose size precludes its attachment at the window site in X- and Y-zeolites.

The ^2H NMR spectrum of polycrystalline phenanthrene-d_{10} is a Pake doublet with a splitting between the peak maxima of 128 kHz. This splitting is smaller than the accepted value for an aromatic C-D bond (138 kHz) [73] and results from a slight asymmetry of the electric-field gradient tensor. The spectrum can be simulated with an asymmetry parameter, η, of 0.057 and a quadrupole coupling constant of 179($\pm$1) kHz.

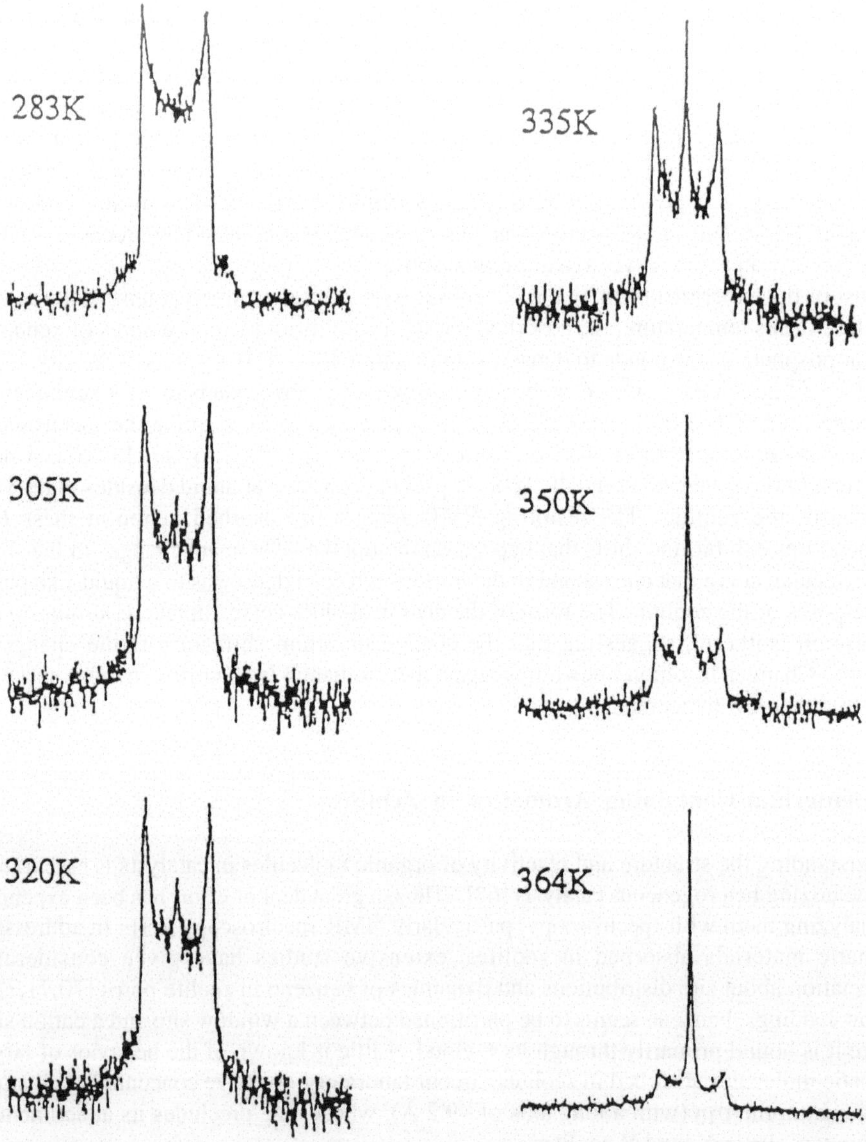

Figure 5. Temperature dependent ^{2}H NMR spectra of phenanthrene-d_{10} included in K-X zeolite. Each two-component spectrum contains a Lorentzian line superimposed upon a static Pake pattern.

The spectra of phenanthrene-d_{10} in Na- and Cs-X zeolites and Cs-Y zeolite have the same general appearance as those of phenanthrene-d_{10} included in K-X (Figure 5). They are composed of a Lorentzian line superimposed on a static Pake pattern with the same splitting as the polycrystalline sample. The relative amounts of narrow and static components in these spectra vary with temperature for every sample. Inversion-recovery experiments performed on phenanthrene-d_{10} in the Na-, K-, and Cs-X zeolites show that the two components of the spectrum have significantly different T_1's and that the relaxation times are in the range of 1 to 100 ms. This observation demonstrates, for these samples, that there are two distinct spin reservoirs with different motional time constants. Since the T_1 values are in the millisecond range, the exchange rate must be less than 1 kHz. For phenanthrene-d_{10} included in Rb-X and Cs-Y zeolites, only a single T_1 is observed across the line, indicating that exchange between the two states in these samples is rapid enough to equilibrate these two spin reservoirs. Phenanthrene-d_{10} in Rb-X zeolite gives a spectrum that cannot be deconvoluted into two distinct sub-spectra. The lack of hysteresis observed in the line-shape, when the temperature is changed, distinguishes this case from that for a distribution of correlation times, where similar line-shapes are obtained [74].

The nature of the two sites in which phenanthrene may exist when absorbed in these faujasites near room temperature may be addressed by ^2H NMR spectroscopy. Virtually no exchange occurs during the quadrupole-echo experiment, and the deconvolution of the observed spectrum into two distinct sub-spectra is justified. In one state, it is solid-like as detected by NMR spectroscopy. In the other, it is rapidly tumbling, effectively averaging the effects of static quadrupolar couplings to zero. We attribute the observed solid-like state to an ion-bound molecule and the motionally-averaged state to a molecule with enough energy to be free of the ion site. Phenanthrene-d_{10} at these two sites undergoes chemical exchange with a rate constant less than 1 kHz in all but the Rb-exchanged sample. Exchange between two sites has also been inferred from the ^2H NMR spectra of pyrene and anthracene [75].

If two states are connected by chemical exchange, then, at any temperature, an equilibrium is established. The equilibrium constant may be determined from the ratio of the integrated intensities of the two spectral components. The temperature dependence of the equilibrium constant may be used to recover the enthalpy and entropy of association to form the ion-molecule complex from the free phenanthrene-d_{10}. ΔH and ΔS for phenanthrene-ion association in these faujasites, as determined from least-squares fits of van't Hoff plots, are given in Table I. These results agree reasonably well with gas-phase measurements for similar ions interacting with aromatics. For example, the enthalpy and entropy of association for phenanthrene in the K-X zeolite compare well with the results for benzene coordinated to a hydrated potassium ion in the gas phase ($\Delta H = 12.6$ kcal/mol; $\Delta S = 27.6$ cal/K·mol) [76].

The binding of the aromatic ring to the cation is expected to be electrostatic. The binding strength will depend on factors like the electric field or charge density at the binding site and the polarizability and geometry of the interaction. As can be seen in Figure 6, a linear relationship is observed between the heat of dissociation of the complex and the cation radius for phenanthrene-d_{10} in the alkaline-earth exchanged faujasites we have investigated.

Table 1. Enthalpies and Entropies of Association for Ion-Phenanthrene-d_{10} Complexes in Faujasites.

Zeolite	-ΔH (kcal/mol)	ΔS (cal/K·mol)
Sodium-X	14.9±3.5	52.5±12
Potassium-X	11.0±1.8	29.9±5.2
Cesium-X	7.9±2.0	22.7±7.0
Cesium-Y	6.2±2.6	20.3±9.0

The errors shown are the 90% confidence interval of the fitted parameter.

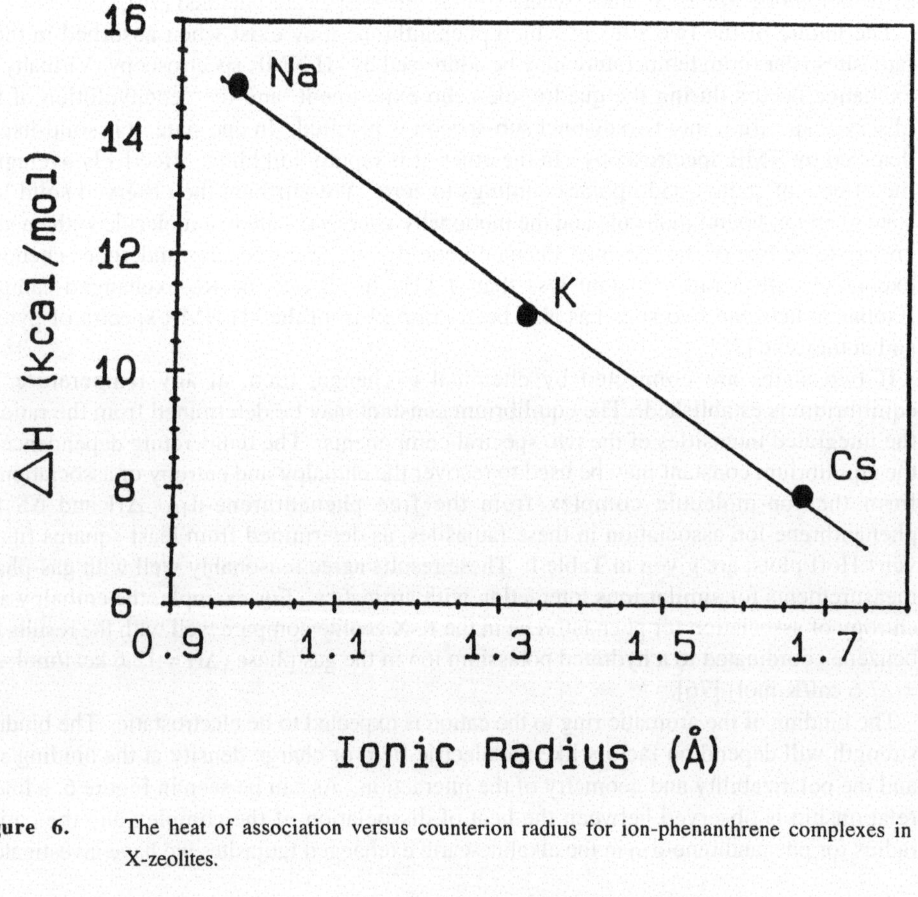

Figure 6. The heat of association versus counterion radius for ion-phenanthrene complexes in X-zeolites.

5. Summary

NMR spectroscopy of xenon is sensitive to the local structure and dynamics of the xenon atom. Because of this sensitivity and the relative ease of performing the ^{129}Xe experiment, xenon is a convenient probe of the state of microporous materials. A number of competing effects influence the shift: interactions with the walls of the micropores, interactions with other xenon atoms, interactions with other absorbed material including direct electrical interactions with ions and the framework, and magnetic effects of ions. Sorting out these various effects can be quite difficult and is a major concern for routine use of xenon NMR spectroscopy as a probe of the environment; it is, in effect, almost too sensitive to many different characteristics of the local environment. Superimposed on these effects is an averaging attendant to the investigation of a highly mobile species. In cases in which one (or a few) effect(s) is (are) dominant, simple models suffice to explain the observed trends in NMR parameters such as the shift. Except for measurements of the diffusion of xenon in the zeolites, few experiments have directly addressed the dynamic state of xenon in these materials.

Acknowledgments

I gratefully acknowledge the financial support of the National Science Foundation (grant CHEM-9013926) during the writing of this review. Students are the lifeblood of an academic research scientist, and I gratefully acknowledge the work of Navin Bansal, Chih-ji Tsiao, Matthew L. Smith, Mark A. Hepp, and Eric Scharpf, without whom this would not have been written.

References

1. R. K Harris, *Nuclear Magnetic Resonance Spectroscopy*, Pitman, London (1983).
2. T. M. Duncan and C. Dybowski, *Surf. Sci. Rep.*, **1**, 157 (1981).
3. J. L. Kaplan and G. Fraenkel, *NMR of Chemically Exchanging Systems*, Academic Press, Boca Raton, Florida (1980).
4. G. J. Schrobilgen, in *NMR and the Periodic Table*, R. K. Harris and B. E. Mann (Eds), Academic Press, New York, Chapter 14 (1978).
5. G. Malli and C. Foroese, *Internat. J. Quantum Chem.*, **1**, 95 (1967).
6. J. A. Pople, W. G. Schneider, and H. J. Bernstein, *High Resolution Nuclear Magnetic Resonance*, McGraw-Hill, New York (1959).
7. D. Brinkman and H. Y. Carr, *Phys. Rev.*, **150**, 174 (1966).
8. (a) R. L. Streever and H. Y. Carr, *Phys. Rev.*, **121**, 20 (1961);
   (b) E. R. Hunt and H. Y. Carr, *Phys. Rev.*, **130**, 2302 (1963);
   (c) W. M. Yen and R. L. Norberg, *Phys. Rev.*, **131**, 269 (1963);

(d) W. W. Warren and R. L. Norberg, *Phys. Rev.*, **148**, 402 (1966).
9. D. Brinkmann, E. Brun, and H. H. Staub, *Helv. Phys. Acta*, **35**, 431 (1962).
10. E. Kanegsberg, B. Pass, and H. Y. Carr, *Phys. Rev. Lett.*, **23**, 572 (1969).
11. A. K. Jameson, C. J. Jameson, and H. S. Gutowsky, *J. Chem. Phys.*, **53**, 2310 (1970).
12. (a) F. J. Adrian, *Phys. Rev. A*, **136**, 980 (1964);
(b) F. J. Adrian, *ibid.*, **138**, 403 (1965).
13. C. J. Jameson, A. K. Jameson, and S. M. Cohen, *J. Chem. Phys.*, **59**, 4540 (1973).
14. (a) A. K. Jameson and C. J. Jameson, *J. Am. Chem. Soc.*, **95**, 8559 (1973);
(b) C. J. Jameson, A. K. Jameson, and S. M. Cohen, *J. Magn. Reson.*, **19**, 385 (1975).
15. A. D. Buckingham and P. A. Kollman, *Mol. Phys.*, **23**, 65 (1972).
16. (a) C. J. Jameson, A. K. Jameson, and S. M. Cohen, *J. Chem. Phys.*, **62**, 4224 (1975);
(b) C. J. Jameson, A. K. Jameson, and S. M. Cohen, *ibid.*, **65**, 3401 (1976);
(c) C. J. Jameson, A. K. Jameson, and S. M. Cohen, *ibid.*, **66**, 5226 (1977).
17. T. R. Stengle, N. V. Reo, and K. L. Williamson, *J. Phys. Chem.*, **85**, 3772 (1981).
18. K. W. Miller, N. V. Reo, A. J. Schoot Uiterkamp, T. R. Stengle, D. P. Stengle, and K. L. Williamson, *Proc. Natl. Acad. Sci.*, **81**, 4946 (1981).
19. F. H. A. Rummens, *Chem. Phys. Lett.*, **31**, 596 (1975).
20. L. Onsager, *J. Am. Chem. Soc.*, **58**, 1486 (1936).
21. B. Linder, *J. Chem. Phys.*, **33**, 668 (1960).
22. J. A. Ripmeester and D. W. Davidson, *J. Mol. Struct.*, **75**, 67 (1981).
23. T. R. Stengle and K. L. Williamson, *Macromolecules*, **20**, 1428 (1987).
24. S. K. Brownstein, J. E. L. Roovers, and D. J. Worsfold, *J. Magn. Reson.*, **26**, 392 (1988).
25. (a) J. A. Ripmeester, *J. Am. Chem. Soc.*, **104**, 289 (1982);
(b) D. Davidson, Y. P. Handa, and J. A. Ripmeester, *J. Phys. Chem.*, **90**, 6549 (1986);
(c) J. A. Ripmeester, J. S. Tse, C. I. Ratcliffe, and B. M. Powell, *Nature*, **325**, 135 (1987).
26. T. Shibanuma, H. Asada, S. Ishi, and T. Matsui, *Jap. J. Appl. Phys.*, **22**, 1656 (1983).
27. G. Neue, *Z. Phys. Chem.*, **152**, 13 (1987).
28. B. C. Gates, J. R. Katzer, and G. C. A. Schuit, *Chemistry of Catalytic Processes*. McGraw-Hill, New York (1979).
29. T. Ito and J. P. Fraissard, *J. Chem. Phys.*, **76**, 5225 (1982).
30. R. C. Weast, (Ed), *Handbook of Chemistry and Physics*, CRC Press, Boca Raton, Florida (1982).
31. J. A. Ripmeester and C. I. Ratcliffe, *J. Phys. Chem.*, **94**, 7652. (1990)

32. W. Heink, J. Kaerger, H. Pfeifer, and F. Stallmach, *J. Am. Chem. Soc.* **112**, 2175 (1990).
33. E. W. Scharpf, R. Crecely, B. C. Gates, and C. Dybowski, *J. Phys. Chem.*, **90**, 9 (1986).
34. J. Fraissard and T. Ito, *Zeolites*, **8**, 350 (1988).
35. N. Bansal and C. Dybowski, *J. Phys. Chem.*, **92**, 2333 (1988).
36. A. Gedeon, J. L. Bonardet, and J. Fraissard, *J. Chim. Phys.*, **85**, 871 (1988).
37. For example, see T. T. P. Cheung, C. M. Fu, and S. J. Wharry, *J. Phys. Chem.*, **92**, 5170 (1988).
38. D. W. Johnson and L. Griffiths, *Zeolites*, **7**, 484 (1987).
39. A. L. R. Bug, *J. Thermophysics*, **10**, 469 (1989).
40. Q. Chen and J. P. Fraissard, Presented at the Fifth International Symposium on Magnetic Resonance in Colloid and Interface Science, Newark, Delaware, 1989.
41. C. J. Tsiao, D. R. Corbin, and C. Dybowski, *J. Phys. Chem.*, **94**, 4195 (1990).
42. J. A. Ripmeester, *J. Magn. Reson.*, **56**, 247 (1984).
43. T. Ito, M. A. Springuel-Huet, and J. P. Fraissard, *Zeolites*, **9**, 68 (1989).
44. R. Benslama, J. P. Fraissard, A. Albizane, F. Fajula, and F. Figueras, *Zeolites*, **8**, 196 (1988).
45. (a) T. Ito and J. P. Fraissard, *Zeolites*, **7**, 554 (1987);
 (b) M. L. Smith, D. R. Corbin, and C. Dybowski, *J. Phys. Chem.*, **97**, 9045 (1993).
46. N. Dumont, T. Ito, and E. G. Derouane, *Appl. Catal.*, **54**, L1 (1989).
47. (a) L. C. deMenorval, J. P. Fraissard, and T. Ito, *J. Chem. Soc., Farad. Trans. I*, **78**, 403 (1982);
 (b) T. Ito, L. C. deMenorval, J. P. Fraissard, and M. Primet, *J. Chem. Soc.,Farad. Trans. I*, **81**, 2855 (1985).
48. (a) B. F. Chmelka, R. Ryoo, S. B. Liu, L. C. deMenorval, C. J. Radke, E. Petersen, and A. Pines, *J. Am. Chem. Soc.*, **110**, 4465 (1988);
 (b) B. F. Chmelka, L. C. deMenorval, R. Csencsits, R. Ryoo, S. B. Liu, C. J. Radke, E. E. Petersen, and A. Pines, *Stud. Surf. Sci. Catal.*, **48**, 269 (1989).
49. M. Boudart, L. C. deMenorval, J. P. Fraissard, and G. P. Valenca, *J. Phys. Chem.*, **92**, 4033 (1988).
50. T. Ito, L. C. deMenorval, E. Guerrier, and J. P. Fraissard, *Chem. Phys. Lett.*, **111**, 271 (1984).
51. J. Demarquay and J. P. Fraissard, *Chem. Phys. Lett.*, **136**, 314 (1987).
52. R. L. Cotterman, D. A. Hickson, S. Cartlidge, C. Dybowski, C. J. Tsiao, and A. F. Venero, *Zeolites*, **11**, 27 (1991).
53. E. G. Derouane and J. B. Nagy, *Chem. Phys. Lett.*, **137**, 341 (1987).
54. E. G. Derouane and M. E. Davis, *J. Mol. Catal.*, **48**, 37 (1988).
55. W. Heink, J. Kaerger, H. Pfeifer and F. Stallmach, *J. Am. Chem. Soc.*, **112**, 2175 (1990).

56. R. Shoemaker and T. M. Apple, *J. Phys. Chem.*, **91**, 4024 (1987).
57. C. J. Tsiao, D. R. Corbin, and C. Dybowski, *J. Phys. Chem.*, **94**, 867 (1990).
58. J. J. Pluth and J. V. Smith, *J. Am. Chem. Soc.*, **195**, 1192 (1983).
59. A. Gedeon, T. Ito and J. P. Fraissard, *Zeolites*, **8**, 376 (1988).
60. (a) R. Ryoo, S. B. Liu, L. C. deMenorval, K. Takegoshi, B. Chmelka, M. Trecoske, and A. Pines, *J. Phys. Chem.*, **91**, 6575 (1987);
    (b) L. C. deMenorval, D. Raftery, S. B. Liu, K. Takegoshi, R. Ryoo, and A. Pines, *ibid.*, **94**, 27 (1990).
61. T. Ito, J. L. Bonardet, J. P. Fraissard, J. B. Nagy, C. Andre, Z. Gabelica, and E. G. Derouane, *Appl. Catal.*, **43**, L5 (1988).
62. C. J. Tsiao, C. Dybowski, A. M. Gaffney, and J. Sofranko, *J. Catal.*, **128**, 520 (1991).
63. C. J. Tsiao and R. Botto, personal communication.
64. W. C. Conner, E. L. Weist, T. Ito, and J. P. Fraissard, *J. Phys. Chem.*, **93**, 4138 (1989).
65. C. J. Tsiao, C. Dybowski, D. Walker, V. Durante, and D. R. Corbin, *Langmuir*, **4**, 1219 (1988).
66. N. Bansal and C. Dybowski, *J. Magn. Reson.*, **89**, 21 (1990).
67. (a) T. T. P. Cheung and C. M. Fu, *J. Phys. Chem.*, **93**, 3740 (1989);
    (b) T. T. P. Cheung, *J. Phys. Chem.*, **94**, 376 (1990).
68. A. de Mallmann, S. Dwizigaj, and D. Barthomeuf, in *Zeolites, Facts, Figures and Future*, P. A. Jacobs and R. A. van Santen, (Eds), Elsevier, Amsterdam, p 935 (1989).
69. H. Pfeifer, W. Meiler, and D. Deininger, *Ann. Rep. NMR Spectrosc.*, **15**, 291 (1983).
70. M. Primet, E. Garbowski, M. Mathieu, and B. Imelik, *J. Chem. Soc., Faraday Trans. 1*, **76**, 1942 (1990).
71. H. Lechert and K. Wittern, *Ber. Bunsenges. Phys. Chem.*, **82**, 1054 (1978).
72. A. N. Fitch, H. Jobic, and A. Renouprez, *J. Chem. Soc., Chem. Commun.*, 284 (1985).
73. J. H. Ok, R. R. Vold, R. L. Vold, and M. C. Etter, *J. Phys. Chem.*, **93**, 7618 (1989).
74. H. A. Riesing, J. K. Thompson, and J. J. Krebs, *J. Phys. Chem.*, **68**, 1621 (1964).
75. Mark A. Hepp, University of Delaware, personal communication
76. R. G. Keesee and A. W. Castleman, *J. Phys. Chem. Ref. Data*, **15**, 1011 (1986).

Abstract. The investigation of xenon and large aromatic molecules adsorbed in zeolitic materials using NMR spectroscopy is discussed. The results suggest a system in which the molecules are often exchanging among adsorption sites. Careful analysis of the spectroscopic results allow one to determine parameters of the magnetic properties of the adsorption sites, as well as factors affecting the exchange.

Journal of Inclusion Phenomena and Molecular Recognition in Chemistry **21**: 137–158, 1995.

Absorbed Molecules in Microporous Hosts - Computational Aspects

NEIL J. HENSON and ANTHONY K. CHEETHAM

1. Introduction

During the last decade, the use of computational methods to study microscopic processes has grown significantly. This is especially noticeable in the field of molecular biology [1], where pharmaceutical companies have a considerable investment in computational resources. In the area of materials science, important advances in the development of computational techniques have also been made in the simulation of materials themselves to both reproduce experimental data and to make predictions [2]. The application of computational methods to zeolitic systems is becoming an increasingly important tool in studying the relation between microscopic properties and their relation to both sorption and catalytic applications [3]. A detailed understanding of important processes at the microscopic level is an essential prerequisite in providing a rational method for computer-aided catalyst and molecular sieve design. The routine application of such methodologies would allow candidate zeolite structures to be screened prior to experimental study, their catalytic activity assessed using a computational algorithm, and the most promising candidates selected for experimental trials. Although the development of such screening techniques still lies in the future, the wide range of computational methods that are currently available can provide a useful insight into the interpretation of both experimental data and the identification of the important fundamental processes at the microscopic level. In this chapter, we will review current progress in this field, with illustrations of recent work and examples in the modeling of sorbates in zeolites of catalytic interest.

2. Theoretical Techniques

Many detailed reviews of the computational techniques available to the solid-state chemist have been written [3,4,5]. In this section, we provide only a short description of each method and highlight the analogous experimental data that is relevant for comparative and predictive purposes. The theoretical approaches can be divided into two main areas. Firstly, there are those methods having a quantum mechanical basis that consider the electronic structure of the atoms in the system explicitly by solving the Schrödinger equation. In addition, there are those methods which make use of interatomic potentials to represent the energies and forces acting with the systems in question.

2.1 QUANTUM MECHANICAL METHODS

Quantum mechanical methods have the advantage that they allow the calculation of properties which directly depend on the electronic configuration of a system such as the bond-breaking and bond-making that occurs during a chemical reaction. However, such calculations tend to be computationally expensive, and the approach is mainly limited to clusters with fewer than 200 atoms. Quantum mechanical calculations have been performed by a number of workers both at the semi-empirical [6] and the ab initio Hartree-Fock level [7,8,9]. The calculations have focused on the structure and properties of the zeolite acid site and its interaction with molecules of catalytic interest. The results of calculations can be usefully compared with crystal structure determinations, infra-red spectra, and acidity measurements of protonated zeolites [10]. One of the problems with cluster calculations is the question of how the cluster is terminated so as to create the correct electrostatic potential in the region of the acid site. This effect was recently investigated by Brand and co-workers [11]. Several methods have been proposed to circumvent this problem. One approach is to embed the quantum mechanical cluster in an infinite array of point charges [12,13], whereas perhaps a more rigorous approach is to use periodic Hartree-Fock techniques. To date, periodic calculations have been performed on several zeolite crystal structures including silicalite [14], mordenite [15], and chabazite [16]. Although these studies do not include guest molecules directly, by means of analyzing of the electrostatic potential, the most favorable binding sites for sorbed species can be postulated. Recently, Nicholas and Hess have performed a periodic Hartree-Fock calculation on the acid sites in sodalite [17] looking at cases in which there are both one and two acid sites in the same β-cage. They found considerable differences between periodic and cluster calculations on the same system.

2.2 FORCEFIELD METHODS

There are several simulation methods that employ interatomic potentials. However, before employing these methods, an accurate set of forcefield parameters must be defined to correctly represent the interactions of interest. Historically, there have been several approaches to forcefield development. The first method uses an empirical procedure to fit the forcefield parameters to experimental data [18,19]. In cases where experimental data are not available, electron gas calculations can be performed to obtain parameters [20]. However, the most promising method is to fit the forcefield to high quality, quantum-mechanically determined ab initio potential energy surface derived calculations on either clusters or a periodic system. Forcefields for both silicates [21] and aluminophosphates [22] have recently been derived by this method.

The numerous forcefields that have been developed to model zeolite structures can be divided into two general categories. Firstly, there are those relying on Born-model type potentials [23]. In this form, the system is presumed to be predominantly ionic with the ions carrying either formal or partial charges interacting via a Coulombic potential, combined with pair potentials (usually of Lennard-Jones or Buckingham form) representing the dispersive interaction. The periodic nature of the system requires a systematic method such as the Ewald summation [24] for dealing with the long range Coulombic interactions. Often, three-body terms are added to correctly represent the bond angles about the tetrahedral atoms. An additional refinement to the forcefield may be provided by the inclusion of shell-model potentials to treat atomic polarizability [25]. In addition, there are forcefields that take a more covalent approach involving potential functions representing the bonding directly with harmonic functions for bonds, angles and torsional terms [26,27]. The interactions between the zeolite lattice and intercalated molecules are presented using similar potential functions as for the zeolite, with some degree of Coulombic and dispersive interaction included in the forcefield. Lennard-Jones functions are often used for the dispersive contribution :

$$E_{ij} = \frac{A_{ij}}{r_{ij}^{12}} - \frac{B_{ij}}{r_{ij}^{6}}$$

where E_{ij} is the interaction energy between species i and j separated by the distance r_{ij}, and A_{ij} and B_{ij} are fitted parameters. Early work by Kiselev and co-workers [28] derived interactions parameters from expressions such as the Slater-Kirkwood formula [29] relating quantities for the polarizabilities and van der Waals radii to the Lennard-Jones parameters :

$$B_{ij} = \frac{3}{2} e^2 \sqrt{a_0} \frac{\alpha_i \alpha_j}{\left(\frac{\alpha_i}{n_i}\right)^{\frac{1}{2}} + \left(\frac{\alpha_j}{n_j}\right)^{\frac{1}{2}}}$$

where α_i and α_j are polarizabilities, e is the electronic charge, a_0 is the Bohr radius, and n_i and n_j are the effective number of electrons in atoms i and j. Since values of polarizabilities of ions inside a zeolite lattice are not directly measurable by experimental methods and there is some ambiguity over the choice of n, various assumptions must be made. A better method, which is similar to that described above, is to fit parameters to experimental data. We note that although more accurate quantum mechanical calculations would be desirable, system size and the often considerable effects of long-range Coulombic interaction on the sorption characteristics of the system generally make the calculation of ab initio potential energy surfaces intractable. Suitable experimental data include structural information from crystallographic studies of zeolites with sorbed guest molecules, thermochemical data, and sorption isotherms. This approach was recently applied to the simulation of benzene in silicalite [30] and in siliceous faujasite [31].

Once a forcefield has been derived that can reproduce the experimental quantities of interest to the desired accuracy, a number of different types of calculations can be considered [32]. The simplest experimentally interesting calculation that can be performed is the determination of the most favorable binding sites for sorption in a zeolite. In early work, this was carried out by manually moving the guest molecule of interest through the zeolite [33]. This methodology has been replaced by static-lattice energy minimization techniques, which are now quite efficient for periodic systems consisting of several hundred atoms. In this approach, all structural degrees of freedom of the system are allowed to relax under the influence of the forcefield until an energy minimum has been located. With suitable potential energy functions for the zeolite, the importance of lattice relaxation on the sorption energetics can also be assessed. One disadvantage of this approach is that, for systems of this complexity, local minima are often a problem, and the final minimized system configuration is strongly dependent on the initial state. Methods to overcome such difficulties, such as simulated annealing techniques, will be discussed later in this section. Another important consideration when comparing sorption energies derived from this method is that they represent determinations at zero Kelvin. Although there are several ways to include the effects of temperature into these methods, for example, via the quasi-harmonic approximation or by the inclusion of lattice dynamical theory, these have not generally been applied to studies of guest-host interactions.

A much simpler method to obtain temperature dependent properties is to use techniques derived from statistical mechanical approaches to the problem. Molecular dynamics (MD) has been widely used in zeolite problems to study the transport properties of molecules through zeolite pores. In this method, Newton's equations are solved by iterative procedures in order to propagate the system through time. A fundamental component of the iterative procedure is the MD time step, which is usually chosen to be smaller than the fastest important process in the system (for example, a bond vibration) so as to give the correct energy conservation properties. A popular method of integration is the velocity modification [34] of the Verlet method [35], which is based on a Taylor series expansion. The positions are advanced in one step by:

$$r(t + \delta t) = r(t) + v(t)\delta t + \frac{1}{2}\delta t^2 \frac{F(t)}{M}$$

where r(t) is the position at time t, v(t) is the velocity, F(t) is the force acting on the particle as computed from the forcefield, M is the particle mass, and δt is the MD time-step. The velocities are advanced to the mid-time step position:

$$v(t + \frac{1}{2}\delta t) = v(t) + \frac{1}{2}\delta t \frac{F(t)}{M}$$

The forces are then computed at the new position and the velocities updated to the full-time step position:

$$v(t + \delta t) = v(t + \frac{1}{2}\delta t) + \frac{1}{2}\delta t\frac{F(t + \delta t)}{M}$$

The simplest MD simulation that can be performed is in the *NVE* ensemble (constant number of particles, volume, and energy), although it is often more useful for comparison with experiment to run a simulation at constant temperature (*NVT* ensemble). In the limit of an infinitely large system, the two ensembles are equivalent; however, in order to study constant temperature dynamics in commonly sized systems, a number of suitable thermostating algorithms have been developed to obtain the *NVT* ensemble [36,37]. Constant temperature MD offers the opportunity of studying the behavior of the system as a function of temperature and deriving activation energies for the transport process.

Early MD studies in zeolites focused on the study of simple sorbates such as xenon [38,39,40]. For example, Pickett and co-workers examined the diffusion of xenon in silicalite with the zeolite lattice held rigid [38]. This is a particularly interesting sorbate since self-diffusivities can now be measured routinely with ^{129}Xe pulsed-field gradient NMR [41] and directly compared with values obtained from the simulations. Yashonath and Santikary have also recently looked at the effect of varying both the size of a spherical sorbate and its interaction strength with the zeolite on the mechanism of diffusion in zeolite Y and zeolite A [42]. They found a strong correlation between the sorbate diameter and the rate of intercage diffusion. There have also been several studies which considered molecules of more catalytic interest such as small hydrocarbons [43-47] and benzene [48,49]. Demontis and co-workers have made an extensive study of methane in silicalite and have also looked at the effect of adding zeolite lattice dynamics into the model on the transport behavior. They conclude that allowing the lattice to flex not only affects the mobility of the sorbate but also more subtle properties such as the distribution and clustering of the sorbate in the pore structure.

There are, however, cases in which the application of molecular dynamics can be impractical, since the simulation needs to sample sufficient time to obtain statistically meaningful results. This is particularly apparent for systems in which the diffusivities of interest are less than 10^{-10} m^2s^{-1} [44]. Indeed, June and co-workers have estimated that, to study a slow moving molecule such as n-hexane in silicalite by MD, a simulation run of at least 300 ns would have to be performed requiring approximately 4000 CPU hours on a large-scale supercomputer [50]. In addition, the inclusion of lattice dynamics can often cause similar problems due the small time steps required to correctly handle lattice vibrations in the numerical integration of Newton's equations of motion. Possible solutions to this problem will be discussed later in this section.

The final statistical mechanical technique that can usefully be applied to zeolite-guest problems is that of Monte Carlo (MC) simulation. The Monte Carlo method provides a further technique for generating an ensemble of system configurations which can be used to calculate averages and correlation functions that can be compared with experimental data. The widely used algorithm of Metropolis and co-workers [51] uses random numbers to create new configurations based on Boltzmann probabilities. Monte Carlo methods tend to be faster than MD simulations, since the calculation of energy derivatives is not required; however, in most traditional MC methods, there is no relation between a single MC step and a time-scale, so dynamical information is not readily accessible. Several studies have been made investigating the sorption of xenon in zeolites using standard Monte Carlo methods [52,53]. If this approach is extended to more complicated molecules, the internal degrees of freedom of the guest molecule must also be varied, which increases the computational cost to obtain statistically meaningful ensemble averages. June and co-workers studied the sorptive properties of linear alkanes in silicalite and found that the molecules locate preferentially in the main body of the channels rather than at channel intersections [54]. Recently, Smit and Siepmann have developed a configurational-bias Monte Carlo methodology for zeolites which greatly improves the efficiency of such simulations for chain molecules [55].

The method of Grand Canonical Monte Carlo (GCMC) [32] allows, in addition to the standard random moves provided by the Metropolis method, the possibility of creating and destroying particles. One useful application of this method is to fix the chemical potential of the system (and thus the pressure) and use several different GCMC simulations to calculate a sorption isotherm that can be compared directly with experimental sorption data. Woods and Rowlinson used this technique to study the sorption of Xe in faujasite [56] whereas two more recent studies have looked at the behavior of Xe in Na-A [57,58]. Cracknell and Gubbins have also investigated the sorption of Ar in the aluminophosphates $AlPO_4$-5, $AlPO_4$-8, and VPI-5 [59].

The development of hybrid methods which combine two or more of these techniques in a single calculation provides the most promising and powerful tools in forcefield methods. For example, the problem of local minima in static-lattice energy minimizations can be approached by the method of simulated annealing in which the system is subjected to a period of molecular dynamics prior to the minimization stage. The goal here is for the MD phase to perturb the system sufficiently to escape the local minima. The technique tends to produce a series of different minima of which one is the true global minimum on the potential energy surface. Deem and Newsam used this method to generate possible zeolite crystal structures given limited information on structural composition [60]. Another method of dealing with this problem is provided by the docking protocol first demonstrated for zeolites by Freeman and co-workers [61]. Using this approach, the zeolite-guest potential energy surface is probed by first carrying out high temperature dynamics on the guest molecule itself to generate a series of thermally accessible conformations. The guest molecule is then inserted into the zeolite at random locations and each zeolite-guest configuration energy minimized. As with simulated annealing, this technique generates a series of energy minima which can be analyzed to determine the preferred binding sites. This method has been applied by a number of group to the study of zeolite-template interactions. Santilli and co-workers have used this method to study the effect of *inverse shape selectivity* [62]. They concluded that the selectivity of, for example, a hydrocracking reaction can be predicted by calculating the binding energies of the various products in the cavity of the zeolite pores. Cox and co-workers [63] have also extended the docking approach to assess the importance in template-template interactions on the binding energy.

Several methods have been demonstrated to probe the diffusion of molecules through zeolites that occurs too slowly for conventional MD methods. The general principles combine techniques in reaction rate theory with conventional forcefield methods. We can use the docking protocol detailed above to probe the zeolite pores for favorable binding sites and then postulate diffusion as occurring via a series of hopping events. By performing a series of energy minimizations along the hopping path, the transition state and thus the activation energy for the hop can be identified. June and co-workers used a transition-state theory approach similar to this to calculate rate constants for the hopping processes and thus diffusion coefficients for the diffusion of Xe and SF_6 in silicalite [50]. An alternative to using transition state theory is to calculate the rate constants directly from the activation energy making realistic estimates for the pre-exponential factor from the reaction rate expression:

$$\nu = \nu_0 \exp\left(\frac{-E_a}{kT}\right)$$

where ν is the reaction rate, ν_0 is the pre-factor, E_a is the activation energy, k is the Boltzmann constant, and T is the temperature. The rate constants can then be used in a *kinetic* Monte Carlo simulation. The most widely used algorithm is based on the Metropolis method. For each kinetic Monte Carlo step, an attempted hop is either rejected or accepted based on Arrhenius-Boltzmann statistics. A system clock can be updated based on the shortest time-scale process in the system. Standard analysis techniques can be applied to determine the diffusion coefficient. This method was recently applied by Auerbach and co-workers to investigate the transport of benzene in Na-Y [64].

3. Examples

Rare-earth and high silica faujasites are widely used as catalysts in the fluid catalytic cracking process [65], showing both good selectivity and high conversion rates. The development of computational models to study the behavior of aromatic molecules, which are important products of this reaction, is an interesting challenge in this area. In the following examples, we will show how we can use experimental data to derive a forcefield for this system and how it can be applied to rationalize experimental data on this system.

The faujasite framework has two main structural features (Figure 1). There are large spherical supercages which are about 13 Å in diameter. Access to the supercages is afforded by four 12-ring windows about 10 Å in diameter which are tetrahedrally distributed about the center of the supercage.

This work was motivated by recent ^2H spin-lattice relaxation NMR measurements on the transport of benzene in faujasite samples of varying Si/Al ratio [48]. The NMR results are summarized in Table 1. It can be seen that diffusion is quite facile in the purely siliceous system, but once we introduce cations into the zeolite, the diffusivity is reduced by about two orders of magnitude due to the benzene molecule's preferred, strong interaction with the Na^+ cations. However, this effect does not have a simple dependence on sodium ion concentration; since when we move to a system with a lower Si/Al ratio (and hence more cations), the diffusion is found to be enhanced. Clearly, a microscopic picture of the important processes in this system would be of value in rationalizing these experimental trends.

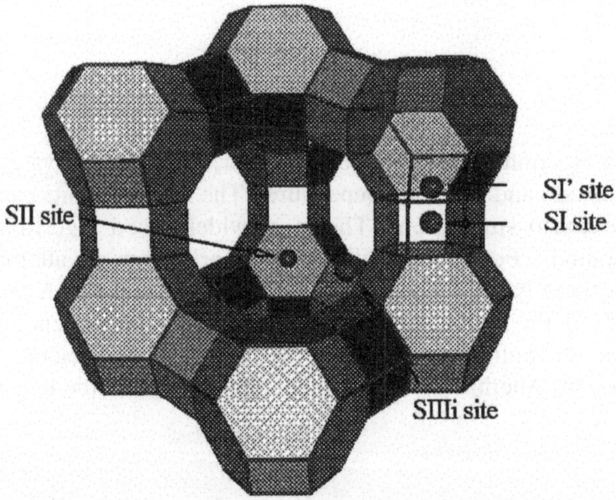

Figure 1. The structure of faujasite showing cation positions.

Table 1. Diffusion coefficients (D) and activation energies (E_a) for benzene absorbed in various faujasite structures (taken from Bull et al. [48]).

	D (298 K) (m^2s^{-1})	E_a ($kJmol^{-1}$)
siliceous faujasite	5×10^{-10}	10
Na-Y (Si/Al = 1.2)	4×10^{-11}	14
Na-X (Si/Al = 1.7)	2×10^{-13}	24

3.1 DEVELOPMENT OF A ZEOLITE-AROMATIC FORCEFIELD

The first step in this work is the development of a forcefield that can be used to calculate the experimental properties of interest. We begin by considering the simplest system, that of the siliceous zeolite and its interaction with benzene. There are two main experimental measurements which have been made on this system that are relevant to our work. Firstly, a low-temperature neutron diffraction study failed to locate a benzene sorption site [66]. This indicates that although the benzene is indeed sorbed, there are a large number of similar binding sites in the structure of similar energy, and a number of them are low-symmetry sites. Also, a recent thermochemical study of this system shows the heat of sorption to be approximately 55 $kJmol^{-1}$ [31].

Since there are a number of forcefields that have been already developed for zeolites, we chose simply to adopt one of these and concentrate on fitting realistic parameters for the zeolite-benzene interaction. We decided to employ the silicate potential derived by Kramer and co-workers [67], since it uses partial charges ($q_{Si} = 2.4$, $q_{Al} = 1.4$, $q_{Na} = 1.0$, $q_O = -0.7$) which numerous ab initio quantum mechanical calculations on clusters have shown to be more realistic than the formal values. In addition, the Kramer forcefield provides parameters for Al and Na which will be important in looking at the cation-containing faujasites. For our model of siliceous faujasite, we took the structure determination of Hriljac and co-workers [66]. A point-charge model for benzene was obtained from a semi-empirical quantum mechanics calculation (MNDO) using a Mulliken charge population analysis ($q_C = -0.153$, $q_H = 0.153$) and has been shown to approximately reproduce the quadrupole moment of benzene [30].

To obtain reasonable parameters for the zeolite-benzene interactions, we carried out a docking calculation as detailed in the previous section. This located a number of binding sites in the siliceous system, of which a site lying above the four rings on the wall of the supercage was the most favorable. Given this, we made the approximation that this site dominates the heat of sorption. With the Coulombic interactions already fixed by the choice of partial charges on benzene and on the zeolite lattice, we fitted the O-C and O-H dispersive parameters to reproduce the experimental heat of sorption at this site.

In the case of the cation-containing faujasite models, we used the experimental heat of sorption from the measurement of Barthomeuf and Ha [68] (75 $kJmol^{-1}$) as the first parameter in our fit. The crystal structure of Fitch and co-workers [69] determined by low-temperature neutron diffraction shows two sites: the first with the benzene molecule bound facially through its π-electron cloud to the SII cation lying in center of the six-ring in the wall of the supercage; at higher coverages, a window site is also found to be occupied. Taking the parameters fitted to the siliceous system as our basis, we adjusted the C-Na, and C-H dispersive parameters to reproduce the correct minimum at the SII site, and give an interaction energy at this site approximately equal to the experimental heat of sorption. This rather simplified fitting method takes no account of the temperature variation of the heat of sorption and can be considered to be a first approximation to the interactions present in the experimental system.

3.2 DYNAMICS OF BENZENE IN SILICEOUS FAUJASITE

Since the value of the diffusivity for benzene in siliceous faujasite as measured in the NMR experiment is 4.5×10^{-10} m^2s^{-1}, our first attempt to investigate this system was using a traditional molecular dynamics simulation.

As an initial test of our potential, we performed a static-lattice energy minimization on the structure of siliceous faujasite itself, removing strains on the cell by optimizing the lattice vectors as well as the atomic positions. The cell parameters are reproduced to within 2% of the experimental values. Although a better representation of siliceous faujasite could be obtained by using a more sophisticated potential model such as one of the several shell model potentials available for silicates [21,70], the high frequency of oscillation of the core-shell pairs would require the use of such a small time-step in the MD simulation that the total time spanned by the simulation would be vastly reduced. To improve further the validity of the siliceous faujasite model, we fixed the cell parameters at their experimental values using the *NVT* ensemble. An Ewald summation was used to sum the electrostatic interactions, whereas a cutoff of 12 Å was used for the short-range terms. A time-step of 1 fs was used and Newton's equations solved by the velocity Verlet algorithm. The temperature was constrained by the Nosé algorithm [37]. The system was equilibrated for a period of 25 ps and data collected over a time-scale of 100 ps. A series of molecular dynamics simulations were performed at 298 K, 350 K, 400 K, and 450 K. Diffusion coefficients were calculated from the Einstein relation using a multiple time-origin analysis on the center of mass of the benzene molecule as a function of time.

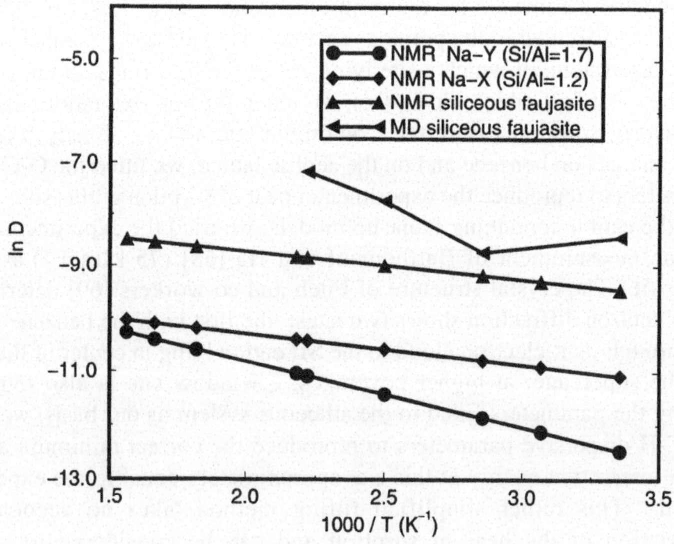

Figure 2. Comparison of the temperature dependence of the diffusion coefficients obtained for Na-Y and siliceous faujasite from the NMR experiments and MD simulations.

An activation energy and pre-exponential factor were determined from fitting the temperature dependence of the diffusion coefficients to an Arrhenius equation. The results are compared to NMR data in Figure 2.

The results are in reasonable agreement with experiment. The molecular dynamics calculation also provides information on possible trajectories. For example, by the use of animation, it can be seen that the benzene molecule spends a large proportion of its time close to the wall of the supercage and little time in the energetically less favorable supercage center. We find that the cage-to-cage hopping process is highly improbable at the temperatures studied and during the limited time-scales of the simulations. Indeed, we observe this process only to occur at the highest temperature studied (450 K).

3.3 DOCKING OF BENZENE IN FAUJASITES

Since the NMR experiment measures the diffusivity in both Na-Y and Na-X to be at least two orders of magnitude lower than in the siliceous system, neither of these systems are accessible to study by a conventional molecular dynamics simulation. With this in mind, we decided to investigate the cation-containing systems by an examination of the zeolite-benzene potential energy surface for possible binding sites. For completeness, the first system studied was that of siliceous faujasite. The docking methodology developed by Freemen and co-workers for zeolites was employed, as detailed in the last section. The initial docking minimizations were performed using a fixed zeolite lattice, which was then allowed to relax once the benzene minimum was located. Several binding sites were located in the siliceous structure, which could be divided into the following broad categories:

(i) sites in the region of the 4-ring on the wall of the supercage;
(ii) sites in the region of the 6-ring in the wall of the supercage;
(iii) 12-ring window site.

The geometry of these binding sites is shown in Figure 3. The relative energetics of the various sites indicate that the potential energy surface around the walls of the supercage is relatively flat, and consists of a number of low symmetry sites; this is consistent with the diffraction data. The effect of lattice relaxation on these binding sites is shown in Table 2 and is most pronounced for the sites lying close to the supercage walls, reflecting the more intimate coordination afforded by those sites.

This procedure was then repeated for the cation-containing systems. Firstly, a model for Na-Y was constructed using the structure of Fitch and co-workers as a basis. There are three cation sites in this structure which are of relevance to our modeling (see Figure 1). The SI cation site is located in the center of the double 6-ring, whereas the SI' cation site is located in the wall of the sodalite cage above the 6-ring window. The separation between these sites is such that they cannot be occupied simultaneously. The final site, which is the most important for its interaction with benzene, is the SII site which is

Table 2. Effect of lattice relaxation on the docked sorption energies for benzene in siliceous faujasite.

site	binding energy (fixed zeolite) (kJmol^{-1})	binding energy (flexible lattice) (kJmol^{-1})
4-ring	55.8	58.4
6-ring	42.9	43.7
window	48.2	50.6

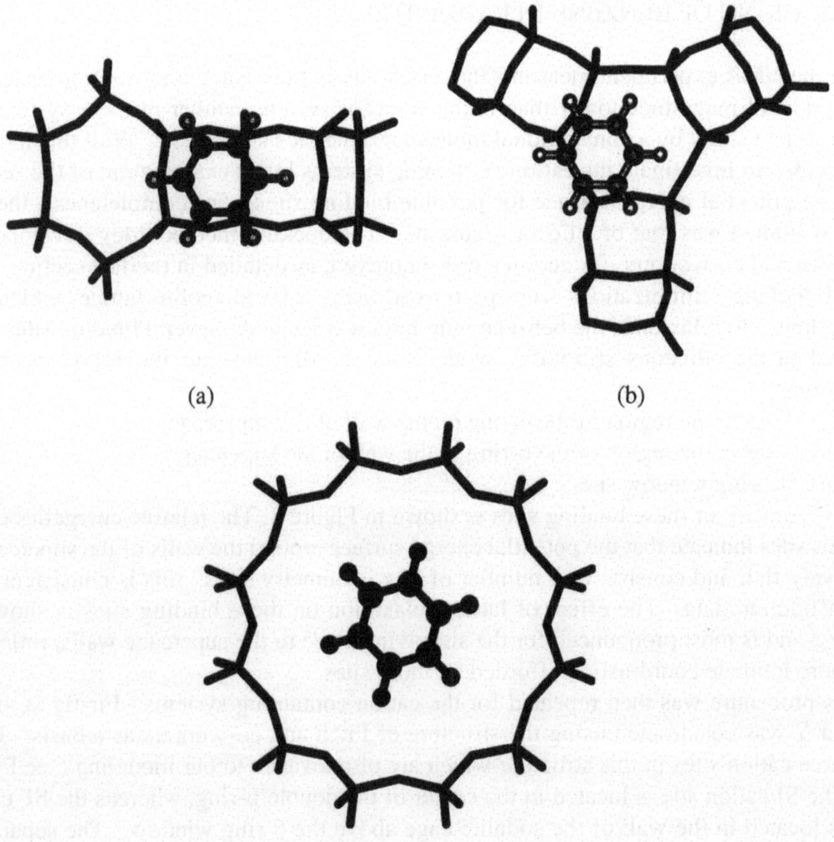

Figure 3. Docked sorption sites for benzene in siliceous faujasite: (a) 4-ring site, (b) 6-ring site , and (c) window site.

located in the wall of the supercage at the center of the 6-ring. To create an ordered model for Na-Y, we filled all the SI' and SII sites to give a Si/Al ratio of 2.0. Since at this Si/Al ratio the Si-Al substitution is known to be disordered, we made no attempt in this initial study to include Al explicitly. Instead, we used an average T-site model in which the charge on Si is reduced to balance the effect of the Na^+ cations. The results of the docking calculation on this system are shown in Figure 4. We locate only two binding sites in this system, both of which are observed in the diffraction experiment; the SII

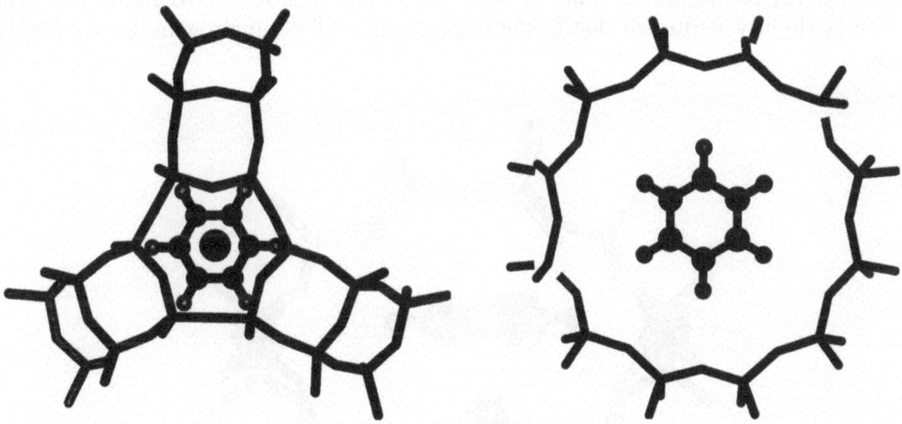

(a) (b)

Figure 4. Docked sorption sites for benzene in Na-Y: (a) SII cation binding site (74.7 $kJmol^{-1}$) and (b) window site (50.2 $kJmol^{-1}$).

binding site is the most favorable by about 25 $kJmol^{-1}$. It can be noted that the presence of cations in the supercage has greatly enhanced the preference of binding in the supercage as compared to the window. Interestingly, the window site is shown to be a stable binding site, even at low dilution, since our docking calculation corresponds to a loading of one benzene molecule per unit cell. Although the presence of nearby benzene cations bound at the SII site undoubtedly perturbs the energetics of the window site, it is not a necessary prerequisite for its stability.

In order to construct models having Si/Al ratios lower than 2.0, further cation sites must be filled. Diffraction studies suggest that these cation sites are located in the region of the 4-rings in the wall of the supercage near to the 12-ring windows. In the absence of any diffraction for benzene in a Na-X sample, we used the crystal structure determination of Hseu [71] as a basis for our Na-X models. In this study, Hseu identifies the SIIIi cation site as the repository for extra Na^+ cations, of which there are 96 per unit cell (see Figure 1). These occur in pairs which are related by a mirror plane, lying above the 4-rings. There are two constraints on the occupation of the SIIIi sites. Firstly, the separation of these pairs of cations precludes the simultaneous occupancy of both sites; this reduces the maximum number of cation sites to 48. Secondly, Löwenstein's rule for this system limits the Si/Al to 1.0 (the structure of this system is referred to Na low-silica X or Na-LSX), which corresponds to 32 SIIIi cation sites filled. With this in mind, we constructed two models, one with 16 cation sites filled (Si/Al =1.4), and a second with 32 cation sites filled (Si/Al =1.0). For the Na-X models, we find the binding sites that occur for Na-Y are also sites in which the benzene is bound directly to the SIIIi cation (Figure 5). The energy of this site is found to be very similar to that of the SII cation site. The SII site is slightly perturbed due to the close contact of the hydrogens with the SIIIi cation.

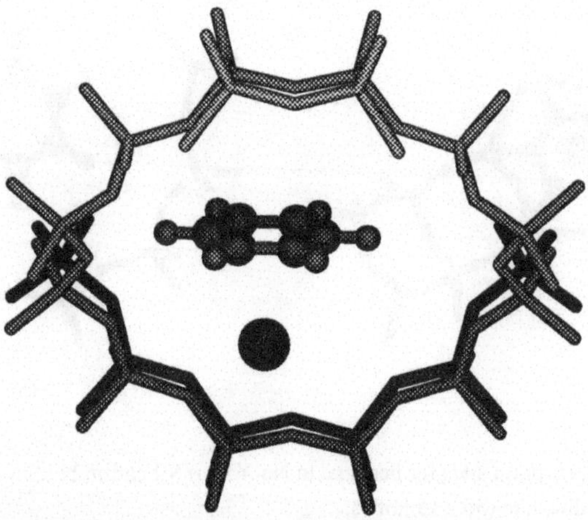

Figure 5. Docked sorption site for benzene at the SIIIi cation in Na-X (68.5 $kJmol^{-1}$).

3.4 DIFFUSION OF BENZENE IN CATION-CONTAINING FAUJASITES

In the previous section, we determined possible binding sites for benzene in various cation-containing faujasite models. While the depth and location give little insight into the diffusion mechanism in these structures, they do provide a basis from which we can determine possible trajectories. If we assume that the transport occurs by hopping between adjacent energy minima, we can analyze the potential energy surface along such pathways. Since the forcefield was fitted to the equilibrium properties of the experimental system, we are making the assumption that the same parameters are also valid close to transition states. Considering the structure of Na-Y which has two binding

(a)

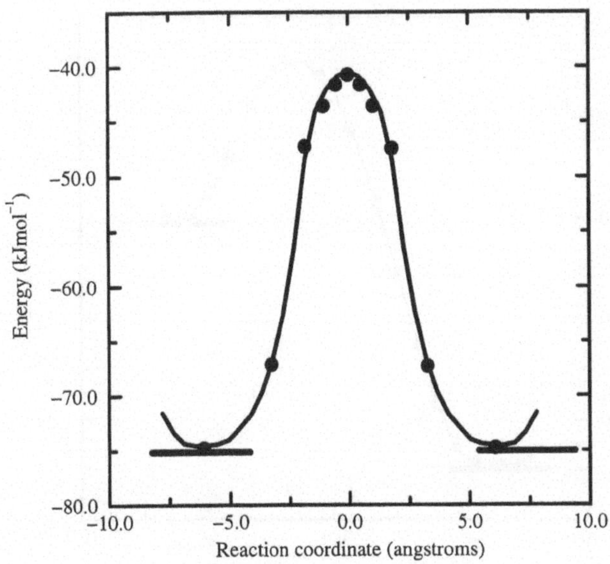

(b)

Figure 6. Calculated hopping trajectories for benzene in Na-Y: (a) SII $\rightarrow$ SII site and (b) energetics.

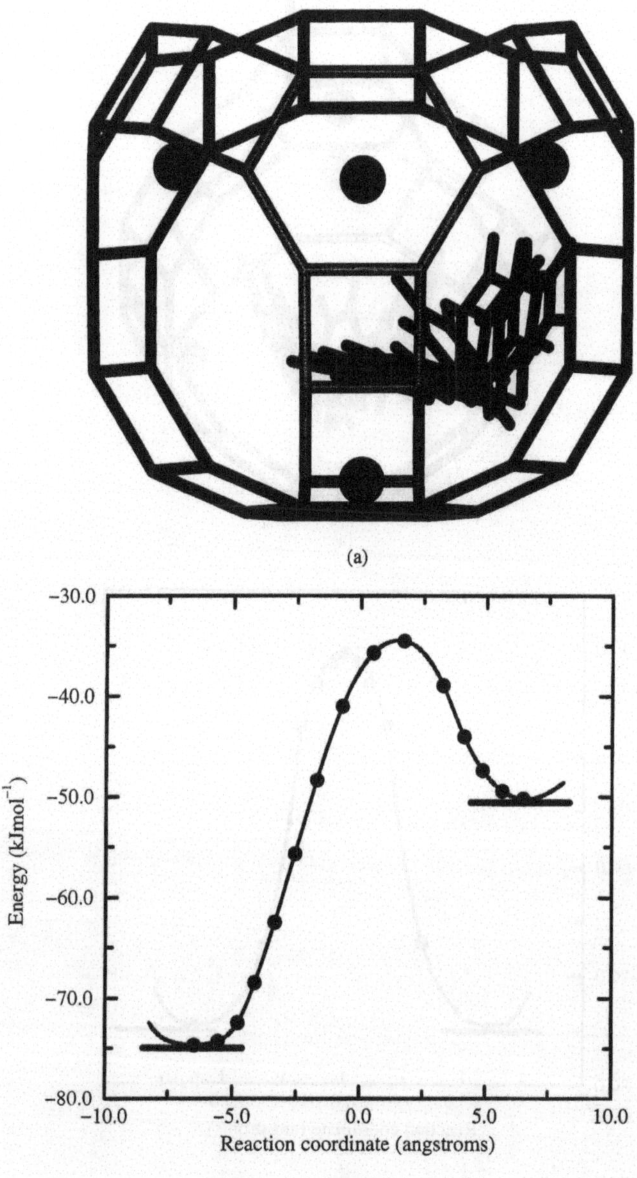

Figure 7. Calculated hopping trajectories for benzene in Na-Y: (a) SII → window site and (b) energetics.

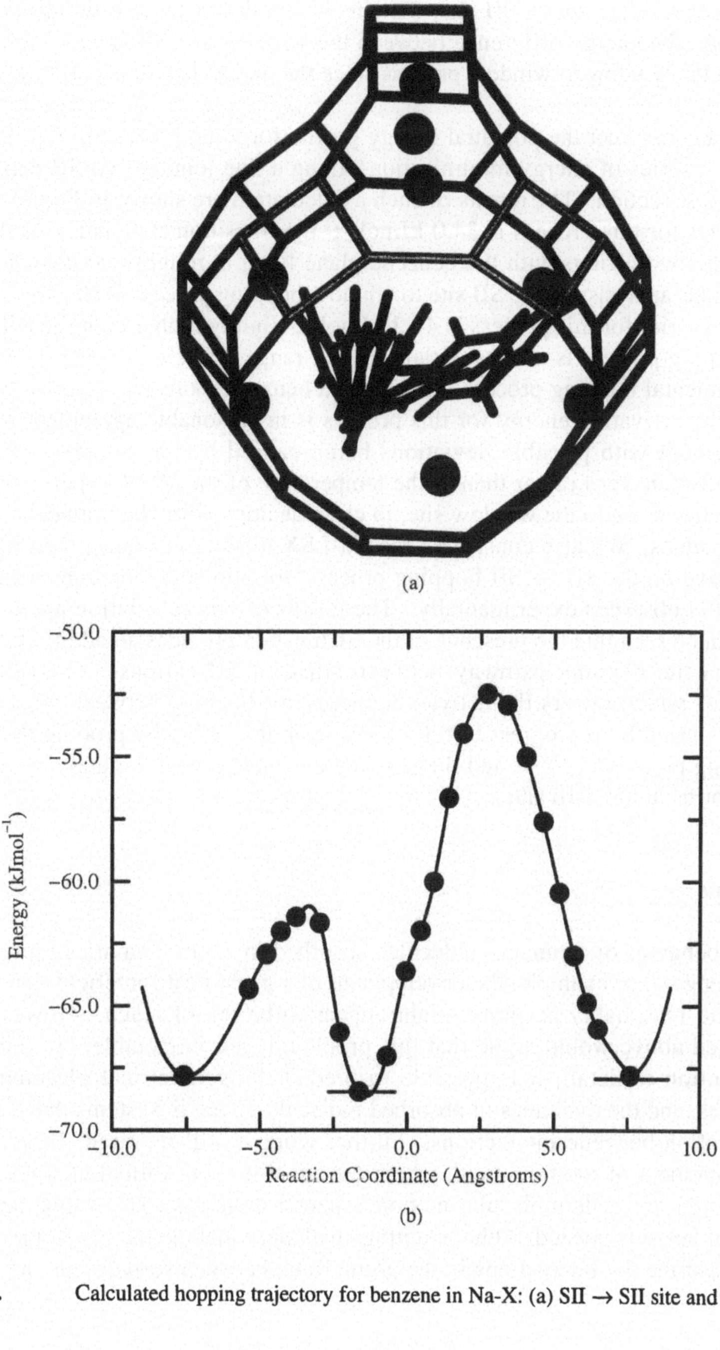

Figure 8. Calculated hopping trajectory for benzene in Na-X: (a) SII → SII site and (b) energetics.

sites of interest, the SII cation site and the window site, we can propose three possible hopping processes: SII site → SII site, SII site → window site, and window site → window site. As the energy difference between the window and SII sites is significant, we can neglect the window to window process since the molecule will spend little time at the window site.

Firstly, we can construct the potential energy profile for a jump between two SII sites by performing a series of energy minimizations along a line joining two SII cations, as detailed in the last section. The results of such a calculation are shown in Figure 6. The activation barrier for this process is 34.0 $kJmol^{-1}$; the transition state is above the four rings joining the two 6-rings with the benzene plane lying at roughly 90 degrees to the 4-rings. A similar analysis for the SII site to window hopping process is shown in Figure 7. The energy barrier for this process is 40.2 $kJmol^{-1}$, slightly higher than the SII → SII hop. This hopping process is also essential for long range diffusion to occur, since it is the only fundamental hopping process that allows a benzene molecule to move between supercages. The activation energy for this process is in reasonable agreement with the NMR measurement with possible deviations being caused by the forcefield fit being performed at absolute zero rather than at the temperature of the NMR experiment. The transition state lies close to the window site, in close analogy with Hammond's postulate in reaction dynamics. We also considered the Na-LSX model to estimate the effect that SIIIi cations have on the SII → SII hopping process, to rationalize the enhancement in diffusivity that is observed experimentally. The results of this calculation are shown in Figure 8. It can be seen that the presence of the SIIIi cation provides an additional strong binding site on the hopping pathway between adjacent SII cations. This acts as a "stepping-stone" which lowers the activation energy for hopping between adjacent SII sites. Work is currently in progress to fully investigate this effect by probing the SII → window hopping process in Na-X and the possible enhancement of the hopping rate due to lattice relaxation at the SIIIi site.

4. Conclusions

Modeling the behavior of aromatic molecules absorbed in zeolitic cavities represents a daunting challenge. For example, the development of a guest-host forcefield that is both transferable and reasonably accurate might appear to be out-of-reach. However, the results presented above would argue that this problem is not intractable, but that, with sufficient attention to detail, it is possible to predict the geometrical placement, the binding energies, and the dynamics of absorbed molecules, even in systems that require a treatment of cation-benzene interactions. Further work is still required, especially in relation to refinement of the forcefield, where we are still using formal charges for the cations, and long time-scale molecular motion, where a more rigorous evaluation of the pre-exponential terms is needed. Other exciting challenges include the development of a forcefield to describe the interactions in the groundstate between sorbates and acid sites.

This will begin to impact our understanding of catalysis at the atomic level, and will lead, in the future, to detailed quantum mechanical treatments of reactivity.

Acknowledgments

The authors would like to thank the Science and Engineering Research Council (UK), Los Alamos National Laboratory, and Biosym Technologies for financial support throughout this work. This work was also supported by the UCSB-MRL under National Science Foundation Award No. DMR-9123048. We also acknowledge Drs. A. Redondo and S. M. Auerbach for useful discussion.

References

1. W. F. van Gusteren and H. J. C. Berendsen, *Angew. Chem. Int. Ed. Engl.*, **29**, 992 (1990).
2. C. R. A. Catlow, J. D. Gale, and R. G. Bell, *J. Mater. Chem.*, **4**, 781 (1994).
3. *Modeling of Structure and Reactivity in Zeolites*, C. R. A. Catlow (Ed), Academic Press, London, San Diego (1992).
4. *Computer Simulation of Solids*, C. R. A. Catlow and W. C. Mackrodt (Eds), Lecture Notes in Physics, Springer-Verlag, Berlin, Vol. 166 (1982).
5. S. C. Parker and G. D. Price, in *Advances in Solid State Chemistry*, C. R. A. Catlow (Ed), JAI Press, London, Vol. 1 (1989).
6. A. Redondo and P. J. J. Hay, *J. Phys. Chem.*, **97**, 11754 (1993).
7. A. E. Alvarado-Swaisgood, M. K. Barr, P. J. Hay, and A. Redondo, *J. Phys. Chem.*, **95**, 10031 (1991).
8. J. B. Nicholas, R. E. Winans, R. J. Harrison, L. E. Iton, L. A. Curtiss, and A. J. J. Hopfinger, *J. Phys. Chem.*, **96**, 10247 (1992).
9. J. Sauer, *Chem. Rev.*, **89**, 199 (1989).
10. G. J. Kramer, R. A. van Santen, C. A. Emeis, and A. K. Nowak, *Nature*, **363**, 529 (1993).
11. H. V. Brand, L. A. Curtiss, and L. E. Iton, *J. Phys. Chem.*, **96**, 7725 (1992).
12. S. J. Cook, A. K. Chakraborty, A. T. Bell, and D. N. Theodorou, *J. Phys. Chem.*, **97**, 6679 (1993).
13. S. P. Greatbanks, P. Sherwood, and I. H. Hillier, *J. Phys. Chem.*, **98**, 8134 (1994).
14. J. C. White and A. C. Hess, *J. Phys. Chem.*, **97**, 8703 (1993).
15. J. C. White and A. C. Hess, *J. Phys. Chem.*, **97**, 6398 (1993).
16. E. Apra, R. Dovesi, C. Freyriafava, and C. Pisani, *Modeling and Simulation in Materials Science and Engineering*, **1** , 297 (1993).
17. J. B. Nicholas and A. C. Hess, *J. Am. Chem. Soc.*, **116**, 5428 (1994).
18. T. S. Bush, J. D. Gale, C. R. A. Catlow, and P. D. Battle, *J. Mater. Chem.*, **4**, 831, (1994).

19. J. D. Gale and N. J. Henson, *J. Chem. Soc., Faraday Trans.*, **90**, 3175 (1994).
20. R. G. Gordon and Y. S. Kim, *J. Chem. Phys.*, **56**, 3122, (1976).
21. K. De Boer, A. P. J. Jansen, and R. A. van Santen, *Chem. Phys.*, **223**, 46 (1994).
22. B. W. K. van Beest, G. J. Kramer, and R. A. van Santen, *Phys. Rev. Lett.*, **64**, 1955, (1990).
23. R. A. Jackson and C. R. A. Catlow, *Mol. Simul.*, **1**, 207 (1988).
24. P. P. Ewald, *Ann. Phys. (Leipzig)*, **64**, 1 (1921).
25. B. G. Dick and A. W. Overhauser, *Phys. Rev.*, **112**, 90 (1958).
26. E. de Vos Burchart, V. A. Verheij, H. van Bekkum, and B. van de Graaf, *J. Chem. Soc., Faraday Trans.*, **88**, 2761 (1992).
27. J. B. Nicholas, A. J. Hopfinger, F. R. Trow, and L. E. Iton, *J. Am. Chem. Soc.*, **113**, 4792 (1991).
28. A. G. Bezus, A. V. Kiselev, A. A. Lopatkin, and P. Q. Du, *J. Chem. Soc., Faraday Trans. II*, **74**, 367 (1978).
29. J. C. Slater and J. G. Kirkwood, *Phys. Rev.*, **37**, 682 (1931).
30. R. Q. Snurr, A. T. Bell and D. N. Theodorou, in *Proc. 9th Int. Zeolite Conference*, Montreal 1992, R. von Ballmoos, J. B. Higgins, and M. M. J. Treacy (Eds), Butterworth-Heinemann, Stoneham, MA, Vol. 2, p 71 (1993).
31. N. J. Henson, M. Stockenhuber, A. K. Cheetham, and J. A. Lercher, to be published.
32. M. P. Allen and D. J. Tildesley, *Computer Simulation of Liquids*, Oxford University Press, Oxford, U. K. (1987).
33. P. A. Wright, J. M. Thomas, A. K. Cheetham, and A. K. Nowak, *Nature*, **318**, 611 (1985).
34. W. C. Swope, H. C. Andersen, P. H. Berens, and K. R. Wilson, *J. Chem. Phys.*, **76**, 637 (1982).
35. L. Verlet, *Phys. Rev.*, **159**, 98 (1967).
36. H. J. C. Berendsen, J. P. M. Postma, W. F. van Gusteren, A. DiNola, and J. R. Haak, *J. Chem. Phys.*, **81**, 3684 (1984).
37. S. Nosé, *J. Chem. Phys.*, **81**, 511 (1984).
38. S. D. Pickett, A. K. Nowak J. M. Thomas, B. K. Peterson, J. F. P. Swift, A. K. Cheetham, C. J. J. den Ouden and B. Smith, *J. Phys. Chem.*, **94**, 1233 (1990).
39. N. J. Henson, A. K. Cheetham, B. K. Peterson, S. D. Pickett, and J. M. Thomas, *J. Comp.-Aided. Mat. Des.*, **1**, 41 (1993).
40. P. Santikary, S. Yashonath, and G. Anathakrishna, *J. Phys. Chem.*, **96**, 10249 (1992).
41. W. Heink, J. Körger, H. Pfeifer, and F. Stallmach, *J. Am. Chem. Soc.*, **112**, 2175 (1990).
42. S. Yashonath and P. Santikary, *J. Phys. Chem.*, **98**, 6368 (1994).
43. E. Hernandez, M. Kawano, A. A. Shubin, C. M. Freeman, C. R. A. Catlow, J. M. Thomas, and K. I. Zamaraev, in *Proc. 9th Int. Zeolite Conference*, Montreal 1992, R. von Ballmoos, J. B. Higgins, and M. M. J. Treacy (Eds), Butterworth-Heinemann, Stoneham, MA (1993).

44. R. L. June, A. T. Bell, and D. N. Theodorou, *J. Phys. Chem.*, **96**, 1051 (1992).
45. P. Demontis, E. S. Fois, G. B. Suffritti, and S. Quartieri, *J. Phys. Chem.*, **94**, 4329 (1990).
46. P. Demontis, G. B. Suffritti, E. Fois, and S. Quartieri, *J. Phys. Chem.*, **96**, 1482 (1992).
47. P. Demontis and G. B. Suffritti, in *Modeling of Structure and Reactivity in Zeolites,* C. R. A. Catlow (Ed), Academic Press, London, U. K. (1992).
48. L. M. Bull, N. J. Henson, A. K. Cheetham, J. M. Newsam, and S. J. Heyes, *J. Phys. Chem.*, **97**, 11776 (1993).
49. P. Demontis, S. Yashonath, and M. L. Klein, *J. Phys. Chem.*, **93**, 5016 (1989).
50. R. L. June, A. T. Bell, and D. N. Theorodou, *J. Phys. Chem.*, **95**, 8866 (1991).
51. N. Metropolis, H. W. Rosenbluth, M. N. Rosenbluth, A. H. Teller, and G. Teller, *J. Chem. Phys.*, **21**, 1087 (1953).
52. P. R. van Tassel, H. T. Davis, and A. V. McCormick, *Mol. Phys.*, **73**, 1107 (1991).
53. P. R. van Tassel, H. T. Davis, and A. V. McCormick, *Mol. Phys.*, **76**, 411 (1992).
54. R. L. June, R. L., A. T. Bell, and D. N. Theorodou, *J. Phys. Chem.*, **94**, 1508 (1990).
55. B. Smit and J. I. Siepmann, *J. Phys. Chem.*, **98,** 8442 (1994).
56. G. B. Woods and J. S. Rowlinson, *J. Chem. Soc., Faraday Trans. II*, **85**, 765 (1989).
57. P. R. van Tassel, H. T. Davis, and A. V. McCormick, *J. Chem. Phys.*, **98**, 8919 (1993).
58. C. J. Jameson, A. K. Jameson, H. M. Lim, and B. I. Baello, *J. Chem. Phys.*, **100**, 5965 (1994).
59. R. F. Cracknell and K. E. Gubbins, *Langmuir*, **9**, 824 (1993).
60. M. W. Deem and J. M. Newsam, *Nature*, **342**, 260 (1989).
61. C. M. Freeman, C. R. A. Catlow, J. M. Thomas, and S. Brode, *Chem. Phys. Lett.*, **186**, 137 (1991).
62. D. S. Santilli, T. V. Harris, S. I. Zones, *Microporous Materials*, **1**, 329 (1993).
63. P. A. Cox, A. P. Stevens, L. Banting, and A. M. Gorman, in *Zeolites and Related Microporous Materials: State of the Art 1994*, (*Stud. Surf. Sci. Catal.*, **84**), J. Weitkamp, H. G. Karge, H. Pfeifer, and W. Hoelderich (Eds), Elsevier, Amsterdam, p 2115 (1994)
64. S. M. Auerbach, N. J. Henson, H. I. Metiu, and A. K. Cheetham, submitted for publication, 1995.
65. A. G. Oblad, *Oil and Gas J.*, **84** (1972).
66. J. A. Hriljac, M. M. Eddy, A. K. Cheetham, and J. A. Donohue, *J. Solid State Chem.*, **106**, 66 (1993).
67. G. J. Kramer, N. P. Farragher, B. W. H. van Beest, and R. A. van Santen, *Phys. Rev. B*, **43**, 5068 (1991).
68. D. Barthomeuf and B.-H. Ha, *J. Chem. Soc., Faraday Trans. I*, **69**, 2147 (1973).
69. A. N. Fitch, H. Jobic, and A. Renouprez, *J. Phys. Chem.*, **90**, 1311 (1986).
70. M. J. Sanders, M. Leslie, and C. R. A. Catlow, *J. Chem. Soc., Chem. Comm.*, 1271 (1984).

71. T. Hseu, Ph. D. Thesis, University of Washington 1972, University Microfilm number 73-13835, Ann Arbor, Michigan, USA.

Abstract. This review examines the computational tools that are available for modeling the behavior of absorbed molecules in microporous hosts such as the aluminosilicate zeolites. The role of quantum mechanical methods is briefly discussed, followed by a more detailed examination of approaches that utilize forcefields for both the zeolite host, itself, and the host-guest interactions. These approaches include molecular dynamics and Monte Carlo simulations. Several examples are presented, focusing on the behavior of benzene absorbed in zeolites belonging to the faujasite family. The development of an appropriate forcefield is described, and the utilization of this forcefield to study the structure, energetics, and dynamics of benzene in siliceous faujasite, Na-Y and sodium Na-X is explored. The results are compared with the available experimental data.

Journal of Inclusion Phenomena and Molecular Recognition in Chemistry **21**: 159–184, 1995.

Zeolite Encapsulated Metal Complexes

KENNETH J. BALKUS, Jr. and ALEXEI G. GABRIELOV

1. Introduction

The potential for coupling the shape selectivity associated with the well-defined channels and cages of zeolites with the reactivity of metal complexes makes these molecular sieves particularly attractive as solid supports. Zeolites have a distinct advantage over conventional support materials in that a metal complex can be physically trapped in the pores and not necessarily bound to the oxide surface. A metal complex of the appropriate dimensions might be encapsulated in a zeolite yet be free to move within the confines of a cage or channel. This could be viewed as a bridge between a homogeneous and heterogeneous system. Herron et al. [1] first referred to such zeolite guest molecules as ship-in-a-bottle complexes. This term has become quite popular in describing zeolite-encapsulated metal chelates and cluster complexes. The novel reactivity that was anticipated for zeolite ship-in-a-bottle complexes is gradually being realized as efforts in this area grow. In particular, progress in the area of biomimetic chemistry such as dioxygen binding and oxidation catalysis has elicited the terms inorganic protein [2,3] and zeozyme [4] to describe certain zeolite encapsulated metal complexes. In this context, even silicon-based life has been contemplated [5].

Romanovsky et al. first reported the synthesis of a metal phthalocyanine (MPc) inside zeolite Na-Y in 1977 [6-9]. Since then several groups have studied the encapsulation and reactivity of MPc complexes [3,4,6-52] as well as related porphyrins [23,51,53,54] in zeolite molecular sieves. The MPc complex is trapped in a zeolite cavity that has restricted apertures and provides a good example of a ship-in-a-bottle complex. The tetradentate Schiff base ligand N,N'-bis(salicylidene)ethylenediimine or SALEN has also been complexed in synthetic faujasite type zeolites [2,55-62]. Ship-in-a-bottle complexes are not limited to large chelates. For example, bis(cyclopentadienyl)cobalt(III) ion can be encapsulated in Nonasil, a clathrasil type molecular sieve, during synthesis [63,64]. Therefore, the only limitation as to the nature of the zeolite and metal complex is the method of encapsulation. In this chapter, we hope to review the synthetic strategies and characterization techniques for zeolite encapsulated metal complexes as well as give a brief summary of emerging applications. In many cases, parallel studies of simple intrazeolite coordination complexes have also been shown to be quite promising but this area has been more frequently reviewed [65,66] and will not be addressed here. Likewise, zeolite supported bulk metal particles [67,68] as well as metal oxides [69] have been omitted from this review.

2. Synthesis of Zeolite Encapsulated Complexes

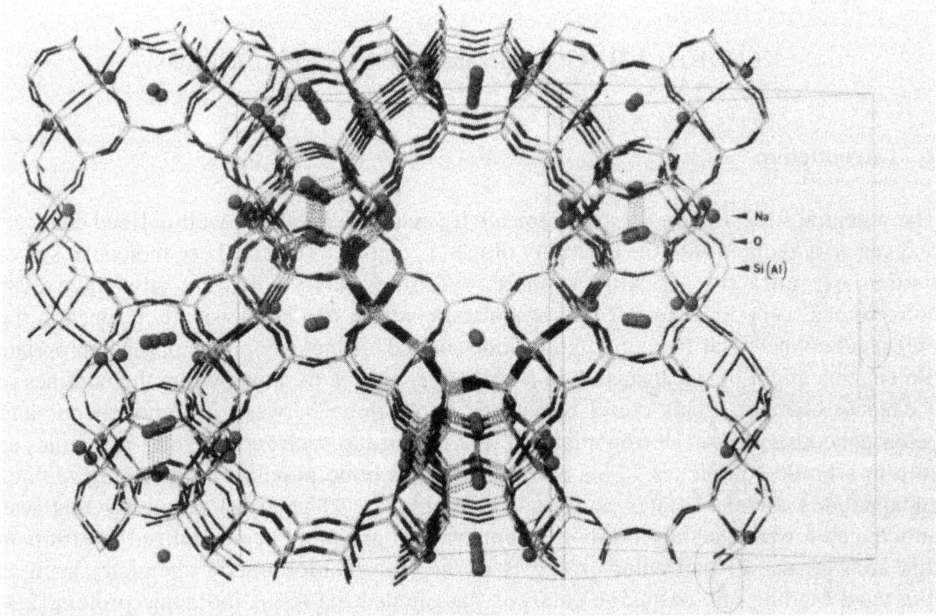

Figure 1. A [110] view of zeolite Na-Y.

There are four general strategies for the encapsulation of metal complexes in molecular sieves that include,

(i) the in situ preparation of intrazeolite metal carbonyl clusters;
(ii) the flexible ligand method;
(iii) the template synthesis method;
(iv) the zeolite synthesis method.

The first three methods generally involve large pore zeolites and molecular sieves. The synthetic faujasite type (FAU) zeolites X and Y have a three-dimensional pore system with ~7.4 Å channels that open up to supercages 12 Å in diameter [70] as shown in Figure 1. From this structure one can envision an intrazeolite metal complex that is small enough to fit inside the supercage but is effectively larger than the cavity openings. Metal complexes have also been encapsulated in VPI-5 (VFI) type molecular sieves shown in Figure 2 which have a one-dimensional channel system with a fixed diameter of 12.1 Å. In the case of metal phthalocyanines, these large complexes are probably wedged in the slightly smaller channels.

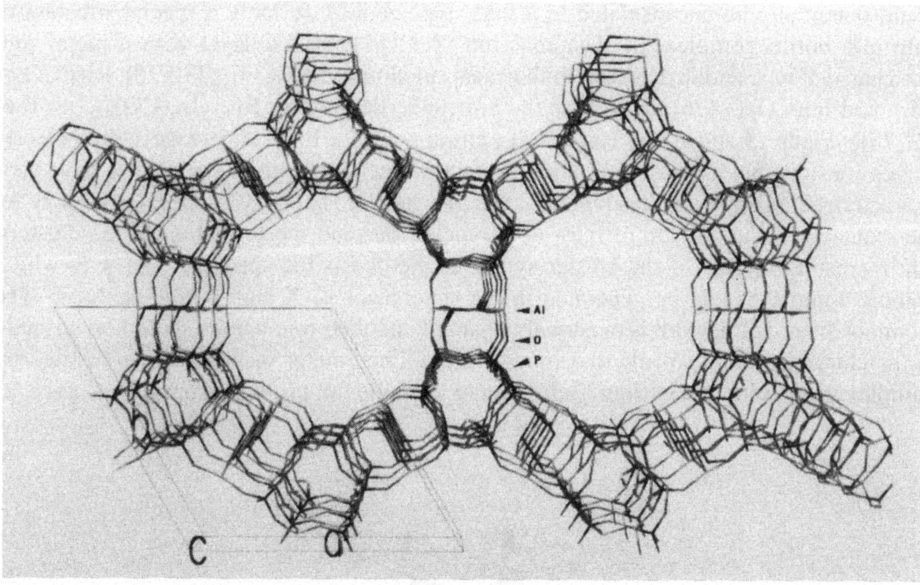

Figure 2. A [001] view of the aluminum phosphate molecular sieve VPI-5.

The encapsulation methods that involve preparing a complex inside a zeolite are limited by the ability to diffuse reagents into the crystalline molecular sieve. Zeolites such as Na-A, which has a large cavity 11.9 Å in diameter but openings that are only 4.1 Å, can not be modified with most ligands or ligand precursors because of the size restrictions of the cage windows. However, if the zeolite is synthesized around the metal complex, then only the cage dimensions are important as far as ligand size is concerned. In this case, the metal complex must meet the requirement of stability during crystallization but this is not a major restriction since zeolites/molecular sieves can be synthesized under many conditions that include low pH fluoride syntheses, high pH hydroxide mediated synthesis, and crystallizations from non-aqueous solvents. The zeolite synthesis approach has been employed to encapsulate metal complexes in a variety of molecular sieves ranging from the small pore clathrasils to the large pore FAU type zeolites. We next briefly review key examples of encapsulated metal complex syntheses before discussing their characterization and reactivity. For a more detailed discussion of synthesis techniques see Chapter 7 by Schoonheydt et al. in this volume.

2.1 METAL CLUSTER SYNTHESIS

Metal carbonyl adducts probably constitute the most widely studied group of intrazeolite coordination complexes [71]. Metal carbonyl clusters with a nuclearity greater than three could potentially be encapsulated in a FAU type zeolite, forming a special subclass of ship-in-a-bottle complexes. The reaction of CO/H_2 or CO/H_2O with a metal ion-exchanged Y type zeolite has lead to the encapsulation of $Rh_6(CO)_{16}$ [72-76], $Rh_4(CO)_{12}$ [76], and $Ir_4(CO)_{12}$ [76], as well as the bimetallic derivatives $Rh_{6-x}Ir_x(CO)_{16}$ (x= 0-6) [77,78]. Figure 3 shows the $Rh_6(CO)_{16}$ entrapped in a FAU supercage which clearly leaves considerable room in the cavity. In fact, clusters as large as $Pd_{13}(CO)_x$ have been characterized in Na-Y zeolites [79,80]. The presence of H_2 or H_2O appears necessary for the reductive carbonylation of intrazeolite metal ions and the formation of the clusters. An interesting feature of the cluster synthesis method is the apparent ease with which anionic complexes can be generated in the supercages of X and Y type zeolites. The aluminosilicate framework is negatively charged and therefore anions should be unstable in the channels. Some of the first intrazeolite anionic metal species were generated by complexation with the strong field ligand cyanide to produce complexes such as

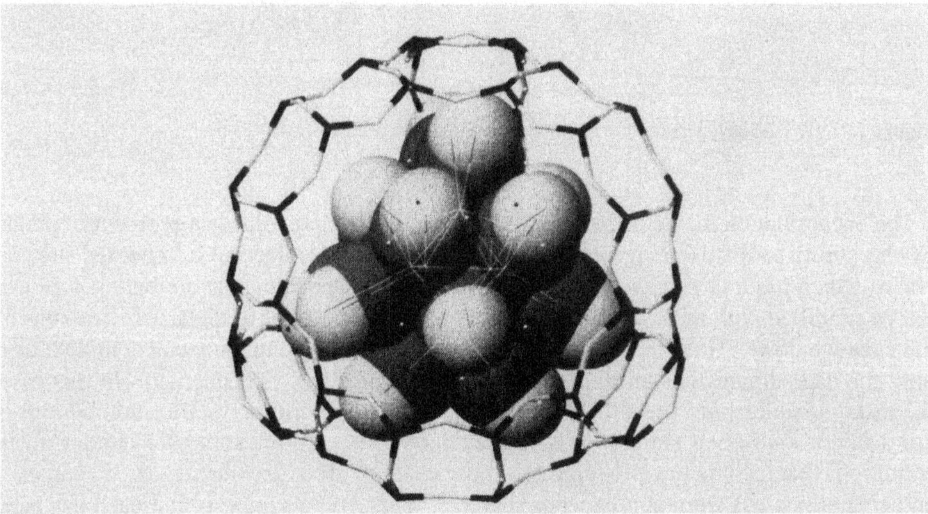

Figure 3. A space filling model of $Rh_6(CO)_{16}$ in a Na-Y supercage.

$Co(CN)_4^{2-}$ [81-83] and $Fe(CN)_4^{2-}$ [84]. Although, charge formalism is not as straightforward, anionic clusters of the type $[HFe_3(CO)_{11}]^-$ [77,78,85], $[HOs_3(CO)_{11}]^-$ [86,87], $[Ir_6(CO)_{15}]^{2-}$ [88,89], $[Co_6(CO)_{15}]^{2-}$ [90], and the Chini-type complexes $[Pt_3(CO)_3(\mu_2\text{-}CO)_3]_n^{2-}$ (n =3,4,5) [91-94] have been characterized in Na-Y zeolites. A bimetallic cluster, $[Rh_4Fe_2(CO)_{16}]^{2-}$, has been prepared in Na-Y by the reaction of

CO/H_2 with Rh(III)/Fe(II) exchanged zeolite or by the reaction of $Rh_4(CO)_{12}$ with encapsulated $[HFe_3(CO)_{11}]^-$ [77,78]. The question becomes why are these anionic clusters stable within the anionic zeolite lattice? In fact, the trinuclear species are not physically precluded from exiting the supercage. Also, the migration and aggregation of "$Pt_3(CO)_6$" species have been mechanistically implicated in the formation of encapsulated Chini complexes [94]. Therefore, certain trinuclear clusters may not be entrapped. However, the 12-ring windows to the supercage are defined by oxygens that bear partial negative charge. Passage of anions through this aperture is unfavorable so that the smaller anionic clusters may be electrostatically trapped. Additionally, the bifunctional activation of carbon monoxide, which arises from the interaction of a metal bound carbonyl with a charge balancing cation, may stabilize the intrazeolite cluster.

Molecular precursors such as $Ir(CO)_2(acac)$ may be absorbed into the zeolite and then reacted with CO to form encapsulated $Ir_4(CO)_{12}$ [95,96] or with CO/H_2 to give $Ir_6(CO)_{16}$ [97]. $Co_2(CO)_8$ absorbed in Na-Y leads to encapsulated $Co_4(CO)_{12}$ [98]. The numerous possibilities for molecular precursors to carbonyl clusters dramatically expands the potential for new and interesting intrazeolite species.

2.2 FLEXIBLE LIGAND METHOD

The flexible ligand method involves the diffusion of a multidentate ligand into the zeolite pores where, upon complexation with a metal ion, it becomes too large to exit. This approach is typified by the encapsulation of metal-SALEN complexes in synthetic faujasite type zeolites. Figure 4 illustrates how the free SALEN ligand might diffuse through the 12-ring windows and complex a supercage metal ion. The encapsulated MSALEN complex is sterically constrained to the supercage. However, computer modeling suggests that a MSALEN species might be able to squeeze itself back out [58]. Nevertheless, encapsulated SALEN complexes of Co [2,59,60], Fe [61], Mn [55,61], Rh [56], Ru [62], and Pd [57,58] have been reported. Typically, a dehydrated or partially dehydrated metal ion exchanged X or Y type zeolite is slurried with the melted SALEN free base ligand. Subsequently, the SALEN modified zeolite is solvent extracted to remove surface species.

The SALEN ligand generally binds in a tetradentate fashion. However, SALEN may also bind one or more metals in a bidentate fashion. Additionally, the ligand may bind in other than a square planar geometry depending upon what else is bound to the metal ion. The nature of the encapsulated SALEN complexes is not clear from spectroscopic data. This is further complicated by the presence of free base ligand as well as uncomplexed metal ions. In the case of Co(II)SALEN encapsulated in Na-Y, cyclic voltammetry has revealed the presence of two intrazeolite complexes [60]. One complex binds added base, such as pyridine, while the other is unreactive. This may be a consequence of different structural isomers or more likely different modes of coordination. Not all the MSALEN complexes behave in this fashion. For example, the higher oxidation state Mn(III)SALEN exhibits only one form inside Na-Y [61].

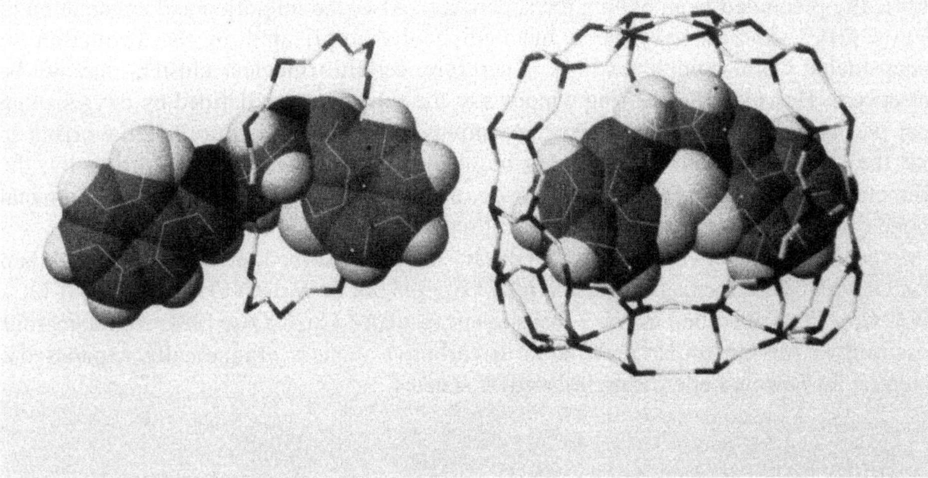

Figure 4. The flexible ligand SALEN is depicted diffusing through a 12-ring to form an FeSALEN complex in a Na-Y supercage.

The flexible ligand method requires a ligand(s) of such dimensions that intrazeolite complexation precludes diffusion back out of the zeolite. This places severe limitations on the number and type of ligands that might be employed. Additionally, the intrazeolite complexation must occur from the outer parts of the crystal to the inner parts. One would anticipate a heterogeneous distribution of complexes as a consequence of pore blockage. Nevertheless, this approach is probably the easiest in practice.

2.3 TEMPLATE SYNTHESIS METHOD

The template synthesis method involves the diffusion of ligand precursors into the zeolite pores where they assemble around an intrazeolite metal ion that acts as a template. This approach is typified by the synthesis of intrazeolite phthalocyanine complexes. The large pore zeolites X, Y, and the aluminum phosphate VPI-5 modified with metal ions, metallocenes (Cp_2Ni, Cp_2Ru, Cp_2Fe, $CpMn(CO)_3$) and metal carbonyl complexes ($Ni(CO)_4$, $Fe(CO)_5$, $Fe_2(CO)_9$, $Co_2(CO)_8$, $Os_3(CO)_{12}$) have been reacted with dicyanobenzene (DCB) to form the intrazeolite MPc complexes. Figure 5 illustrates how 4 DCB molecules would have to approach a FAU type supercage containing an $Fe(CO)_5$ molecule to form the FePc guest complex. The template synthesis of zeolite-encapsulated phthalocyanine complexes of Mn [23,32,42,51], Fe [3,4,21-25,33,47-49,51],

Co [6-18,23,33,35,36,43,47-49,51], Ni [6-10,12,13,15-17,36,38,47-49,51], Cu [6,7,9,10,12-17,30,33,36], Ru [23,36,37,47-49,51], Rh [31], Os [43,47-49], and Ti [51] as well as Li_2Pc and H_2Pc [51] have been reported. The synthesis generally involves heating an intimate mixture of partially dehydrated zeolite with an excess of DCB in a bomb reactor between 150 and 350° C. There has been one report of a solvent mediated reaction at 180° C which appears to lower the reaction temperature [23]. The condensation of four DCB molecules around a metal ion to form a phthalocyanine requires two reducing equivalents, the source of which is the subject of much debate. These electrons may originate from water or the metal ions in the case of the organometallic precursors. This is consistent with the fact that intrazeolite MPc complexes form most easily with zero valent metal carbonyls, followed by metallocenes and metal ions. Better results for MPc formation in ion-exchanged zeolites are obtained when the zeolite is only partially dehydrated.

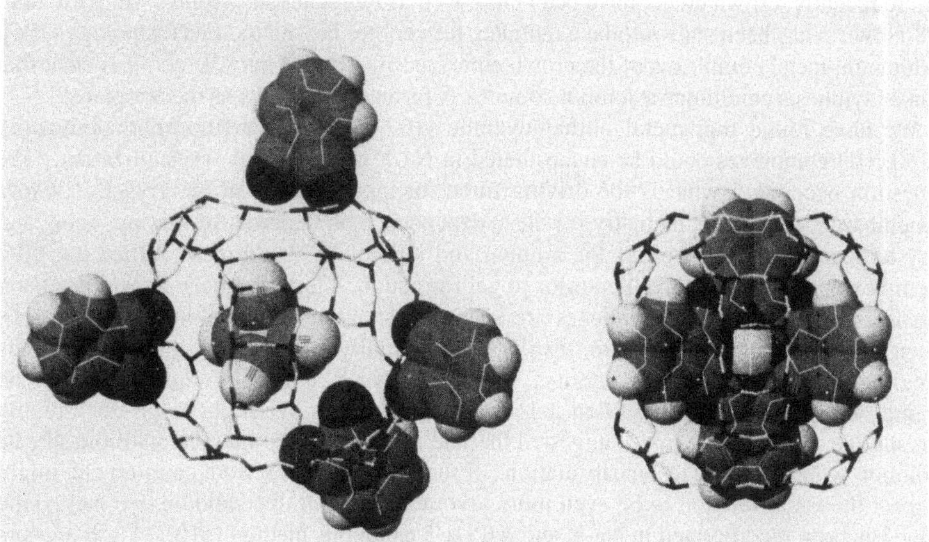

Figure 5. Dicyanobenzene molecules are depicted diffusing into an $Fe(CO)_5$ modified Na-Y supercage to form the encapsulated FePc complex.

Several MPc derivatives have been encapsulated in X and Y type zeolites including perhalogenated phthalocyanine complexes [100,101], *t*-butylphthalocyanines [26,27], and nitrophthalocyanines [51]. There have also been claims of Co-porphyrin [54], Fe- and Mn- tetramethylporphyrin complexes [51,53], as well as tetraphenylporphyrin complexes [23] encapsulated in Na-Y. The porphyrin template synthesis usually requires two reagents, pyrrole and either formaldehyde, acetaldehyde, or benzaldehyde.

2.4 ZEOLITE SYNTHESIS METHOD

The limitations and disadvantages associated with the flexible ligand and template synthesis methods for encapsulating metal complexes in zeolites encouraged us to explore the possibility of crystallizing a zeolite around a metal complex, i.e., build the bottle around the ship. This would afford the advantage of encapsulating a well-defined intrazeolite complex without contamination from free ligand as well as uncomplexed and partially complexed metal ions. There were some early efforts that employed simple chelate complexes such as bis(ethylenediamine)cobalt(II) during zeolite synthesis [102,103]. Unfortunately, these complexes are not stable during zeolite synthesis and may lead to partially exchanged or occluded metal. There have been claims of ZSM-5 and mordenite being synthesized in the presence of MPc complexes [104,105]. However, the 13 Å diameter of the Pc ligand far exceeds the pore dimensions of these zeolites. More recently, crown ethers have been employed as templates for zeolites. In particular, 18-crown-6 has been shown to be a template for zeolites having the EMT topology [106]. Although, metal complexes of the crown ethers are introduced into the gel, it is clear that under synthesis conditions a sodium complex is formed which acts as the template.

We have found that metal phthalocyanine [107-109] and perfluorophthalocyanine [100,101] complexes could be encapsulated in Na-X during zeolite crystallization. The question becomes - what is the driving force for incorporation in the crystals? Close examination of the gel chemistry reveals a dependence on the order of mixing, aging and crystallization time as well as the amount and type of MPc complex. When the MPc complex is mixed with the silica prior to gel formation, a fairly homogeneous dispersion results. However, if the complexes are added to the aluminate or aluminosilicate gels, then a heterogeneous mixture results and virtually no metal complex becomes incorporated in the resulting zeolites. A normal Na-X synthesis requires little or no aging, whereas the MPc modified gels may require aging overnight to produce highly crystalline zeolites. We have suggested that the metal complex interacts with the silicate solution which encourages encapsulation. If the metal complex were cationic one might expect the encapsulation to be even more favorable. In fact, the cationic dye methylene blue has been incorporated in Na-Y and $AlPO_4$-5 using this method [110,111]. It appears that the zeolite synthesis method is a viable approach to preparing ship-in-a-bottle complexes.

Na-X does not require a template so it is not clear what, if any, structure-directing properties the MPc complexes might have. More promising results have been obtained with bis(cyclopentadienyl)cobalt(III) ion, Cp_2Co^+, where ZSM-51 [112], nonasil [63,64], and $AlPO_4$-5 [113] have been crystallized in the presence of this complex. All of these molecular sieves require a structure directing agent. Therefore, cobalticinium ion acts as a template. The zeolite ZSM-51 and the all silica nonasil both have the NON topology [70] which belongs to a family of structural types known as the clathrasils. These molecular sieves are characterized by the complete entrapment of the template during crystallization. The nature of the anion for these syntheses is also important. For example, Cp_2CoPF_6 was claimed as a template for ZSM-51 [112], while we found this complex resulted in only amorphous materials for the nonasil synthesis. Cp_2CoOH, on the other hand, produced highly crystalline nonasil in both hydroxide and fluoride media after only a few days, while the organic templates typically require several weeks. The encapsulation of Cp_2Co^+ during synthesis resulted in nearly one complex per $[5^86^{12}]$ cage (see Chapter 1, Table 3b, p 8 for details regarding this nomenclature), the only cavity large enough to accommodate the complex. This ellipsoidal cavity, shown in Figure 6, is a 20-hedron cage accessible only through six-membered rings. The Cp_2Co^+ complex easily fits in this cage and appears to have a similar shape. In contrast to the NON cage structure, $AlPO_4$-5 has a one-dimensional channel system which easily accommodates the Cp_2Co^+ complex (Figure 6) but structure-shape relationships are less obvious. Rigid templates such as cobalticinium ion and derivatives may provide information on effects of template symmetry during the synthesis process as well as assist in the rational design of new materials.

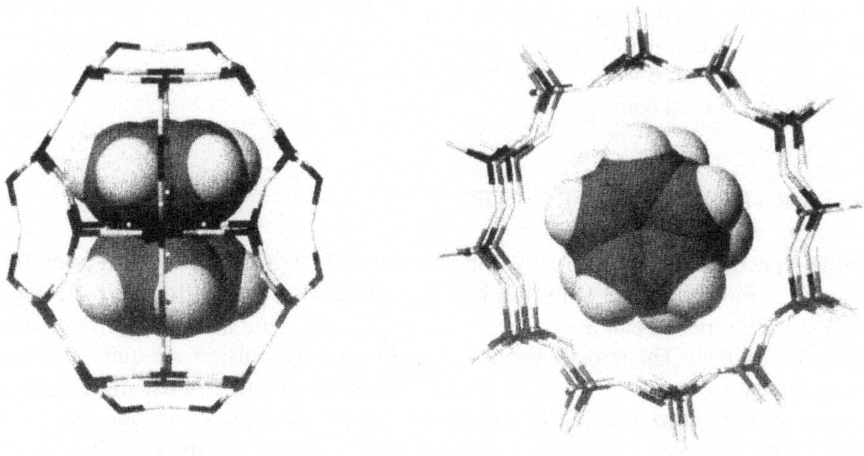

Figure 6. Bis(cyclopentadienyl)cobalt(III) ion is depicted in the nonasil $[5^86^{12}]$ cage (left) and inside a channel of $AlPO_4$-5 (right).

3. Characterization of Intrazeolite Complexes

The characterization of zeolite encapsulated metal complexes is quite difficult and generally requires a battery of techniques in order to convincingly argue for the intrazeolite location of the metal complexes. The first step in this procedure is to remove surface species by Soxhlet extraction with a series of solvents and/or vacuum sublimation. One has to be particularly careful with metal complexes such as the metallophthalocyanines, that strongly adsorb to the metal oxide surface. We have found that in the case of some zeolites modified with MPc complexes that even after solvent extraction and sublimation some residual metal complex could be removed electrochemically. When a negative potential is applied to freshly prepared zeolite encapsulated MPc complex/graphite composite electrodes, a small amount of reduced metal complex is released from the surface as evidenced by a faint blue color in the electrolyte solution. With subsequent cycles and fresh solvent, there was no further evidence of surface MPc species. This is not the case with zeolites modified with MSALEN complexes where solvent extraction is sufficient to remove surface species. Even though the outer crystal surface area of most zeolites may be several orders of magnitude less than the internal surface area, it is clear that one cannot completely rule out the possibility that there are some complexes adsorbed on the outer surface or partially occluded in mesopores. Therefore, zeolites with very low loadings of metal complex should be considered carefully. In addition to traditional elemental analyses, encapsulated metal complexes such as the phthalocyanines may be recovered after digestion of the zeolite and quantified by UV-Vis spectroscopy. This is preferred for MPc complexes prepared by the template method because the complexation may not be quantitative and elemental analysis does not necessarily reflect the amount of metal complex. Otherwise, a variety of techniques are required to characterize the intrazeolite complexes as detailed below.

3.1 VIBRATIONAL SPECTROSCOPY

Infrared spectroscopy is probably the most widely applied analytical tool for the characterization of zeolite ship-in-a-bottle complexes. Either diffuse reflectance or transmission IR spectra can provide information on the zeolite lattice as well as the nature of guest molecules. The mid-IR spectra of metal carbonyl clusters are quite useful given the sensitivity of the CO ligand to its environment. In fact, most of the structural evidence of such intrazeolite clusters is based on CO [72,74-80,85-97] and ^{13}CO [92,94] stretching frequencies. The weak coordination of a supported metal carbonyl oxygen with the zeolite host (eqn. 1) may affect the stability and reactivity of encapsulated clusters.

$$M\text{-}C{\equiv}O\cdots\cdots Na^{+}Z^{-} \qquad (1)$$

A shift for both bridging and terminal carbonyl IR stretching frequencies relative to the free clusters, is consistent with this type of interaction. Such shifts have been noted for several encapsulated metal carbonyl clusters [72,78,79,85,88,89,91,97,98]. The zeolite lattice effects are also influenced by water. Zeolite encapsulated $Ir_4(CO)_{12}$ upon dehydration exhibits bridging carbonyl bands in the IR which were interpreted as a consequence of removing the protective sheath of water between the cluster and the electrostatics of the cage [76].

The mid-IR spectroscopic results for intrazeolite MPc [25,28,31,33,45,47-49,51], substituted MPc [26,27,100,101] and MSALEN [55,56,58,61] complexes have been reported. The various T-O vibrational modes associated with the zeolite dominate most of the region between 1150-450 cm^{-1}, however, at reasonable loadings, the complexes can be detected. The bands associated with C=N and C=C stretching modes for the Pc ligand are easily observed and may show a slight shift to lower wavenumbers inside the zeolite. Kimura et al. have interpreted this shift as a result of ligand distortion in the supercage [28]. However, intrazeolite MPc complexes prepared by the template method may contain free Pc ligand which could account for lower frequency bands [51]. The complexation of SALEN should result in a blue shift of a strong C=C stretch at 1504 cm^{-1}. Other bands not associated with free ligand or complex have been observed [56,61] and may be the result of coordination modes other than tetradentate.

The resonance Raman spectra for Li_2Pc, TiPc and FePc in Na-Y suggest that the intrazeolite complexes are not distorted from D_{4h} symmetry [51]. Such a distortion might be expected to result in the generation of new bands as well as a loss of others. At this point it is not clear where the anticipated spectral changes would be observed. However, these results are in agreement with NMR results obtained in the same study.

3.2 ELECTRONIC SPECTROSCOPY

Electronic spectra can be useful in evaluating intrazeolite complexation as well as any structural perturbations that might occur as a consequence of encapsulation. The aromatic phthalocyanines are intensely blue and green with extinction coefficients on the order of 10^5. Therefore, the UV-Vis spectra of MPc modified zeolites, even those containing low loadings of MPc, are readily obtained as Nujol mulls or via reflectance techniques. The most intense band for the complexes is a Pc based $\pi \rightarrow \pi^*$ transition known as the Q band which is typically in the range of 600-900 nm. MPc complexes encapsulated in FAU type zeolites generally exhibit a Q band that is red-shifted relative to the free or surface physisorbed complex. Shultz-Ekloff first observed these spectral changes [35] and Herron later suggested that this arises from distortion of the planar Pc ligand as a result of steric interactions within the supercage [22]. Molecular modeling predicts a saddling of the MPc complex which is in agreement with the fact that the Pc ligand is slightly larger than the 12 Å diameter of the supercage. XPS [14,36,37], EXAFS [25-28], and IR [28] results might be interpreted as consistent with a deformation of the ligand from planarity. However, Raman and NMR results [51] suggest there is no distortion. If the intrazeolite MPc complexes are not distorted, then a bathochromic shift for the Q band might arise from zeolite solvent effects. FePc in sulfuric acid exhibits an ~100 nm red-shift of the Q band relative to organic solvents which is attributed to protonation of the peripheral nitrogens. It has been suggested that intrazeolite phthalocyanines might be protonated [122]. Evidence for protonated intrazeolite Pc species has been reported by Parton et al. [45]. Additionally, XPS results could be interpreted as a result of protonated nitrogens [36,37]. It is also quite amazing that Na-X can be synthesized around a MPc complex that has to be distorted. Nevertheless, the saddle structure for a zeolite encapsulated phthalocyanine is possible and probable given the constraints of the supercage. Similarly, the Q band for FePc(t-Bu)$_4$ (19 x 19 Å) prepared in Na-Y [27] is red shifted. Additionally, zeolite encapsulated tetramethylporphyrin complexes exhibit a suppressed Q band which has been interpreted as an effect of ligand distortion in the supercage [53].

The UV-Vis spectra of intrazeolite SALEN complexes are not quite as sensitive to structural perturbations. The bright yellow SALEN ligand exhibits bands at 323 and 409 nm which shift upon complexation. Both Mn(II) and Mn(III)SALEN [55] as well as Rh(III) [56], Pd(II) [58] and Co(II)SALEN [59] have been characterized by UV-Vis spectroscopy.

The color of the intrazeolite Chini-type carbonyl clusters vary with nuclearity such that $[Pt_9(CO)_{18}]^{2-}$ (orange-brown), $[Pt_{12}(CO)_{24}]^{2-}$ (dark green), and $[Pt_{15}(CO)_{30}]^{2-}$ (yellow-green) are readily characterized by their UV-Vis spectra [90-92]. Differences between solution and encapsulated clusters have been noted but not explained. Perhaps the interaction of coordinated CO with the zeolite lattice contributes to these differences, however, this needs to be studied further.

3.3 ELECTRON PARAMAGNETIC RESONANCE/MÖSSBAUER

Both EPR and Mössbauer spectroscopy provide information on the states of iron-based zeolite ship-in-a-bottle complexes [25-28,39-41]. The Mössbauer spectra of Fe(II)Pc encapsulated in Y type zeolites indicate the presence of two types of FePc, one that is reactive with pyridine (py) and one that is not [39-41]. The formation of FePc(py)$_2$ depends on the free volume within the supercage. FePc entrapped in Na-Y, K-Y, and Rb-Y showed a decreasing tendency to react with pyridine as the size of the charge balancing cation was increased [41]. There is very little difference in the isomer shift or quadrupole splitting between intrazeolite and free FePc. In contrast, the Mössbauer spectrum for FePc(t-Bu)$_4$ inside the zeolite is quite different from the free complex and indicates reduced electron density for the encapsulated Fe [26,27]. A temperature dependence study of the Mössbauer absorption area of FePc in Na-Y indicated the results do not follow a Debye approximation suggesting the complexes are well-dispersed [39]. Complimentary EPR experiments indicate the FePc complexes are close enough for the spins to interact [41].

A dioxygen adduct of Na-Y encapsulated CoSALEN has been studied by EPR [2]. The intrazeolite CoSALEN prepared by the flexible ligand method is EPR silent and presumably high-spin Co(II). Exposure to pyridine still does not generate an EPR signal, however, subsequent absorption of oxygen produces a classic EPR spectrum for a Co-O$_2$ adduct. The signal is axial which indicates the complex is not rotating in the supercage. Additionally, the EPR parameters are quite similar to those obtained in solution such that the intrazeolite complex is probably not distorted.

3.4 NUCLEAR MAGNETIC RESONANCE

There has been only one report of zeolite ship-in-a-bottle complexes being characterized by solid-state NMR [51]. Na-Y encapsulated Li_2Pc, FePc, Fe(nitro)$_4$Pc, and TiPc were studied using cross-polarization magic-angle-spinning ^{13}C-NMR. There are 8 different carbons that can be differentiated by this technique, but only 4 were resolved for the intrazeolite species. One might expect the carbon resonances to shift as the hybridization or aromaticity is altered by distortion of the Pc macrocycle from planarity inside the supercage. However, the peaks observed for zeolite encapsulated TiPc were nearly the same as for the free complex. This would seem to argue against any significant deformation of the intrazeolite Pc ligand. The fact that four signals are well-resolved and have narrow line widths suggests the complexes are fairly symmetric. The FePc in Na-Y was shown to be largely free H_2Pc ligand which again is a consequence of the template method of encapsulation.

The ^{7}Li-NMR spectra of Na-Y modified with Li_2Pc indicate the surface and entrapped complexes can be resolved. The signal for the intrazeolite complex was shifted ~1.8 ppm upfield from the surface complex which was attributed to the effects of the electrostatic field inside the zeolite [51].

Zeolites containing paramagnetic metal ions have been shown to be effective contrast agents for magnetic resonance imaging (MRI) [114]. A diagnostic pharmaceutical for this purpose known as GADOLITE®, which is based on a Gd(III) modified Na-Y zeolite, is currently in clinical trials. As part of this effort, Gd(III) complexes of 8-hydroxyquinoline were prepared in zeolites Na-X and Na-A by the zeolite synthesis method [115]. In the case of the Na-A zeolite, the 4.1 Å apertures to the large cage preclude the 8-hydroxyquinoline ligand from entering. Therefore, the only way to prepare such ship-in-a-bottle species is to crystallize the zeolite around the complexes. A measure of potential in vivo efficacy as an MRI contrast agent is the relaxivity which represents the relationship between water proton T_1 shortening and gadolinium concentration. Preliminary results suggest the intrazeolite complexes exhibit higher relaxivities than the ion-exchanged zeolites.

3.5 ELECTROCHEMISTRY

There is a well-established effort to include zeolites in the design of new electrocatalytic surfaces as evidenced from recent reviews [116,117]. The application of cyclic voltammetry to characterize intrazeolite metal complexes has also proven to be quite useful. Both intrazeolite metal SALEN [60-62] and phthalocyanine [50,62] as well as perfluorophthalocyanine [62,100,101] complexes have been characterized electrochemically.

For the zeolite ship-in-a-bottle complexes, electron transfer must occur inside the zeolite, presumably by an extended electron-hopping mechanism [118]. Therefore, communication between zeolite cavities depends on the complex loading. The outermost supercages of the zeolite crystals in contact with graphite will be redox accessible. However, the presence of vacant supercages hinders the electron transfer process between complexes in the bulk. Additionally, occupied cages may prevent solvent and electrolyte from reaching inner crystal redox sites. As a result, only a few percent of the encapsulated metal complexes are electroactive. The addition of a surface mediator, such as a metal complex too large to enter the zeolite, can improve the electrochemical response of the entrapped complexes [118]. However, electrochemical access to all of the encapsulated complexes remains a challenge. For those intrazeolite metal complexes that are redox accessible, the possibility exists for developing new types of electrocatalysts. The catalytic reduction of alkyl halides was demonstrated for CoSALEN in Na-Y [60] and the electrochemical formation of MnSALEN superoxo and oxo species [61] presents an opportunity for exploring oxidation electrocatalysis.

3.6 X-RAY METHODS

X-Ray powder diffraction is certainly an indispensable technique for the zeolite synthesis method where phase identification and purity are critical. XRD also provides valuable information on crystallinity as well as any change in unit cell parameters that might arise from the in situ generation of intrazeolite ship-in-a-bottle complexes. Variations in peak intensities resulting from a redistribution of charge balancing cations after inclusion of cationic complexes in the supercage have been noted. However, for the most part, ship-in-a-bottle complexes are neutral. Intrazeolite MPc complexes, including the perfluorinated analogs, have never shown an expansion of an X or Y zeolite unit cell in spite of the fact that the complex has a tight fit inside the supercage [22,31,33,35,45, 51,100,101]. Zeolite crystallinity is largely preserved after the encapsulation of MPc complexes by the template method. Complexes prepared from ion-exchanged zeolites might be more susceptible to dealumination as the formation of protons is likely. The preparation of RhPc complexes in Y or X zeolites results in a loss of crystallinity, however, this system also involves partial reduction of the Rh(III) [31]. The zeolite crystallinity was reported to be intact after entrapped MnSALEN was prepared by the flexible ligand method [55]. Na-Y was also found to be unaffected by the generation of intrazeolite osmium carbonyl clusters [87].

The intrazeolite carbonyl clusters are more often characterized by extended X-ray absorption fine structure spectroscopy (EXAFS). This technique provides valuable information on the type and number of atoms as well as the distance between them. Unfortunately, access to the required synchrotron radiation may be limited. X-Ray absorption near edge spectra (XANES) are often obtained with EXAFS and generally are used to evaluate oxidation states. EXAFS spectroscopy has provided evidence of $Ir_4(CO)_{12}$ [89,94], $[Ir_6(CO)_{15}]^{2-}$ [89], $Rh_6(CO)_{16}$ [72,78], $[Ni(CO)_4]_2$ [97], and $[Pt_3(CO)_6]_n^{2-}$ [91,92,94]. Both the Rh_6 and Chini-type clusters appear to interact with the zeolite framework - with the latter complexes possibly being distorted. The formation of Na-Y encapsulated FePc was confirmed by XANES and EXAFS [25,28]. The Fe-N distance was determined to be slightly longer than for the free complex which is consistent with distortion of the complex in the supercage. Similarly, Na-Y modified with Fe(*t*-butyl)$_4$Pc exhibited a longer Fe-N distance and a lower coordination number for Fe (3.6) [26,27]. These results do not confirm encapsulation but agree with other spectroscopic results. For a more detailed description of X-ray techniques used in characterization see Chapter 3 by Parise in this volume.

3.7 SURFACE AREA/ADSORPTION EXPERIMENTS

Since zeolite surface area is largely internal, the inclusion of guest molecules should dramatically reduce the absorption capacity of such molecular sieves. The BET surface area of X and Y zeolites containing MPc complexes (M=Mn,Fe,Co,Cu) [32,33] and Co porphyrin [54] are typically less than 100 m^2g^{-1} which is a reduction of at least 600 m^2g^{-1}. Interestingly, the nitrogen absorption capacity of Na-X with entrapped CoPc was reduced by only half [35]. This indicates very little pore blockage at the surface and may attest to the distribution of complexes. The absorption capacity for water in X and Y zeolites containing FePc was reported to be reduced [20]. The absorption results for O_2 at -78 and -196° C by zeolites modified with MPc complexes (M=Co,Cu,Ni) indicated a high dispersion of complexes [7,9]. Pre-absorption of pyridine by these zeolites reduces O_2 absorption in the order NiPc>CoPc>CuPc [12]. Probe molecules such as bulky phosphines have not been applied to the intrazeolite clusters, MPc, or MSALEN complexes but have shown promise in discriminating surface versus internal species, as well as generating ship-in-a-bottle complexes [1,97].

3.8 THERMAL METHODS

Differential Scanning Calorimetry (DSC) has been used to study the energetics of intrazeolite MPc formation by the template synthesis method [29,30]. In the case of CuPc, it was found that the exotherm for CuPc formation occurs at increasing temperature in the order $CuCl_2$<CuNa-Y<CuNa-X<CuNa-A (surface only). The lower peak temperature for CuNa-Y is attributed to the higher mobility of Cu^{2+} in the Y zeolite. The amount of intrazeolite CuPc complex formed in Na-Y increased with both increasing copper loading as well as pressure. At low pressure and high copper loadings, the exotherms for CuPc formation both inside and outside the zeolite can be resolved, where the intrazeolite complex is formed at higher temperatures.

Differential Thermal Analysis (DTA) [33,35] and Thermogravimetric Analysis (TGA) [32,51] have been used to characterize intrazeolite MPc complexes prepared by the template method. The amount of intrazeolite MPc can be estimated from the weight loss and is generally more accurate than the spectrophotometric analysis after zeolite digestion. In the case of iron(II) phthalocyanine and tetranitrophthalocyanine complexes in Na-Y, several exotherms are observed during the decomposition which have been ascribed to free ligand and organic residues [51]. The encapsulated complexes appear more thermally stable than the free complexes. These results are in contrast with those observed by Gudkov et al. [14,15] where zeolite Y encapsulated CoPc, CuPc and NiPc were deemed less stable than the free complexes. This difference was ascribed to the distortion of the Pc ligand in the supercage.

3.9 SURFACE TECHNIQUES

X-Ray photoelectron spectroscopy (XPS) has been applied to the characterization of intrazeolite MPc [14,18,22,33,36,37,47-49] and MSALEN [2,56,59] complexes. XPS can provide information on the relative concentration of elements and oxidation states in the first ~50 Å of the zeolite which may actually reflect less than 1% of the crystal. Nevertheless, shifts in binding energies for both the intrazeolite metals and nitrogen in SALEN or phthalocyanines indicate complexation. Shpiro et al. noted two types of nitrogens for Na-Y encapsulated Ni, Cu, Co, and RuPc [14,36,37]. This was interpreted as either the result of Pc ligand distortion or strong interactions with the zeolite lattice. A close match between surface concentrations, estimated from metal/Si or metal/Al ratios, and bulk elemental analyses has been used as evidence for a homogeneous distribution of intrazeolite complexes. A higher concentration of metal at the surface does not necessarily imply extrazeolite complexes since ions and molecules may migrate close to the surface as a consequence of the thermal treatment associated with the flexible ligand or template syntheses.

Scanning and transmission electron microscopy (SEM or TEM) provides evidence for metal complex or metal oxide depositions on the outer zeolite crystal surface. SEM images of zeolites modified with MPc complexes before and after solvent extraction exhibit significant changes in crystal surface morphology [18,33,35]. For comparative purposes, Meyer et al. prepared a Na-Y covered with a two-dimensional phthalocyanine polymer derived from tetracyanobenzene which showed a uniform but roughened surface [18]. Whereas the solvent extracted MPc/Na-Y samples had smooth, clean surfaces. Zeolite encapsulated Co porphyrin was also studied by SEM with similar results [54]. TEM has provided evidence of uniform dispersions of osmium carbonyl clusters in Y type zeolites [87]. These techniques, in concert with the others described above, provide compelling evidence for the intrazeolite location and distribution of metal complexes.

3.10 COMPUTER MODELING

Molecular modeling is rapidly becoming an indispensable tool for predicting zeolite guest/host interactions as well as corroborating empirical results. Both the phthalocyanine [22,45,51,100,119,120] and SALEN [2,58,61] complexes have been modeled in the FAU supercage. Several important questions have been posed including, are the intrazeolite MPc complexes distorted and is MSALEN truly entrapped? Herron first reported a zeolite Y encapsulated FePc model where a saddle distortion of $<20°$ from planarity allowed the phenyl rings to poke through the supercage windows without any close van der Waals contacts [22]. Subsequently, Parton et al. calculated the minimum Pc ligand distortion to be 37.27° from planarity using Chem-X software [45]. It was also shown that FePc in the 12.1 Å cylindrical channels of VPI-5 is not distorted if the complex is positioned between two opposite 6-rings. There is no such possibility of steric relief in the 12 Å FAU supercage given the spherical nature of the cavity. The lowest energy configuration of $FeF_{16}Pc$ in Na-Y was found to be distorted 34.5° from planarity using CAChe molecular mechanics and MM2 parameters [100]. The consistency of these calculations in predicting a saddle shape complex argues for the distortion used to explain many spectroscopic results.

CoSALEN and the pyridine adduct were modeled inside Na-Y and shown to be entrapped [2]. However, De Vos et al. claimed that by moving the CoSALEN in small steps (0.2 Å and 3°), the complex could pass through the supercage windows [58]. This was accomplished using Chem-X software and crystallographic parameters for CoSALEN. This calculation does not take into account other absorbed molecules or the lowest energy configuration for encapsulated MSALEN. A possible bidentate SALEN complex with iron at an S_{II} site was also modeled using CAChe software [61]. Since the nature of the ligand coordination appears to vary with the metal ion, it may be appropriate to refer to intrazeolite SALEN complexes as sterically hindered.

4. Reactivity

A number of reviews have appeared which specifically address catalytic advances in the area of zeolite guest/host chemistry, especially ship-in-a-bottle complexes [120-127]. The largest effort is directed towards the development of oxidation catalysts, although, a number of new applications are beginning to emerge. In addition to the potential for new catalyst development, reactivity studies can provide convincing evidence for the encapsulation of metal complexes.

4.1 DIOXYGEN BINDING

Some of the major challenges in the development of a commercially viable oxygen carrier include capacity and reversibility. A target capacity of 10 cc O_2/g of carrier is amenable to the high surface area zeolites modified with metal complexes. Additionally, the zeolite ship-in-a-bottle complexes are effectively site-isolated which precludes formation of peroxo-bridged dimers that could occur on conventional supports or in solution. Zeolite Na-Y encapsulated CoSALEN which is EPR silent and presumably high-spin Co(II), is unreactive towards O_2. However, intrazeolite CoSALEN readily binds O_2 after the complex is exposed to pyridine [2,59]. A negative cooperativity was deduced from Hill plots which was interpreted as the ability of the zeolite to maintain an

O_2 pressure within the pores that was effectively greater than the external pressure [2]. This system may actually be more complicated since electrochemical studies have shown it is possible to have at least two types of intrazeolite CoSALEN complexes with only one that binds pyridine [60].

4.2 OXIDATION CATALYSIS

Herron et al. first reported the oxidation of alkanes by iodosylbenzene, catalyzed by FePc encapsulated in zeolites Na-X and Na-Y [3]. Substrate selectivity was reported where, in a competitive experiment, cyclohexane was preferentially oxidized over cyclododecane. The oxidation of n-octane favored the 2-position near the end of the molecule which is presumably the result of orientation effects in the zeolite. Norbornane was oxidized to norborneol with an exo:endo ratio of 6 compared with the ratio of 9 observed with the homogeneous catalyst. Na-Y encapsulated FePc catalyzed methylcyclohexane oxidation with a turnover number of 5.6 and a yield of *trans/cis* alcohols at the 4 position in a 2:1 ratio whereas the free FePc complex gave a 1.1:1 mixture and 1.1 turnovers. It was later shown that the selectivity could be further modified by changing the charge balancing cation of the zeolite [22]. Although the activity of intrazeolite FePc was greater than the free complex, it was still rather poor and this was attributed to pore blockage by iodoxybenzene [3]. Subsequently, Parton et al. reported that using *t*-butylhydroperoxide and Na-Y zeolites containing 1 FePc complex per 77 supercages for the oxidation of alkanes resulted in turnovers as high as 6000 [45,46]. Not only was the lower FePc loading deemed significant in the prevention of pore blockage, but the original source of iron, ferrocene, was also important in reducing the amount of uncomplexed iron. The encapsulated FePc eventually deactivates, presumably from ligand degradation. In fact, the peroxide must be slowly added because the catalyst can be easily bleached with excess oxidant. Interestingly, FePc occluded inside VPI-5 appears more stable than the FAU zeolite analogs, with no signs of deactivation even after 2000 turnovers [46]. There are two major differences between these ship-in-a-bottle systems. Firstly, inside VPI-5 the complex is predicted to be planar and secondly, the one-dimensional channels of VPI-5 may limit activity to the ends of the pores.

Both MnPc [32,51] and MnSALEN [55] encapsulated in Na-Y have been reported to catalyze the oxidation of alkenes using iodosylbenzene. In contrast to FePc, the activity of intrazeolite MnPc is considerably lower than for the free complex but the selectivity is altered [32]. For example, the conversion of 1-undecene is greater than for 2-octene which is the opposite of that obtained with the free complex. Similarly, the conversion of *trans*-2-octene is greater than *cis*-2-octene which again is opposite from the homogeneous catalyst. Na-Y encapsulated MnSALEN also exhibited much lower activity than the free complex which was attributed to substrate diffusion and competition with solvent oxidation. However, shape selectivity was observed with the order of reactivity, cyclohexene>styrene>*trans*-stilbene. This is in contrast to free MnSALEN which exhibits the opposite selectivity. Substitution of the Pc ligand periphery with four nitro groups enhances the catalytic activity of the Mn and Fe complexes [51]. This is believed to be the result of their electron withdrawing properties. A perfluorinated phthalocyanine complex of iron, which we anticipate will exhibit further enhanced catalytic activity and oxidative stability, has been encapsulated and reactivity studies are in progress [100].

There are relatively few examples of O_2 oxidations catalyzed by zeolite-encapsulated MPc complexes. Na-X encapsulated CoPc was found to facilitate the oxidation of propene to formaldehyde and acetaldehyde, while the free complex was unreactive and Co-Y produced acetone [35]. The oxidation of methane at 375° C catalyzed by Na-Y encapsulated Co, Fe, Mn and Ru phthalocyanines as well as tetraphenylporphyrin complexes produced primarily CO_2 and H_2O [23]. The intrazeolite RuPc, CoTPP, and MnTPP produced some methanol. The oxidation of cumene catalyzed by FePc [21] and cysteine by CoPc [11] in Na-Y have also been reported.

Zeolite-encapsulated MPc complexes are beginning to attract interest as eco-catalysts. Romanovsky et al. have studied the oxidation of CO and the reduction of NO [7,12,47-49]. CO can be converted to CO_2 over the temperature range -78 to 60° C. At room temperature, CO_3^{2-} forms and poisons the catalyst. However, pre-absorption of pyridine prevents formation of carbonate over these catalysts. NO was reduced by both CO and H_2 over MPc (M=Co,Fe,Ni,Ru,Os) complexes entrapped in Na-Y zeolites between 200 and 300° C. In all cases, CoPc/Na-Y was the most active, producing primarily NH_3 during H_2 reduction and N_2O during CO reduction of the NO. Zeolite-encapsulated $[Pt_3(CO)_6]_n^{2-}$ clusters have also been shown to be effective catalysts for NO reduction by CO [91,92].

4.3 HYDROGENATION CATALYSIS

Palladium(II)SALEN is an efficient catalyst for the room temperature hydrogenation of olefins but poorly selective [128]. Encapsulation of the PdSALEN complex in zeolites Na-X or Na-Y dramatically enhances the substrate selectivity. For example, a mixture of cyclohexene and 1-hexene are converted to nearly equal amounts of cyclohexane and n-hexane by the free PdSALEN complex whereas the zeolite encapsulated PdSALEN hydrogenates only the terminal olefin in the same mixture [57]. For a mixture of cyclooctadiene and 1-octene, the hydrogenation of 1-octene was suppressed [58]. It was also shown in this study that the zeolite encapsulated PdSALEN was more selective for the hydrogenation of cyclooctadiene to cyclooctene than Pd on carbon or PdY. The results for hydrogenation of the diene and mono olefins in these two studies require further study; however, at this point it is clear that the complexation of intrazeolite Pd(II) imparts reactivity not observed in solution or with supported Pd.

The catalytic hydrogenation of toluene to methylcyclohexane in the presence of iron(II) phthalocyanine encapsulated in Na-Y has been reported [34]. Benzene was unreactive under the same conditions so it was proposed that the methyl group of toluene interacts with the zeolite lattice favoring reaction. FePc/Na-Y was reported to be unreactive towards butadiene hydrogenation [28]. However, treatment of the intrazeolite FePc with sodium naphthalide induces catalytic activity. At low conversions, higher selectivity was observed towards *trans*-2-butene compared with the extra zeolite $(Na^+)_4(FePc^{4-})$ donor-acceptor complex.

The hydrogenation of CO catalyzed by several intrazeolite carbonyl clusters has been reported [72,86,87,89]. The encapsulated $Ir_4(CO)_{12}$ [89], $[Ir_6(CO)_{15}]^{2-}$ [89], and $HOs_3(CO)_{11}^-$ [86,87] produce hydrocarbons but in a non-Schultz-Flory distribution where the selectivity is towards alkene rich C_2-C_4 molecules.

4.4 OTHER REACTIONS

There are a variety of reactions where the metal carbonyl clusters are either precursors to metal particles or the clusters are simply implicated in the mechanism [71]. There is proportionately less information available on intact intrazeolite carbonyl clusters involved in catalysis. The water gas shift reaction has been reported to be catalyzed by $HFe_3(CO)_{11}^-$ [85] and $[Pt_3(CO)_6]_n^{2-}$ (n=3,4) [129] in Na-Y. The iron cluster catalyst exhibited high activity at low temperatures (60-180° C) and atmospheric pressure. The intrazeolite Chini clusters, which were also quite active at low temperatures (50-150° C), displayed enhanced activity (38 times higher) during illumination from a Xe lamp.

Zeolite encapsulated metal phthalocyanines have been reported to be effective catalysts for the dehydrogenation of cyclohexane [7,15] and isopropanol [7,8]. It was noted that pre-absorption of aromatics such as pyridine enhanced the activity of intrazeolite NiPc and CoPc but not that of CuPc. The conversion of cyclohexene oxide catalyzed by zeolite encapsulated CuPc was found to be nearly 100% selective for cyclohexadiene [10]. Finally, the conversion of butylmercaptan to butylenes, dibutylsulfide, and hydrogen sulfide over Na-Y zeolites containing various metal phthalocyanines was reported [16] and more recently, Na-X encapsulated CoPc was shown to be an efficient catalyst for the oxidation of mercaptans to disulfides [44].

5. Future Trends

It should be apparent that a large effort has been directed towards the preparation and characterization of zeolite ship-in-a-bottle complexes. The synthesis of molecular sieves around metal complexes is probably the most versatile strategy discussed because there is no predetermined size restrictions on the complex. Since the complexes are well-defined before encapsulation, the characterization of the intrazeolite species should be easier. Additionally, there are no limitations as to the pore dimensions or structural composition of potential host molecular sieves. Moreover, the possible structure directing properties of a metal complex presents exciting opportunities for the generation of new molecular sieves.

The bulk of this review has involved the modification of synthetic faujasite type zeolites, partly because of the pore structure. The diffusional limitations of the X or Y zeolite containing metal complexes casts some doubt as to the commercial viability of such catalysts. However, there are many new porous materials such as MCM-41 [130] or cloverite [131] that would be amenable to the encapsulation methods outlined above. One could imagine that structural and compositional variance might impart new and interesting properties to a ship-in-a-bottle complex. In addition to the many catalytic transformations that will undoubtedly be elucidated in the near future, we expect new directions in areas ranging from pharmaceuticals to materials chemistry will be forthcoming.

References

1. N. Herron, G. D. Stucky, and C. A. Tolman, *Inorg. Chim. Acta*, **100**, 135 (1985).
2. N. Herron, *Inorg. Chem.*, **25**, 4714 (1986).
3. N. Herron, C. A. Tolman, and G. D. Stucky, *J. Chem. Soc., Chem. Commun.*, 1521 (1986).
4. R. Parton, D. De Vos, and P. A. Jacobs, in *Proceedings of the NATO Advanced Study Institute on Zeolite Microporous Solids: Synthesis, Structure and Reactivity*, E. G. Derouane, F. Lemos, C. Naccache, and F. R. Ribeiro (Eds), Kluwer, Dordrecht, pp 555 - 578 (1992).
5. N. Herron, *Chemtech,* **19**, 542 (1989).
6. V. Yu. Zakharov and B. V. Romanovsky, *Vestn. Mosk. Univ., Ser. Khim.*, **18**, 142 (1977) [Eng. Trans. in *Sov. Mosc. Univ. Bull.*, **32**, 16 (1977)].
7. B. V. Romanovsky, R. E. Mardaleishvili, V. Yu Zakharov, and O. M. Zakharova, *Vestn. Mosk. Univ., Ser. 2: Khim.*, **18**, 232 (1977).
8. V. Yu. Zakharov and B. V. Romanovsky, *Vestn. Mosk. Univ., Ser. 2: Khim.*, **18**, 348 (1977).
9. V. Yu. Zakharov, O. M. Zakharova, B. V. Romanovsky, and R. E. Mardaleishvili, *React. Kinet. Catal. Lett.*, **6**, 133 (1977).
10. M. V. Gusenkov, V. Yu. Zakharov, and B. V. Romanovsky, *Neftekhimiya*, **18**, 105 (1978).
11. O. M. Zakharova and B. V. Romanovsky, *Vestn. Mosk. Univ., Ser. 2: Khim.*, **20**, 284 (1979).
12. B. V. Romanovsky, O. M. Zakharova, and V. Yu. Zakharov, *Vestn. Mosk. Univ., Ser. 2: Khim.*, **20**, 43 (1979).
13. V. Yu. Zakharov and B. V. Romanovsky, *Vestn. Mosk. Univ., Ser. 2: Khim.*, **20**, 78 (1979).
14. S. V. Gudkov, E. S. Shpiro, and B. V. Romanovsky, *Izv. Akad. Nauk. SSSR, Ser. Khim.*, 2448 (1980).
15. S. V. Gudkov, B. V. Romanovsky, E. S. Shpiro, G. V. Antoshin, and Kh. M. Minachev, *VINITI Deposited Doc.,* 2472 (1980), [CA **95**:61069n (1981)].
16. O. M. Zakharova and B. V. Romanovsky, *Neftekhimiya*, **21**, 924 (1981).

17. B. V. Romanovsky, in *Proceedings of the 8th International Congress on Catalysis,* Verlag Chemie, Weinheim, p 657 (1984).
18. G. Meyer, D. Wohrle, D. Mohl, and G. Schultz-Ekloff, *Zeolites*, **4**, 30 (1984).
19. A. N. Zakharov and B. V. Romanovsky, *J. Inclus. Phenom.*, **3**, 389 (1985).
20. N. Herron, G. D. Stucky, and C. A. Tolman, *J. Chem. Soc., Chem. Commun.*, 1521, (1986).
21. T. V. Korol'kova and B. V. Romanovsky, *Neftekhimiya*, **26**, 546 (1986).
22. N. Herron, *J. Coord. Chem.*, **19**, 25 (1988).
23. Y. W. Chan and R. B. Wislon, *Preprint Papers - ACS, Div. Fuel Chem.*, **33**, 453 (1988).
24. T. Kimura, A. Fukuoka, and M. Ichikawa, *Shokubai*, **31,** 357 (1988).
25. T. Kimura, A. Fukuoka, and M. Ichikawa, *62th CATSJ Meeting Abstr.: No. 1D203*, **30**, 444 (1988).
26. T. Kimura, A. Fukuoka, and M. Ichikawa, *64th CATSJ Meeting Abstr.: No. 1A09*, **31**, 357 (1989).
27. M. Ichikawa, T. Kimura, and A. Fukuoka, *Stud. Surf. Sci. Catal.*, **60**, 335 (1991).
28. T. Kimura, A. Fukuoka, and M. Ichikawa, *Catal. Lett.*, **4**, 279 (1990).
29. K. J. Balkus, Jr. and J. P. Ferraris, *J. Phys. Chem.*, **94**, 8019 (1990).
30. J. P. Ferraris, K. J. Balkus, Jr., and A. Schade, *J. Inclus. Phenom. Molec. Recog. Chem.*, **14**, 163 (1992).
31. K. J. Balkus, Jr., A. A. Welch, and B. E. Gnade, *J. Inclus. Phenom. Molec. Recog. Chem.*, **10**, 141 (1991).
32. Z. Jiang and Z. Xi, *Fenzi Cuihua*, **6**, 467 (1992), [CA **118**: 212554 (1992)].
33. W. Zhang, X. Ye, and Y. Wu, *Fenzi Cuihua*, **5**, 168 (1991), [CA **115**: 143584x (1991)]
34. A. N. Zakharov, *Mendeleev Commun.*, 80 (1991).
35. H. Diegruber, P. J. Plath, and G. Schultz-Ekloff, *J. Mol. Catal.*, **24**, 115 (1984).
36. E. S. Shpiro, G. V. Antoshin, O. P. Tkachenko, S. V. Gudkov, B. V. Romanovsky, and Kh. M. Minachev, *Stud. Surf. Sci. Catal.,* **18**, 31 (1984).
37. A. G. Gabrielov, A. N. Zakharov, B. V. Romanovsky, O. P. Tkachenko, E. S. Shpiro, and Kh. M. Minachev, *Koord. Khim.*, **14**, 821 (1988).
38. A. G. Gabrielov, A. N. Zakharov, and B. V. Romanovsky, *Koord. Khim.*, **14**, 214 (1988).
39. M. Tanaka, Y. Sakai, T. Tominaga, A. Fukuoka, T. Kimura, and M. Ichikawa *J. Radioanal. Nucl. Chem., Lett.*, **137**, 287 (1989).
40. M. Tanaka, Y. Minai, T. Watanabe, and T. Tominaga, *J. Radioanal. Nucl. Chem., Lett.*, **154**, 197 (1991).
41. M. Tanaka, Y. Minai, T. Watanabe, and T. Tominaga, *J. Radioanal. Nucl. Chem., Lett.*, **164**, 255 (1992).
42. A. N. Zakharov, B. V. Romanovsky, D. Luka, and V. I. Sokolov, *Organomet. Chem. USSR,* **1**, 64 (1988).
43. A. N. Zakharov, A. G. Gabrielov, B. V. Romanovsky, and V. I. Sokolov, *Vestn. Mosk. Univ., Ser. 2: Khim.*, **30**, 234 (1989).

44. G. Schultz-Ekloff, D. Wohrle, and A. Andreev, *Wiss. Zeitschr. Leuna-Merseburg,* **32**, 649 (1990).
45. R. F. Parton, L. Utytterhoeven, and P. A. Jacobs, *Stud. Surf. Sci. Catal.*, **59**, 395 (1991).
46. R. F. Parton, D. R. C. Huybrechts, Ph. Buskens, and P. A. Jacobs, *Stud. Surf. Sci. Catal.*, **70**, 47 (1991).
47. B. V. Romanovsky and A. G. Gabrielov, *Mendeleev Commun.*, 14 (1991).
48. B. V. Romanovsky and A. G. Gabrielov, *Stud. Surf. Sci. Catal.*, **72**, 443 (1992).
49. B. V. Romanovsky and A. G. Gabrielov, *J. Mol. Catal.*, **74**, 293 (1992).
50 F. Bedioui, E. DeBoysson, J. Devynck, and K. J. Balkus, Jr., *J. Electroanal. Chem., Interfac. Electrochem.,* **315**, 313 (1991).
51. R. Parton, Dissertation, Katholieke Universiteit Leuven (1993).
52. E. Paez-Mozo, N. Gabriunas, F. Lucaccioni, D. D. Acosta, P. Patrono, A. La Ginestra, P. Ruiz, and B. Delmon, *J. Phys. Chem.*, **97**, 12819 (1993).
53. M. Nakamura, T. Tatsumi, and H. Tominaga, *Bull. Chem. Soc. Jpn.*, **63**, 3334 (1990).
54. X. Wang, Y. Liang, Y. Lui, L. Yu, Y. Li, F. Li, and X. Cao, *Gaodeng Xuexiao Hiaxue Xuebao*, **14**, 19 (1993), [CA **119**: 107854j (1993)].
55. C. Bowers and P. K. Dutta, *J. Catal.*, **122**, 271 (1990).
56. K. J . Balkus, Jr., A. A. Welch, and B. E. Gnade, *Zeolites*, **10**, 722 (1990).
57. S. Kowalak, R. C. Weiss, and K. J. Balkus, Jr., *J. Chem. Soc., Chem. Commun.*, 57 (1991).
58. D. E. DeVos and P. A. Jacobs, in *Proceedings from the 9th International Zeolite Conference,* R. Von Ballmoos, J. B. Higgins, and M. M. J. Treacy (Eds), Butterworth-Heinemann, Boston, Vol. 2, pp 615-622 (1992).
59. K. Putyera, G. Plesch, L. Benco, J. Dobrovodsky, A. V. Tchuvaev, V. I. Nefedov, and M. Zikmund, *Proc. 12th Conf. Coord. Chem.*, 295 (1990).
60. F. Bedioui, E. DeBoysson, J. Devynck, and K. J. Balkus, Jr., *J. Chem. Soc., Farad. Trans.*, **87**, 3831 (1991).
61. L. Gaillon, N. Sajot, F. Bedioui, J. Devynck, and K. J. Balkus, Jr., *J. Electroanal. Chem., Interfac. Electrochem.,* **345**, 157(1993).
62. F. Bedioui, L. Roue, L. Gaillon, J. Devynck, S. L. Bell, and K. J. Balkus, Jr., *Petrol. Preprints*, **38**, 529 (1993).
63. K. J. Balkus, Jr., and S. Shepelev, *Micropor. Mater.*, **1**, 383, (1993).
64. K. J. Balkus, Jr., and S. Shepelev, *Petrol. Preprints*, **38**, 512 (1993).
65. G. A. Ozin and C. Gil, *Chem. Rev.*, **89**, 1749 (1989).
66. N. Jaeger, P. Plath, and G. Schultz-Ekloff, *Acta Phys. Chem.*, **31**, 189 (1985).
67. W. M. H. Sachtler, *Acc. Chem. Res.*, **26**, 383 (1993).
68. P. Gallezot, in *Metal Clusters*, E. M. Moskovits (Ed), J. Wiley, New York, pp 219-247 (1986).
69. G. A. Ozin, S. Ozkar, and R. A. Prokopowicz, *Acc. Chem. Res.*, **25**, 553 (1992).
70. W. M. Meier and D. H. Olson, *Atlas of Zeolite Structure Types*, Butterworth-Heinemann, Boston (1992).

71. M. Ichikawa, *Adv. Catal.*, **38**, 283 (1992).
72. L. F. Rao, A. Fukuoka, and M. Ichikawa, *J. Chem. Soc., Chem. Commun.*, 458 (1988).
73. L. F. Rao, A. Fukuoka, N. Kosugi, H. Kuroda, and M. Ichikawa, *J. Phys. Chem.*, **94**, 5317 (1990).
74. N. Takahashi, A. Mijin, H. Suematsu, S. Shinohara, and H. Matsuoka, *J. Catal.*, **177**, 348 (1989).
75. B. E. Hanson, M. E. Davis, D. Taylor, and E. Rode, *Inorg. Chem.*, **23**, 52 (1984).
76. P. Gelin, C. Naccache, and Y. Ben Taarit, *Pure Appl. Chem.*, **60**, 1315 (1988).
77. M. Ichikawa, L. -F. Rao, and A. Fukuoka, *Catal. Sci. Tech.*, **1**, 111 (1991).
78. A. Fukuoka, L. -F. Rao, N. Kosugi, H. Kuroda, and M. Ichikawa, *Appl. Catal.*, **50**, 295 (1989).
79. L. L. Sheu, H. Knozinger, and W. M. H. Sachtler, *Catal. Lett.*, **2**, 129 (1989).
80. L. L. Sheu, H. Knozinger, and W. M. H. Sachtler, *J. Am. Chem. Soc.*, **111**, 8125 (1989).
81. R. S. Drago, I. Bresinska, J. E. George, K. J. Balkus, Jr., and R. J. Taylor, *J. Am. Chem. Soc.*, **110**, 304 (1988).
82. R. J. Taylor, R. S. Drago, and J. E. George, *J. Am. Chem. Soc.*, **111**, 6610 (1989).
83. R. J. Taylor, R. S. Drago, and J. P. Hage, *Inorg. Chem.*, **31**, 253 (1992).
84. I. Bresinska and R. S. Drago, *Stud. Surf. Sci. Catal.*, **69**, 101 (1991).
85. M. Iwamoto, S. -I. Nakamura, H. Kusano, and Sh. Kagawa, *J. Phys. Chem.*, **90**, 5244 (1986).
86. P. -L. Zhou and B. C. Gates, *J. Chem. Soc., Chem. Commun.*, 347 (1989).
87. P. -L. Zhou, S. D. Maloney, and B. C. Gates, *J. Catal.*, **129**, 315 (1991).
88. S. Kawi and B. C. Gates, *J. Chem. Soc., Chem. Commun.*, 702 (1992).
89. S. Kawi, J. -R. Chang, and B. C. Gates, *J. Catal.*, **142**, 585 (1993).
90. R. Nakamura, N. Okada, A. Oomura, and E. Echigoya, *Chem. Lett.*, **119** (1984).
91. A. De Mallmann and D. Barthomeuf, *Catal. Lett.*, **5**, 293 (1990).
92. G. J. Li, T. Fujimoto, A. Fukuoka, and M. Ichikawa, *J. Chem. Soc., Chem. Commun.*, 1337 (1991).
93. G. J. Li, T. Fujimoto, A. Fukuoka, and M. Ichikawa, *Catal. Lett.*, **12**, 171 (1992)
94. G. J. Li, M. Ichikawa, T. Fujimoto, and X. Guo,. *Petrol. Preprints*, **30**, 523 (1993).
95. S. Kawi, J. -R. Chang, and B. C. Gates, *Catal. Lett.*, **10**, 263 (1991).
96. S. Kawi, J. -R. Chang, and B. C. Gates, *J. Phys. Chem.*, **97**, 10599 (1993).
97. T. Bein, S. J. McLain, D. R. Corbin, R. D. Farlee, K. Moller, G. D. Stucky, G. Woolery, and S. Sayers, *J. Am. Chem. Soc.*, **110**, 1801 (1988).
98. S. Kawi and B. C. Gates, *J. Chem. Soc., Chem. Commun.*, 994 (1991).
99. C. Bremard, E. Denneulin, C. Depecker, and P. Legrand, *Stud. Surf. Sci. Catal.*, **48**, 219 (1989).
100. A. G. Gabrielov, K. J. Balkus, Jr., F. Bedioui, and J. Devynck, *Micropor. Mater.*, **2**, 119 (1994).
101. K. J. Balkus, Jr., A. G. Gabrielov, F. Bedioui, and J. Devynck, *Inorg. Chem.*, **33**, 67 (1994).

102. V. J. Frilette and R. W. Maatman, *U. S. Patent 3,373,109* (1968).
103. C. A. Rundell and C. V. McDaniel, *U. S. Patent 3,769,386* (1973).
104. L. A. Rankel and E.W. Valyocsik, *U. S. Patent 4,500,503* (1985).
105. L. A. Rankel and E.W. Valyocsik, *U. S. Patent 4,388,285* (1983).
106. F. Delprato, L. Demotte, J. L. Guth, and L. Huve, *Zeolites*, **10**, 564 (1990).
107. K. J. Balkus, Jr. and S. Kowalak, *U. S. Patent 5,167,942* (1992).
108. K. J. Balkus, Jr., C. D. Hargis, and S. Kowalak, *ACS Symp. Ser.*, **499**, 347 (1992).
109. K. J. Balkus, Jr., S. Kowalak, K. T. Ly, and C. D. Hargis, *Stud. Surf. Sci. Catal.*, **69**, 93 (1991).
110. R. Hoppe, G. Schultz-Ekloff, D. Wöhrle, M. Ehrl, and C. Bräuchle, *Stud. Surf. Sci. Catal.*, **69**, 199 (1991).
111. S. Wohlrab, R. Hoppe, G. Schultz-Ekloff, and D. Wöhrle, *Zeolites*, **12**, 862 (1992).
112. E. W. Valyocsik, *U. S. Patent 4,568,654* (1986).
113. K. J. Balkus, Jr. and S. Shepelev, Unpublished results.
114. K. J. Balkus, Jr., A. D. Sherry, and S. W. Young, *U.S. Patent 5,122,363* (1992).
115. K. J. Balkus, Jr., I. Bresinska, and S. W. Young, in *Proceedings from the 9th International Zeolite Conference,* R. Von Ballmoos, J. B. Higgins, and M. M. J. Treacy (Eds), Butterworth-Heinemann, Boston, Vol. 2, pp 193-201 (1992).
116. D. R. Rolisson, *Chem. Rev.*, **90**, 867 (1990).
117. A. J. Bard and T. E. Mallouk, in *Molecular Design of Electrode Surfaces*, R. W. Murray (Ed), J. Wiley, New York, p 271 (1992).
118. F. Bedioui, L. Roue, E. Briot, J. Devynck, S. L. Bell, and K. J. Balkus, Jr., *J. Electroanal. Chem., Interfac. Electrochem.*, in press.
119. M. G. B. Drew in *Spectroscopic and Computational Studies of Supramolecular Species,* J. E. D. Davies (Ed), Kluwer, Dordrecht, pp 207-237 (1992).
120. P. C. H. Mitchell, *Chem. Ind.* 308, (1991).
121. B. V. Romanovsky, *Acta Phys. Chem.*, **31**, 215, (1985).
122. B. V. Romanovsky, in *Homogeneous and Heterogeneous Catalysis*, I. Y. Ermakov and V. A. Likholobov (Eds), VNU Sci. Press, Utrecht, pp 343-365 (1986).
123. B. V. Romanovsky and A. G. Gabrielov, *Neftekhimiya*, **30**, 617 (1990).
124. C. A. Tolman and N. Herron, *Catal. Today,* **3**, 235 (1988).
125. N. Herron, in *Inclusions Compounds*, J. L. Atwood, J. E. D. Davies, and D. D. MacNicol (Eds), Oxford University Press, Oxford, Vol. 5, pp 90-103 (1991).
126. D. R. C. Huybrechts, R. F. Parton, and P. A. Jacobs, *Stud. Surf. Sci. Catal.*, **60**, 225 (1991).
127. J. Weitkamp, in *Proceedings from the 9th International Zeolite Conference,* R. Von Ballmoos, J. B. Higgins, and M. M. J. Treacy (Eds.), Butterworth-Heinemann, Boston, Vol. 2, pp 13-45 (1992).
128. G. Henrici-Olive and S. Olive, *J. Mol. Catal.*, **1**, 121 (1975-76).
129. R. -J. Wang, T. Fujimoto, T. Shido, and M. Ichikawa, *J. Chem. Soc., Chem. Commum.*, 962 (1992).

130. J. S. Beck, J. C. Vartuli, W. Roth, M. E. Leonowicz, C. T. Kresge, K. D. Schmitt, C. T. -W. Chu, D. H. Olson, E. W. Sheppard, S. B. McCullen, J. B. Higgins, and J. L. Schlenker, *J. Am. Chem. Soc.*, **114**, 10834 (1992).
131. M. Estermann, L. B. McCusker, C. Baerlocher, A. Merrouche, and H. Kessler, *Nature*, **352**, 320 (1991).

Abstract. The various strategies for the preparation of zeolite ship-in-a-bottle complexes are reviewed. Key examples of the cluster, flexible ligand, template and zeolite synthesis methods are described. Evidence for the intrazeolite location of the metal complexes as well as the prevailing interpretation of these results have been organized in terms of the various methods of characterization. A brief summary of known reactivity studies of zeolite encapsulated metal complexes, including catalytic oxidations and hydrogenations, is also presented.

Journal of Inclusion Phenomena and Molecular Recognition in Chemistry **21**: 185–213, 1995.

Coordination Chemistry in Zeolites

DIRK E. De VOS, PETER. P. KNOPS-GERRITS, RUDY F. PARTON, BERT M. WECKHUYSEN, PETER A. JACOBS and ROBERT A. SCHOONHEYDT

1. Introduction

Zeolites are relatively rigid inorganic matrices with cavities and channels of molecular dimensions and of different sizes and shapes. Transition metal ion (TMI) complexes in molecular sieves have three main characteristics: (1) the complexes have strongly reduced mobility with respect to their solution mobility and deactivating aggregations may be avoided; (2) complexation with extra-lattice ligands is in competition with complexation to lattice oxygens. Unusual complexes can be synthesized, depending on the relative coordinating strength of the structural oxygens and the extra-lattice ligands; and (3) shape-selectivity is not only operative on the size and shape of complexes which can be synthesized, but is also evident in the catalytic properties of the occluded complexes. Research on coordination chemistry in zeolites started around 1970 and most of the early work has been summarized by Lunsford [1].

The field can be subdivided into three areas: (1) Complexes which are smaller than the free diameters of the channels and windows giving access to the cavities. These complexes can be absorbed from a solution phase into the zeolite or they can be synthesized via absorption of the ligand into a TMI-exchanged zeolite; (2) Complexes which are larger than the free diameters of the channels and windows of the zeolite. Such complexes must be synthesized in situ, i.e., by absorption of the ligands into the TMI-loaded zeolite, or by synthesis of the ligand in the zeolite in the presence of the TMI; (3) Complexes which serve as fingerprints of the oxidation state, the location, and the coordination environment of the TMI. An example of the latter group are the CO and NO complexes. In this chapter, the first two areas will be covered with an emphasis on the synthesis and spectroscopic signatures of the encapsulated complexes.

The overwhelming majority of reported complexes have been immobilized in the supercages of zeolite Y; the site nomenclature of which is recalled in Figure 1. In the supercages two types of sites are of interest: site II (on the oxygen six-ring) and site III or III' (on the four-ring of oxygens). The former are also the preferential sites of TMI in zeolite A.

The spectroscopic signatures of Cu^{2+}, Co^{2+}, Ni^{2+}, and Cr^{2+}, coordinated to the lattice oxygens, are given in Table 1 [2,3]. The ions coordinated to six-rings have one free

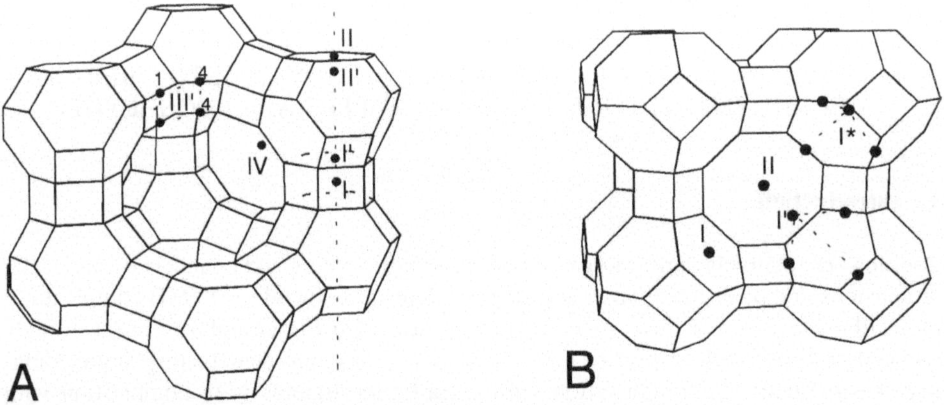

Figure 1. Structures of (A) zeolite Y and (B) zeolite A; different cation sites are indicated with Roman numerals. Arabic numerals indicate positions of structural oxygen atoms.

Table 1. Spectroscopic signatures of transition metal ions coordinated to lattice oxygens in zeolites.*

	site	*d-d* absorption bands	$g_{\parallel}$	$A_{\parallel}$	$g_{\perp}$	$A_{\perp}$
zeolite A						
Cr^{2+}	I	12000, 17000				
Co^{2+}		6000, 15000-20000, 25000				
Ni^{2+}		4500, 9100, 22800				
Cu^{2+}		10400, 12200, 15000	2.40	12.6	2.07	0.2
zeolites X,Y						
Ni^{2+}	I	6100-6400, 19000-20000				
	I',II	4900, 9300, 21500-23500				
Cu^{2+}	I'	10700, 12900, 14800	2.34	15.7	2.07	1.9
	II	10400, 12400, 14700	2.39	12.4	2.07	1.3

* Absorption frequencies in cm^{-1}; hyperfine constants in mT.

coordination site and readily accept one extra-lattice ligand to form pseudo-tetrahedral complexes.

Such is the case for weak ligands such as olefins, as shown by Klier and co-workers for Co-A [3]. Stronger ligands favorably compete with lattice oxygens for coordination sites. If enough ligands are offered in the cavities and channels, and if space is available, full coordination with these extra-lattice ligands is possible, giving (pseudo-)octahedral complexes. In such cases, the zeolite acts as an anionic solvent. Complexes may undergo distortion to match the size of the complex to the size and shape of the zeolite cavities.

2. Synthesis of Complexes in Zeolites

Figure 2 summarizes the different methods used for synthesis of complexes in zeolites. Two main routes can be followed: (1) synthesis starting from a preformed complex and (2) complex assembly inside the zeolite. The choice of the route depends on the zeolite, TMI, and on the ligands.

Synthesis with a preformed complex	↔	Complex assembly in the zeolite
→ Exchange of a preformed complex		→ with solvent : non-aqueous medium aqueous medium
→ Template synthesis		→ without solvent : in situ ligand synthesis gas phase absorption impregnation with liquid ligand
→ Anchorage of a complex precursor		

Figure 2. Overview of the different synthesis routes for complexes in zeolites.

2.1 SYNTHESIS WITH A PREFORMED COMPLEX

2.1.1 Exchange of Preformed Complexes

Direct ion-exchange of preformed cationic complexes depends on the resistance of the zeolite to acid or base and on the dimensions of the ring of oxygen atoms defining the zeolite windows and channels. Therefore, this method is only appropriate if: (a) the complexes are stable; (b) the solutions have no extreme pH; and (c) the complexes are small enough to enter into the zeolite. Thus, only (octahedral) complexes with ligands such as H_2O, NH_3, and some aliphatic amines may be exchanged without steric limitations. For acid solutions, e.g., of Cr^{3+} and Fe^{3+}, the contact time must be very short to avoid zeolite lattice destruction.

2.1.2 Template Synthesis

Preformed complexes can also be used as templates during the synthesis of zeolites [4,5] (see also Chapter 6 in this volume). There are two important requirements which must be fulfilled for this route: (i) The complexes must be stable under synthesis conditions. For example, in the synthesis of zeolite Y, the complex must be stable in basic conditions and at high temperature, while for $AlPO_4$ synthesis it must be stable in acidic conditions. (ii) The complex must be soluble and randomly dispersed in the synthesis mixture.

2.1.3 Anchorage of a Complex Precursor

When heterobinuclear organometallics are absorbed into zeolites, an oxophilic metal may anchor the complex through reaction with the zeolite surface. For example, $Me_3SnMn(CO)_5$ and $Cl_2(THF)GeMo(CO)_5$ can react with acidic zeolites at elevated temperatures; volatile products such as methane or hydrogen chloride are removed [6]. Sample preparation involves mixing of a dehydrated acidic zeolite with the complexes in an organic medium under an inert atmosphere.

2.2 COMPLEX ASSEMBLY INSIDE THE ZEOLITE

Complexes, unstable under ion-exchange or zeolite synthesis conditions, can be synthesized by assembly in the zeolite. This method usually starts by ion-exchange of the zeolite with the desired TMI, followed by introduction of the ligand. This can be done in the absence or in the presence of a solvent. A condition for complex formation is that the complexation constant of the TMI with the lattice oxygens be lower than that with the added ligand.

2.2.1 Assembly without Solvent

Three synthesis routes can be envisaged: (1) gas phase absorption of the ligand; (2) impregnation of the zeolite by the liquid ligand; and (3) in situ ligand synthesis. Gas phase absorption of ligands, like NH_3, H_2O, CO, and NO can be done easily. The only restrictions are that the ligands be small enough to enter the pores or channels and have a vapor pressure at room temperature. For impregnation, the ligand must have a low enough melting point. In situ ligand synthesis needs diffusion of the ligand precursor into the zeolite to form, in situ, a stable ligand, which is able to chelate the TMI.

2.2.2 *Assembly with Solvent*

Complex assembly inside zeolites using solvents can be performed both in aqueous and non-aqueous media. The choice of solvent depends on the solubility of the ligand, the pore architecture of the zeolite material, and the affinity of the solvent for the zeolite. The latter is determined by the Si/Al ratio. Indeed, Si-rich zeolites are hydrophobic and an increasing Al content raises the hydrophilicity.

3. Coordination of Transition Metal Ions with Monodentate Ligands

The state of the art of complexes between TMIs and monodentate ligands in zeolites was reviewed in 1985 by Mortier and Schoonheydt [2]. In this chapter, we take this review as a starting point to give a general overview of the fundamental early work and recent progress. A whole series of monodentate ligands have been used as complexants of TMI in zeolites: H_2O, RNH_2 (R=H, CH_3, or $CH_3CH_2CH_2$), pyridine, picoline, PR_3 (R=CH_3, Ph), CH_3NC, and CN^-, and results are now reviewed below.

3.1 WATER COMPLEXES

TMI-H_2O complexes are easily introduced into zeolites by ion-exchange and can be characterized by their typical *d-d* absorption bands. Upon drying, one (or more) H_2O ligands in the first coordination sphere of the TMI may be replaced by surface oxygens. High temperature treatment can result in complete dehydration and migration of the TMI to the cation exchange sites in the smaller zeolite cages.

Re-absorption of H_2O into a dehydrated zeolite causes back-migration of the cations to the large cages, where hexa-aquo complexes can be re-formed. However, this process may be slow and incomplete.

The absorption-desorption scheme can be represented as:

$$[TMI(H_2O)_6]^{n+} \rightleftarrows [TMI(O_{lat})_p(H_2O)_q]^{n+} \rightleftarrows [TMI(O_{lat})_r]^{n+} \quad (1)$$

where both (p+q), r $\leq$ 6. In zeolites A, X, and Y, controlled absorption and desorption of water occurs via an intermediate pseudo-tetrahedral complex [7]. In this complex, the TMI is coordinated to three oxygens of a six-ring (site II) and one water molecule (p=3, q=1). Coordination of two water molecules, one on either side of the six-ring, has also been suggested.

3.2 ISONITRILE COMPLEXES

Vansant and Lunsford prepared Co(II)-isonitrile complexes in zeolite Y [8]. The gaseous ligand was absorbed into the Co,Na-Y zeolite after a high-temperature dehydration. This treatment makes the Co(II) migrate to sites in the small cages of the zeolite. Although the CH_3NC certainly has no access to these sites, the Co cation smoothly returns to the faujasite supercages to form various isonitrile complexes. The symmetries of the low-spin Co(II) species, tetragonal six-coordinated, and square-pyramidal five-coordinated, were deduced from the X- and Q-band EPR spectra.

3.3 N-BASE COMPLEXES

A whole range of monodentate ligands, such as primary amines, pyridine, and picolines, has been used for designing TMI-complexes in zeolites. Especially for the aliphatic amines, complexation is in competition with protonation of the ligand by residual acidic protons of the TMI-loaded zeolites; however, protonation can be minimized by careful exchange of TMI and by gradual dehydration prior to ligand absorption.

3.3.1 Complexes with Ammonia and Alkylamines

Only in the case of 4*d* and 5*d* TMI is direct exchange of, e.g., $[Pd(NH_3)_4]^{2+}$ and $[Pt(NH_3)_4]^{2+}$ possible [9]. Ammonia complexes of 3*d* TMI are prepared by gas-phase absorption of NH_3 into ion-exchanged and dehydrated zeolites. This may lead to six-coordinated complexes, but sometimes unsaturated $[TMI(NH_3)_m]^{2+}$ complexes are formed, with $m < 6$. For example, square-planar $[Cu(NH_3)_4]^{2+}$ is smoothly formed upon absorption of ammonia into copper zeolites, as revealed by the analysis of the EPR spectra [10,11]. The chemistry of these ammonia complexes is somewhat similar to that of the aquo complexes. Thus, upon controlled desorption, intermediate $[TMI(O_{lat})_3(NH_3)]^{n+}$ complexes are formed, similar to the pseudo-tetrahedral mono-aquo complexes [7].

In an analogous way, Co(II)ammonia-complexes were made in zeolite Y [12]. These complexes are themselves EPR invisible, but their oxygen adducts show characteristic EPR spectra. Thus, zeolite occluded $[Co(NH_3)_5]^{2+}$ interacts with molecular oxygen to form mononuclear $[Co(III)(NH_3)_5O_2^-]$, which is slowly converted into the dinuclear-peroxo complex.

Similar complexes are formed on Co,Na-Y zeolite with CH_3NH_2 and $CH_3CH_2CH_2NH_2$ [13]. The larger dimensions of these complexes restrict their mobility. As a consequence, formation of dimeric complexes, which occurs with NH_3, is not observed for complexes of alkylamines.

3.3.2 Complexes with Aromatic Amines

Furthermore, TMI-complexes with aromatic amines, such as pyridine and picoline, have been prepared and characterized [14-18]. For example, planar $[Cu(pyridine)_4]^{2+}$ was prepared by absorption of pyridine vapor into a Cu,Na-Y zeolite [14,15]. The pyridine ligands have also been attached covalently to the zeolite surface by a silylation procedure. Such immobilized Cu-complexes have been used as catalysts for the oxidative coupling of 2,6-dimethylphenol [16].

Instead of pyridine, α-, β-, or γ-picoline may be used. Yamada used $[Cu(acac)_2]$-impregnated zeolites as the precursor material in this type of synthesis (acac = acetylacetonate) [17,18]. Formally, the complexation reaction can be viewed as an intrazeolitic exchange of non-framework ligands. Surface acidity is thought to facilitate the replacement of the anionic acetylacetonate by picoline.

3.4 PHOSPHINE COMPLEXES

Small monodentate phosphines, such as PMe_3, are very reactive molecules. Any complexation of TMI in zeolites with PMe_3, via gas-phase absorption of the latter, is in competition with reaction with surface oxygen atoms, hydroxyl groups, or residual water [19]. Phosphines are strongly chemisorbed on zeolites, with formation of P-O bonds. Heating is necessary to accelerate the reaction with the TMI. Mostly, the pseudo-tetrahedral high-spin $[TMI(O_l)_3(PMe_3)]^{n+}$ is formed [20,21]. Only in the presence of excess PMe_3, low-spin complexes of Co^{2+} or Ni^{2+} are detected, such as the trigonal bipyramidal $[Ni(PMe_3)_5]^{2+}$ [22,23]. With more bulky ligands, such as $P(Me)_2(C_6H_5)$, only pseudo-tetrahedral complexes can be synthesized [21].

In view of the importance of sorption, it is not surprising that the Si/Al ratio affects the result of the complexation. In Co,Na-Y zeolites, pseudo-tetrahedral Co^{2+} complexes are formed, but phosphine chemisorption on CoNa-X zeolites is so strong that tertiary phosphine complexes are not even detected [21].

3.5 CYANIDE COMPLEXES

The $[TMI^{n+}(CN^-)_m]^{(m-n)-}$ complexes are the only anionic coordination compounds yet occluded in zeolite cages [24-27]. The negative charge of the CN^- ligand necessitates special synthesis procedures. The best results were obtained by impregnation of dehydrated zeolites with methanolic NaCN solutions, although irreversible oxidation of, e.g., Co and Fe to the trivalent-state, is a major side-reaction [24,25].

The formation of low-spin Co(II) complexes and Co(II)-superoxo species on Co,Na-Y zeolites was evidenced by the EPR spectra [25,26]. There is, however, uncertainty concerning the number of CN^- ligands surrounding Co^{2+}. Both four- and five-coordination occurs and the monovalent zeolitic cations, such as Cs^+, seem to influence the complexation stoichiometry [27].

4. Coordination of Transition Metal Ions with Bi- or Polydentate Ligands

4.1 COMPLEXES OF ALIPHATIC POLYAMINES

4.1.1. Complexes of ethylenediamine (en)

a) Complexes with Cu(II)

The formation of complexes between zeolitic Cu(II) and en was studied in both aqueous media and in dehydrated zeolites [28]. As discussed previously, direct aqueous exchange of preformed cationic complexes requires high stability constants. This permits one to work in an aqueous solution with a minimal ligand excess. $[Cu(en)_2]^{2+}$ is an ideal case, as $K_1 = 10^{10.7}$ and $K_2 = 10^{9.3}$, giving an overall stability constant $\beta_2 = 10^{20}$ for :

$$[Cu(H_2O)_6]^{2+} + 2\ en \rightleftarrows [Cu(en)_2(H_2O)_2]^{2+} + 4\ H_2O. \quad (2)$$

In the supercages of zeolites Na-X and Na-Y, the equilibrium is shifted to the right for the *mono*-complex and to the left for the *bis*-complex, giving a stabilization and a destabilization, respectively. This effect is more pronounced on Na-X with its lower Si/Al ratio.

The prevalence of the *mono*-complex on Na-X may be a consequence of the higher negative surface charge density of this zeolite. It is, however, doubtful whether the lattice oxygen atoms can really compete with en for Cu(II). On the dehydrated zeolites Cu,Na-X and Cu,Na-Y, en absorption results in formation of *tris*-complexes. Apparently, even on X zeolites, the affinity of the lattice oxygens for Cu(II) is substantially lower than that of en. The larger apparent stability of the *mono*-complex on zeolite Na-X, therefore, is more likely due to a change in the protonation equilibrium for en :

$$en + H_2O \rightleftarrows enH^+ + OH^-. \quad (3)$$

Inside a Na-X supercage, the H_2O/Na^+ ratio is only about 3. The resulting strong polarization of the water molecules results in a higher acidity. Therefore, less en is available in its unprotonated form and the complexation equilibrium is shifted towards the *mono*-adduct.

b) Complexes with Co(II)

En uptake by dry Co,Na-Y zeolite is fast, but formation of the *tris*-complex may require heating and depends on the availability and mobility of the Co(II) ions. Low loadings of Co(II) tend to disfavor full complexation, while blocking of hidden sites with La^{3+}, accelerates full chelation [29].

The complexation of Co(II) by en is much stronger than that by monoamines. This implies that, if one intends to use the system for O_2 binding, an evacuation treatment is needed to create a vacancy in the Co(II) coordination sphere [29,30]. There is, however, considerable debate on what the actual coordination around the active, O_2 binding Co(II) complex in en loaded Co(II) zeolites may be. Reflectance spectroscopy evidences the simultaneous presence of *mono-*, *bis-* and *tris*-complexes. An average number of two en ligands per Co(II), determined by gravimetry, seems optimal for O_2 binding [29]. The absence of an EPR low-spin Co(II) signal rules out a square-planar, *trans*-geometry for the *bis*-$[Co(en)_2]^{2+}$ complexes. Alternatives are a *cis*-$[Co(en)_2]^{2+}$ complex with four N donors, or a five-coordinate complex containing three en molecules, one of which acts as a monodentate ligand [31].

Oxygen binding results primarily in a $[Co(III)(en)_2{\cdot}O_2^-]$ superoxo complex. The mononuclear nature of this compound was established not only by the EPR spectrum, but also by the energy of the charge transfer bands (Co(III) $\rightarrow O_2^-$ and $O_2^- \rightarrow$ Co(III)) in the reflectance spectrum [29]. In spite of the limited mobility of these complexes, deactivation occurs through formation of dimers. These dimers were identified by the higher energies for the charge transfer bands in the reflectance spectrum [29]. Use of the frequencies of these bands for excitation in resonance Raman spectroscopy results in a strongly enhanced intensity of, e.g., Co-O and O-O vibrations. In this way, the presence of monobridged-peroxo and, after H_2O infusion, of dibridged-peroxo-hydroxo complexes was established [31]:

$$[Co(en)_2]^{2+} + [Co(en)_2{\cdot}O_2]^{2+} \rightarrow [Co(en)_2{\cdot}O_2{\cdot}Co(en)_2]^{4+} \quad (4)$$

$$[Co(en)_2{\cdot}O_2{\cdot}Co(en)_2] + H_2O \rightarrow [Co(en)_2{\cdot}O_2(OH){\cdot}Co(en)_2]^{3+} + H^+. \quad (5)$$

c) Complexes with other TMI

Howe and Lunsford reported complexation of Ni(II) and Mn(II) by en in zeolite Y and proved chelation by IR spectroscopy [32]. Especially $[Mn(II)(en)_2]^{2+}$-Na-Y has an enhanced activity in catalytic catechol oxidation by molecular oxygen.

4.1.2 Complexes with tetren

The use of tetren constitutes an extension of the prior work with amines and diamines. Tetren was absorbed aerobically from a 1 M aqueous solution on hydrated Co,Na-Y and Co,Na-A zeolites [33]. On Na-A, mononuclear superoxo complexes were observed in EPR and Raman spectra, but in Na-Y the reaction proceeds to peroxo Co(III) dimers:

$$\text{(Na-A)} \quad [Co(H_2O)_6]^{2+} + \text{tetren} + O_2 \rightarrow [Co(\text{tetren}){\cdot}O_2]^{2+} + 6\,H_2O \quad (6)$$

$$\text{(Na-Y)} \quad [Co(\text{tetren}){\cdot}O_2]^{2+} + [Co(\text{tetren})]^{2+} \rightarrow [Co(\text{tetren}){\cdot}O_2{\cdot}Co(\text{tetren})]^{4+}. \quad (7)$$

This difference was attributed to the steric constraints imposed by the small α-cage windows of Na-A, which inhibit reaction between the monomeric Co(II) complexes. It is, however, unclear whether the zeolite is stable in the synthesis conditions, as the pH of the aqueous tetren solution exceeds 13.

The oxygen binding affinity of zeolite A confined $[Co(tetren)]^{2+}$ is lower than that of $[Co(bpy)(terpy)]^{2+}$-Y and similar to that of [Co(SALEN)(py)]-Y. This is remarkable as in aqueous solution $[Co(tetren)]^{2+}$ has one of the highest O_2 affinities known for a Co(II) compound. The low O_2 affinity of the zeolite encapsulated complex was ascribed, by Dutta, to the hydrated state of the $[Co(tetren)]^{2+}$-Y zeolite [33]; the filling of the zeolite pores with water strongly decreases the access of O_2 to the active complex (see Chapters 6 and 8 of this volume for more details of this area).

4.2 DIMETHYLGLYOXIME COMPLEXES

The absorption of dimethylglyoxime ($dmgH_2$) into Na-X zeolites, partially exchanged with Co(II) or Ni(II), mostly yields a mixture of different species [34-37]. With Co(II) for instance, proposed speciations include $[Co(II)(dmgH)(O_l)_2]$ (with a distorted tetrahedral geometry), square-planar $[Co(II)(dmgH)_2]$, $[Co(II)(dmgH)_2(H_2O)_2]$ (a tetragonally distorted octahedron), and $[Co(III)(dmgH)_2(H_2O)O_2^-]$ (an octahedral superoxo complex) [34]. The first step in the complexation is the formation of a mixed chelate:

$$[Co(O_l)_3(H_2O)] + dmgH_2 \rightarrow [Co(O_l)_2(dmgH)] + H_3O^+ \qquad (8)$$

where dmgH is the *mono*-anion of $dmgH_2$. The assignment is mainly based on UV-Vis spectroscopy [35] and is logical in view of the pronounced tendency of the Na-X surface to retain the cations or produce mixed chelates, as noted above for en and phosphine complex formation. XPS measurements support this proposition, as the absorption of $dmgH_2$ into Co,Na-Y does not change significantly the Co(II) binding energies [36]. However, for at least part of the Co(II), the reaction proceeds directly to the low-spin, four-coordinated Co(II) complex [34]:

$$Co(II) + 2\,dmgH_2 \rightarrow [Co(dmgH)_2] + 2\,H^+. \qquad (9)$$

Two different types of this complex, a planar one and a distorted planar one, were identified by very well-resolved EPR spectra. Exposure of these square-planar Co(II) complexes to O_2 does not result in Co-O_2 formation, as O_2 binding to planar complexes requires axial base coordination *trans* to the O_2 binding site. H_2O infusion is sufficient to change the electronic structure of the complex and thus promotes O_2 binding:

$$[Co(dmgH)_2] + 2\,H_2O \rightarrow [Co(dmgH)_2(H_2O)_2] \quad (10)$$

$$[Co(dmgH)_2(H_2O)_2] + O_2 \rightarrow [Co(dmgH)_2(H_2O){\cdot}O_2] + H_2O. \quad (11)$$

The EPR spectrum of $[Co(dmgH)_2(H_2O)_2]$ is that of a tetragonally distorted Co^{2+} complex. Reversibility of the O_2 sorption on the complex was not demonstrated.

For Ni(II), the complexation also seems to be a multistep process, since only after H_2O infusion on a $dmgH_2$ loaded Ni,Na-Y are appreciable changes in the XPS Ni(II) binding energies observable [36]. The addition of H_2O may be necessary to promote the cation mobility during the complexation process. XPS is also useful in following the spin-state of the Ni(II) cations, as Ni(II) paramagnetism gives rise to satellite peaks in the XPS measurements. A decrease of the intensity of these peaks upon $dmgH_2$ complex formation indicates that at least part of the Ni(II) has become diamagnetic. XPS thus provides an alternative for measurements of magnetic susceptibility, which, in the zeolitic medium, are strongly complicated by multi-speciation and the presence of Fe impurities.

In some cases, it was observed that reaction between $dmgH_2$ and TMI-exchanged zeolites led to cleavage of Na-X crystals, as monitored by electron microscopy [37], or to migration of the cations to the outer crystal surface, as detected by XPS measurements [36].

4.3 COMPLEXES WITH BIPYRIDINE, TERPYRIDINE, AND ANALOGS

This group of materials was originally devised by Lunsford and co-workers [38-44]. Potential applications of these encapsulation compounds include light energy storage and air separation [44,45]. Moreover, zeolite Na-X occluded $[Gd(III)(8\text{-hydroxyquinoline})_2]^{1+}$ has been used as a magnetic resonance imaging contrast agent [46]. Catalytic applications are lacking until now, but optimism is warranted in view of the recent progress in homogeneous catalysis with bpy complexes.

In the patent literature, claims are given for ZSM-5 synthesis with $[Ru(o\text{-phenanthroline})_3Cl_2]$ and $[Fe(bpy)_3(ClO_4)_2]$ [47]. These complexes are stable enough to be used as templates, but, because of their 1.2 nm diameter, they can only reside at the intersection of two perpendicular channels of the MFI lattice. As the obtained ZSM-5 was of low crystallinity, it is doubtful whether this approach merits further investigation. Better results are obtained when the complexes are assembled inside the zeolites.

4.3.1 Copper complexes

Formation of $[Cu(bpy)]^{2+}$ in zeolite Y was reported by De Wilde and Lunsford [38]. The 1:1 stoichiometry is a direct result of the synthetic approach. Zeolite Y was stirred in an aqueous solution of 2,2'-bipyridinium chloride and was subsequently exchanged with Cu(II). At the very small loading used (1 Cu(II) per 50 supercages), only *mono*

$[Cu(bpy)]^{2+}$ is formed. This was proved by the axially symmetric EPR spectrum with a five line superhyperfine structure, due to two equivalent nitrogen atoms. It was claimed that $[Cu(bpy)]^{2+}$ is not coordinated to lattice oxygens, but the insensitivity of the EPR parameters to dehydration suggests the opposite. Due to the low Cu(II) content a relatively large amount of free bipyridinium cations remain in the zeolite.

This synthetic approach is unique to Cu(II). In all other reports, the ligand is absorbed into metal-containing, dehydrated zeolites, resulting in formation of *bis*- or *tris*-complexes.

4.3.2 Ruthenium and iron complexes

Absorption of bpy into Ru^{3+} or $[Ru(NH_3)_6]^{3+}$ exchanged Na-Y yields *bis*- or *tris*-bpy complexes. The result of the synthesis depends on the metal-ligand stoichiometry and on the choice of the complexation temperature. *Bis*-complexes are selectively formed at 90° C with a 2:1 bpy:Ru stoichiometry [48]; however, at a 3:1 bpy:Ru stoichiometry and at 200° C, *tris*-complexes are formed as depicted in Figure 3 [39,40]. Generally Ru is reduced to its divalent state during synthesis, possibly by one of the following reactions [40]:

$$2\ Ru(III) + 8\ bpy \rightarrow 2\ [Ru(II)(bpy)_3]^{2+} + C_{20}H_{14}N_4 + 2\ H^+ \quad (12)$$

$$2\ [Ru(III)(NH_3)_6]^{3+} + 6\ bpy \rightarrow 2\ [Ru(II)(bpy)_3]^{2+} + N_2H_4 + 2\ NH_4^+ + 8\ NH_3 \quad (13)$$

$$6\ [Ru(III)(NH_3)_6]^{3+} + 18\ bpy \rightarrow 6\ [Ru(II)(bpy)_3]^{2+} + N_2 + 6\ NH_4^+ + 28\ NH_3. \quad (14)$$

The *bis*-$[Ru(II)(bpy)_2]^{2+}$ species is easily identified by its MLCT absorption at 488 nm [48]. A supplementary band at 606 nm was ascribed to lattice coordinated [(Al)Si-O-Ru$(bpy)_2(H_2O)$]. The assignment is based on the spectral similarity with the dinuclear $[(bpy)_2(H_2O)RuORu(H_2O)(bpy)_2]^{4+}$, which absorbs at 640 nm [49]. Apparently, the encapsulation in the zeolite cage prevents formation of the dinuclear complex; the free coordination sites are taken up by the lattice oxygen atoms. The assignment was confirmed by resonance Raman spectroscopy; the only difference between the spectra of the mononuclear zeolitic compound and the dinuclear crystalline compound is observed in the stretching modes of the Si-O-Ru and Ru-O-Ru bonds.

A general problem for *bis*-complexes of bidentate ligands is whether the complexes assume a *cis*- or a *trans*-geometry. For dissolved *bis*-bpy complexes, the *cis*-geometry is favored, but bulky axial ligands induce the *trans*-configuration [49]. The effect of zeolite encapsulation on the geometry is still unclear and may be worth looking at.

In the synthesis of the *tris*-$[Ru(II)(bpy)_3]^{2+}$ species, more than 80% of the total Ru can be transformed into the desired complex [39]. In order to investigate the location of the complexes, a sample, synthesized by the in situ method, was compared to a Na-Y sample which was impregnated with $[Ru(II)(bpy)_3]^{2+}$. XPS data proved that in the latter case the complex surface concentration is much higher than in the former case. This proves that for the in situ sample, complexation is largely confined to the zeolite cages.

Kincaid et al. [48] prepared $[Ru(II)(bpy)_2(bpz)]^{2+}$-Y and $[Ru(II)(bpy)_2(dmb)]^{2+}$-Y by heating a mixture of $[Ru(II)(bpy)_2]^{2+}$-Y and the desired ligand (bpz or dmb) at 200° C. This method constitutes an elegant two-step approach to the preparation of mixed complexes in zeolites. However, the zeolite cage size prevents incorporation of larger

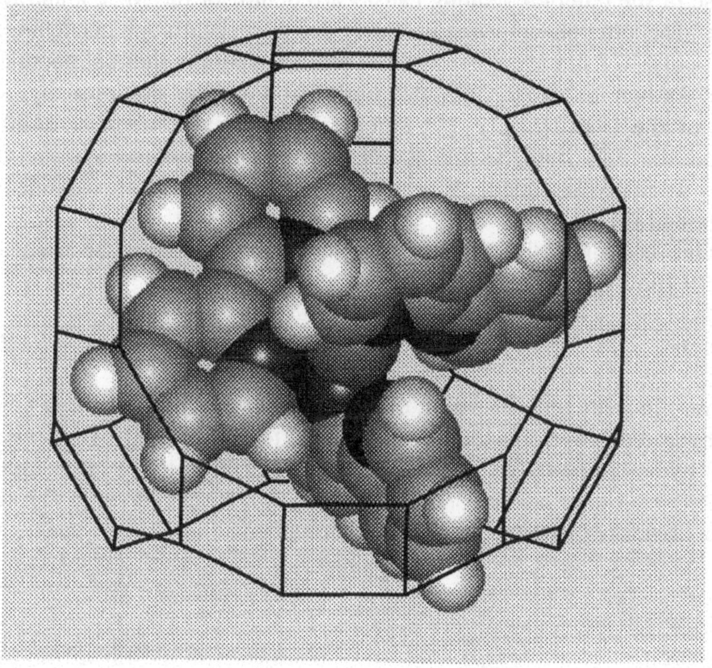

Figure 3. $[Ru(II)(bpy)_3]^{2+}$, occluded inside a zeolite Y supercage.

complexes. For instance, $[Ru(II)(bpz)_2(bpy)]^{2+}$ and $[Ru(II)(dmb)_2(bpy)]^{2+}$ cannot be made by this procedure.

Encapsulation of $[Ru(II)(bpy)_3]^{2+}$ in Na-Y strongly affects the luminescence properties of this complex [39,45,50]. The optimal conditions for emission by zeolite occluded complexes are low-loading, low content of paramagnetic impurities, and absence of oxygen.

In zeolites, *concentration* quenching occurs by resonance energy transfer from the excited complex to its neighbors. *Oxygen* quenching gives singlet oxygen and superoxide ions. *Water* causes quenching, when the ligand-attached hydrogen atoms assist in radiationless transitions. However, at high water loadings, the water molecules form an isolating mantle between the complexes and cause an increase of the emission intensity.

Dehydration of $[Ru(II)(bpy)_3]^{2+}$-Y results in a reduction of the rotational mobility of the complex. The π-π^* bipyridine transition, the MLCT band, and the ^{3}CT emission band are all blue-shifted (from 285, 460 and 620 to 278, 449, and 586 nm, respectively). The ^{3}CT lifetime decay was fitted with a bi-exponential model, with lifetimes of 440 ns and 98 ns. However, this bi-exponential decay model does not exclude a multi-exponential model. Indeed, dehydration results in direct interaction of the complex with the surface; as a result, the complexes are all in slightly different environments.

An increase of the complex loading causes a red shift of the MLCT and reduces the ^{3}CT lifetime. The latter phenomenon is clearly due to concentration quenching.

In zeolites, only ^{3}CT luminescence is observed. In solution, thermal population of the ^{3}dd state is known to cause photodecomposition of the complex by deligation. Entrapment of the $[Ru(II)L_nL'_{3-n}]^{2+}$ complexes (L=bpy; L'=bpz or dpm) prevents population of the ^{3}dd state, as the Ru-N bond elongation (which should lead to deligation) is sterically restricted. As a result, the room temperature lifetime of the ^{3}CT state is increased [48].

An important reaction in the photocatalytic splitting of water is the water oxidation, e.g., by $[Ru(III)(bpy)_3]^{3+}$ [40]:

$$[Ru(III)(bpy)_3]^{3+} + OH^- \rightarrow [Ru(III)(bpy)_3]^{2+} + {}^1/_4\, O_2 + {}^1/_2\, H_2O. \quad (15)$$

The reaction proceeds only in well-defined pH ranges. Under *alkaline* conditions, a covalently hydrated bipyridine ligand is thought to be formed. This results in the formation of hydrogen peroxide, which is further oxidized to O_2 by two equivalents of $[Ru(III)(bpy)_3]^{3+}$. Under *acidic* conditions, H_2O_2 is not a precursor to O_2; the oxygen atoms remain associated with the Ru complex throughout:

$$[Ru(III)bpy_3]^{3+} + H_2O \rightarrow [Ru(II)bpy_3(OH^{\bullet})]^{2+} + H^+ \quad (16)$$

$$[Ru(II)bpy_3(OH^{\bullet})]^{2+} + [Ru(III)bpy_3]^{3+} \rightarrow [Ru(II)bpy_3(OH^{\bullet})]^{3+} + [Ru(II)bpy_3]^{2+} \quad (17)$$

$$[Ru(III)bpy_3(OH^{\bullet})]^{3+} \rightarrow [Ru(III)bpy_3O]^{2+} + H^+ \quad (18)$$

$$[Ru(III)bpy_3(OH^{\bullet})]^{2+} + [Ru(III)bpy_3O]^{2+} \rightarrow 2\,[Ru(II)bpy_3]^{2+} + H^+ + O_2. \quad (19)$$

Upon zeolite entrapment, $[Ru(III)(bpy)_3]^{3+}$ remains reactive toward water. Evolution of oxygen is, however, not observed, although there is some evidence for formation of $OH^{\bullet}$ radicals. Lunsford et al. [40] have ascribed the lack of oxygen production to a pH effect. The effective pH in the zeolite is unknown, but it may well fall inside a range which is not suited for oxygen evolution.

In bpy-loaded Fe(II)-Y, several species were identified by Mössbauer spectroscopy [41]. The characteristics of the most intense signal correspond to those of the well-known octahedral, low-spin $[Fe(II)(bpy)_3]^{2+}$, which is probably situated near the center of the Na-Y supercage. A weaker Mössbauer doublet seems to correspond to a low-spin Fe(II) species with a coordination number greater than 6. In this complex, the Fe cation is probably located at site IV (in the center of the twelve-membered ring of the cage window), where it is π-bonded to the ligands of two $[Fe(II)(bpy)_3]^{2+}$ complexes in adjacent supercages (see Figure 4). Finally, uncomplexed high-spin Fe(III) is observed, especially when the loading exceeds one Fe(II) per supercage.

Upon exposure of bpy-loaded Fe(II)-Y to chlorine, encapsulated $[Fe(II)(bpy)_3]^{2+}$ is largely oxidized to $[Fe(III)(bpy)_3]^{3+}$. The *g* values of the EPR spectrum of the oxidized sample are indeed indicative of a low-spin, axially symmetric $(t_{2g})^5$ complex.

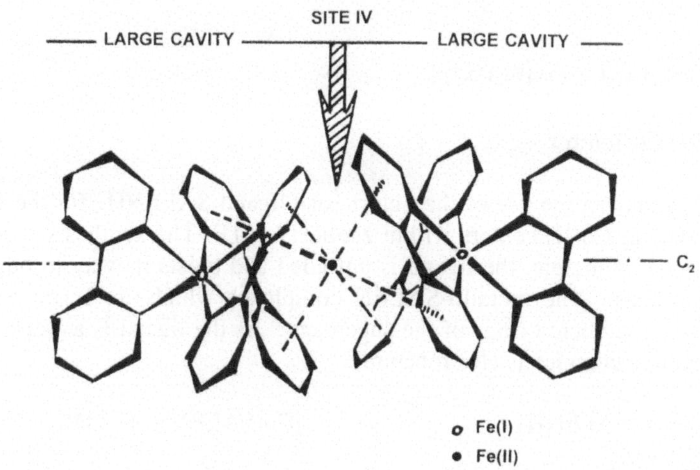

Figure 4. Model of low-spin Fe(II) at site IV in bpy loaded Fe(II)-Y. Fe is π-bonded to the ligands of two $[Fe(bpy)_3]^{2+}$ complexes in neighboring supercages (after Quayle et al., see reference [41] ; copyright *ACS*.).

4.3.3 Cobalt Complexes

$[Co(bpy)_3]^{2+}$-Y and $[Co(terpy)_2]^{2+}$-Y are easily obtained by solid-state reaction between dehydrated Co,Na-Y and bpy or terpy [42,43]. The spin-multiplicity of these complexes is variable. At 77 K, the EPR spectrum shows an 8-line hyperfine structure, which is ascribed to the low-spin complex. As the temperature is increased, the EPR spectrum gradually vanishes, due to the short relaxation time of the high-spin form [42]:

$$^2E \textit{ (low-spin)} \rightleftarrows {}^4T \textit{ (high-spin)}$$

For $[Co(bpy)_3]^{2+}$, this equilibrium is considerably shifted to the left in the zeolite Y supercage. This was ascribed to steric constraints, which force the bpy ligands closer to the Co(II) ion.

These six-coordinated complexes are unable to bind O_2. However, when a bpy-terpy mixture is used for Co,Na-Y impregnation, the mixed, oxygen-binding $[Co(bpy)(terpy)]^{2+}$ can be obtained in fair yields [43]. The zeolitic $[Co(III)(bpy)(terpy)\cdot O_2^-]^{2+}$ adduct is stable in oxygen up to 343 K. The $[Co(bpy)(terpy)]^{2+}$ content of the zeolite was maximized by optimization of the reactant stoichiometry and by variation of the monovalent cations on the zeolite. When Co(II), bpy, and terpy are introduced into Li-Y in a 1:2.5:2 ratio, an efficient system is obtained for separation of dry air into O_2 and N_2 [44].

4.4 SCHIFF-BASE COMPLEXES

4.4.1 Co(II) Complexes

Herron first proposed the use of the Schiff base ligand $SALENH_2$ for the formation of oxygen activating Co(II) centers within zeolite Y [51]. The synthesis is based on the "flexible-ligand" principle: the ligand is unfolded and twists its way through the metal exchanged zeolite. The metallo-SALEN complexes which are formed cannot pass through the 12-membered rings of the supercage. As the ligand is a weak bifunctional acid, the resulting complex is electroneutral:

$$\text{Co(II)} + \text{SALENH}_2 \rightarrow [\text{Co(SALEN)}] + 2\,\text{H}^+. \quad (20)$$

The final Soxhlet extraction only removes the excess ligand. The non-extractability of the neutral chelates indicates that they are retained in the zeolite by spatial restrictions alone; hence, these complexes are truly entrapped species. However, some leaching of Co(II) from the outer zeolite layers was demonstrated by XPS. A molecular graphics picture of [Co(SALEN)] complexes in a Y-type zeolite is shown in Figure 5.

Pyridine treatment of [Co(SALEN)],Na-Y results in the formation of an EPR detectable superoxo species. The observation of this species is a strong proof of the simultaneous, intrazeolitic coordination of Co(II) by SALEN and pyridine:

$$[\text{Co(SALEN)}] + \text{pyridine} + \text{O}_2 \rightarrow [\text{Co(SALEN)}\cdot\text{pyridine}\cdot\text{O}_2]. \quad (21)$$

That a zeolite is a very effective site-isolating solid solvent for this complex was proven by the long lifetime of the oxygen binding species; oxygenation-deoxygenation cycles can be continued for over 24 h without intensity loss.

While the occlusion has little effect on the characteristics of the oxygenated species, the deoxygenated complexes are somewhat different from their solution counterparts. Four-coordinate, square-planar [Co(SALEN)] and five-coordinate, square-pyramidal [Co(SALEN)(pyridine)] normally have an EPR-detectable, low-spin electronic configuration, but [Co(SALEN)]-Y and [Co(SALEN)(pyridine)]-Y are EPR silent. The anomalous spin-multiplicity may be a consequence of the complex entrapment, as in the case of $[Co(II)(bpy)_3]^{2+}$-Y. On the other hand, the high-spin character may, especially for the pyridine free sample, partly be ascribed to a non-planar arrangement of the SALEN donor atoms around Co(II).

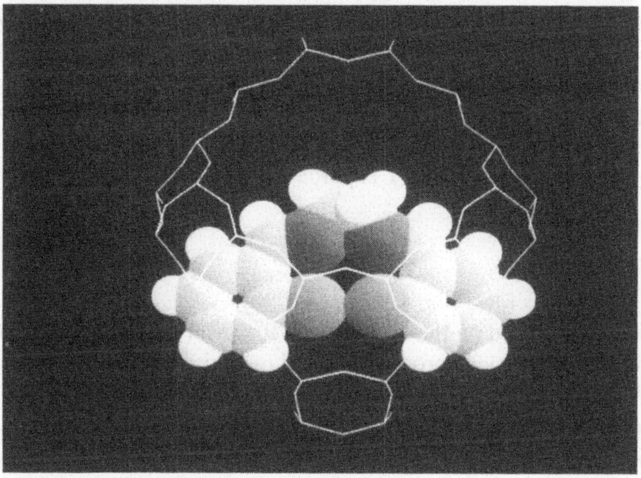

Figure 5. Molecular graphics representation of [Co(SALEN)] in the supercage of faujasite.

Such coordination proposals are mainly based on electrochemical studies of zeolite-entrapped [Co(SALEN)] [52]. In the cyclic voltammograms of [Co(SALEN)]-Y, completely reversible Co^{3+}/Co^{2+} and Co^{2+}/Co^{+} transitions can be observed. This contrasts with the irreversibility of the redox process for ligand-free Co,Na-Y, on which Co metal is precipitated upon reduction. The transitions for [Co(SALEN)]-Y occur at potentials similar to those of [Co(SALEN)] in solution, but there is a supplementary Co^{3+}/Co^{2+} redox couple, with a distinct redox potential. The latter signal is not influenced by pyridine and corresponds hence to a species in which pyridine is unable to coordinate. Proposed coordinations include a partial chelation of Co(II) by the zeolite, or a non-planar organization of the ligand around Co(II).

The problem of various, co-existing Co(II) speciations may be partly resolved by the use of the pentadentate Schiff base ligand smdpt [53]. This ligand already incorporates a potential axial nitrogen base; hence, pyridine absorption is no longer needed for oxygen activation. In [Co(smdpt)], four ligand donor atoms form a plane around Co(II), while the nitrogen base occupies a fifth, axial position, and raises the affinity of Co(II) for oxygen. The oxygen-free [Co(smdpt)] complexes are well-known to be high-spin and EPR-invisible, but the [Co(smdpt)·O_2] oxygen adducts can be readily examined by EPR. The smdpt ligand is quite efficient in providing the correct coordination sphere for oxygen binding on Co(II); in a Co(II) exchanged, smdpt-loaded hexagonal faujasite, up to 25 % of the total Co(II) reversibly binds dioxygen.

An interesting influence of the framework topology on the complex structure was observed. While in Na-Y a very large hyperfine coupling constant was observed on the parallel component of the signal ($g_{||} = 2.100 \pm 0.003$, $A_{||} = 27.0 \pm 0.5$ G), the EPR parameters of the species observed in the hexagonal Na-EMT zeolite resemble closely those of the species in frozen solution ($g_{||} = 2.091 \pm 0.003$ and $A_{||} = 19.5 \pm 0.5$ G). This seems to indicate that the more spacious hypercage of the Na-EMT zeolite, which is accessible through five 12-membered rings, provides enough space for the complex to organize as in solution, whereas the Na-Y supercage causes complex distortion.

4.4.2 Mn-Complexes

Bowers and Dutta first reported on the inclusion of the [Mn(II)SALEN] chelate in zeolite Y [54]. Again, the voltammetric approach is useful in studying the Mn coordination [55]; a reversible Mn^{3+}/Mn^{2+} redox couple is observed. The IR spectrum of the material shows that the extraction, albeit with the rather polar acetonitrile, leaves a considerable amount of free, non-coordinating $SALENH_2$ on the zeolite [54]. The [Mn(II)SALEN]-Y zeolite catalyzes the selective oxidation of olefins with PhIO. Epoxides are formed with [Mn(II)SALEN]-Y, while with Mn,Na-Y, mainly allylic oxidation is observed. This remarkable selectivity difference illustrates the strong influence of the SALEN ligand on the oxygen transfer capacity of the Mn(III). In electrochemical experiments, zeolite Y entrapped [Mn(II)SALEN] reacts with dissolved O_2 in the presence of 1-methylimidazole. This approach opens perspectives for using these materials in electrocatalytic oxidations of, e.g., hydrocarbons.

4.4.3 Complexes of Fe, Rh ,and Pd

Whereas for Mn only one coordination type is detected by electrochemistry, two coordination types are observed for [Fe^{n+}(SALEN)]-Y (n = 2 or 3) [55]. As in the case of [Co(SALEN)]-Y, different tetradentate- or bidentate-coordinations may be involved. The redox activity of the encapsulated Fe complex depends, e.g., on the presence of a coordinating solvent or a coordinating Cl^- anion. The four-coordinate species, which are formed during the synthesis, can be converted into five-coordinate ones, which incorporate, e.g., a DMSO solvent molecule.

An XPS study of Rh chelation by SALEN in Na-Y still corroborates the idea of different coordinations around the TMI [56]. The influence of the SALEN ligand is mainly seen in the higher binding energy of Rh(III) in SALEN loaded Rh,Na-Y, but the electronic spectra suggest several binding modes of SALEN around Rh(III). In [Pd(II)(SALEN)]-Y, the coordination is uncertain as well, but this does not diminish the qualities of this material as a selective hydrogenation catalyst; the SALEN ligand ensures a higher dispersion of Pd(II) than in a Pd,Na-Y zeolite. This results in an enhanced hydrogenation selectivity [57,58].

It should be noted that it is not obvious to synthesize coordinatively unsaturated, four-coordinate SALEN complexes in zeolites. The zeolite tends to function itself as a bi- or tridentate ligand and forces SALEN into a non-planar conformation. Therefore, use of pentadentate ligands instead of the classically used tetradentate SALEN may be promising in further expelling the zeolite from the coordination sphere of the TMI.

4.5 PHTHALOCYANINE AND PORPHYRIN COMPLEXES

4.5.1 Phthalocyanines

The range of zeolites which are suitable as hosts for phthalocyanines is limited, because of the size of these ligands. Generally, faujasite structures are preferred [59-62], but the use of ZSM-5 [47], AlPO-5 [5], AlPO-11 [5], and VPI-5 [63] has also been claimed. Synthesis of phthalocyanines in the internal space of zeolites is exceptional in that mostly, the ligand is itself synthesized in situ before complexation of the metal ion. Moreover, depending on the metal source (salts, carbonyl complexes, or metallocenes), distinct synthesis procedures can be recognized.

When salts are used, the phthalocyanines are synthesized by tetramerization of dicyanobenzene on cation-exchanged zeolites at temperatures between 523 and 573 K [59-65]. Although the complexation of the TMI by the formed phthalocyanine is thermodynamically favorable, it generally proceeds only slowly. The degree of incorporation of TMI in zeolite-hosted phthalocyanines decreases in the sequence Co > Ni > Cu > Fe. Therefore, after synthesis, large quantities of transition metal are still ligated by the lattice or are present as oxides [60,61,64,66]. Lattice-coordinated cations can be removed by exchange with NaCl, but the exchange of the TMI with Na^+ is incomplete [66-69]. Moreover, considerable amounts of TMI may migrate to the surface during dehydration and Pc synthesis, as shown by XPS-analysis of [Rh(Pc)]-Y and [Rh(Pc)]-X [65]. To overcome the unfavorable kinetics of the chelation of TMI by the Pc chelate, procedures using other metal sources have been proposed.

One of the alternative procedures involves absorption of TMI-carbonyl complexes into the zeolite. These complexes may or may not be decomposed prior to the in situ synthesis of the Pc ligand [67,70-72]. Decomposition can be performed thermally or photochemically and results in formation of metal clusters in the zeolite. Photochemical decomposition is preferred, in order to suppress the migration of the metal to the outer surface. As the transition metal is present as metal clusters in the faujasite supercage and as only one phthalocyanine can be synthesized per cage, the presence of unchelated TMI is unavoidable. If, on the other hand, the decomposition step is omitted, the CO ligands are directly replaced by 1,2-dicyanobenzene. However, this procedure also leaves some unchelated TMI in the zeolite, as the rate of decomposition of the carbonyls is higher than the rate of formation of [TMI(Pc)].

Therefore, it is advisable to use more stable complexes, such as metallocenes, as precursors. Zakharov et al. first applied ferrocene and cymantrene in [TMI(Pc)]-Y synthesis [73]. The amount of unchelated TMI can be minimized by selecting a proper synthesis temperature [74]. An analogous synthesis procedure was used by Parton et al. [63], who claimed, based on chemical analysis, that there was almost no residual iron in their [Fe(Pc)]-Y. However, because of the high stability of ferrocene, large amounts of free base phthalocyanines are synthesized. Fortunately, at low loading, these do not interfere with the catalytic activity. Using a metallocene as TMI-source moreover allows the use of molecular sieves without cation exchange capacity as hosts for [TMI(Pc)]. Indeed, when ferrocene is mixed with dry VPI-5 and dicyanobenzene, [Fe(Pc)] complexes are formed inside the channels of this neutral, aluminophosphate molecular sieve [63].

Cyclization of phthalocyanines starting from dicyanobenzene is a two-electron reduction process. When TMI-carbonyl complexes are used, these two electrons are supplied by the metal, which is oxidized to the divalent state:

$$[\mathrm{TMI(CO)_m}] + 4\,\mathrm{DCB} \rightarrow [\mathrm{TMI(Pc)}] + \mathrm{m\,CO}. \quad (22)$$

In the case where salts or metallocenes are used as the metal source, some water must be added as an electron source [61,63]. The subsequent complexation results in the liberation of two protons. These protons can be trapped efficiently by cyclopentadienyl anions, but when salts are used as the metal source, the protons cause surface acidity:

$$4\,\mathrm{DCB} + \mathrm{H_2O} \rightarrow \mathrm{H_2Pc} + {}^1\!/_2\,\mathrm{O_2} \quad (23)$$

$$\mathrm{H_2Pc} + [\mathrm{TMI(Cp)_2}] \rightarrow [\mathrm{TMI(Pc)}] + 2\ \text{cyclopentadiene} \quad (24)$$

$$\mathrm{H_2Pc} + \mathrm{TMI\text{-}Y} \rightarrow [\mathrm{TMI(Pc)}] + \mathrm{H_2\text{-}Y}. \quad (25)$$

The acidic sites can interfere with the catalytic activity of the phthalocyanines and may even cause dealumination or loss of crystallinity of the zeolite [65]. There is also some spectroscopic evidence for protonation of the coordinating nitrogen atoms. This results in a symmetry reduction from D_{4h} to D_{2h} and a splitting of the Q-bands in the Vis-NIR region [75]. In IR spectra of [Co(Pc)]-X, a band at 1020 cm^{-1} was attributed to protonation of the inner nitrogen atoms [76]. XPS measurements on [Ni(Pc)]-Y showed non-equivalency of the chelating nitrogen atoms [66].

Instead of synthesizing the phthalocyanines in situ, they can be used as a template during zeolite synthesis. Phthalocyanines are suitable for such a synthesis as they are thermally and chemically extremely stable. Consequently, no residual TMI is formed during synthesis. A major problem is, however, to keep the phthalocyanines monomolecularly dispersed in the aqueous zeolite synthesis medium. By careful control of the synthesis gel chemistry, Balkus et al. obtained a zeolite X material in which 50% of the unit cells were occupied by Pc, even after severe extractions [4] (see also Chapter 6). Various [TMI(Pc)]s were also successfully used as templates for AlPO-5 and AlPO-11 synthesis [5]. However, in view of the relative size of the Pc ligand and the pore system of these $AlPO_4$'s, it is likely that the complexes are mainly located at structural defects.

As discussed previously, [TMI(Pc)] zeolites often contain free-base H_2Pc. Distinction between free-base and metallated phthalocyanines is facilitated by the symmetry change upon chelation. The symmetry of free-base H_2Pc is D_{2h}, whereas [TMI(Pc)] belongs to D_{4h}. In IR spectroscopy, this symmetry difference results in the splitting of the conjugated isoindole band (at 1332 cm^{-1}) and the C-H in-plane deformation band (at 1287 cm^{-1}) into doublets for the free base H_2Pc's (at 1336 and 1322 cm^{-1}, and at 1304 and 1278 cm^{-1}, respectively; see Figure 6). Moreover, the TMI-N vibrations around 900 cm^{-1} are also typical for metallated Pc [63]. The Q-bands in the electronic spectra should also be sensitive to symmetry changes, but are usually too broad to be of practical use [75,77].

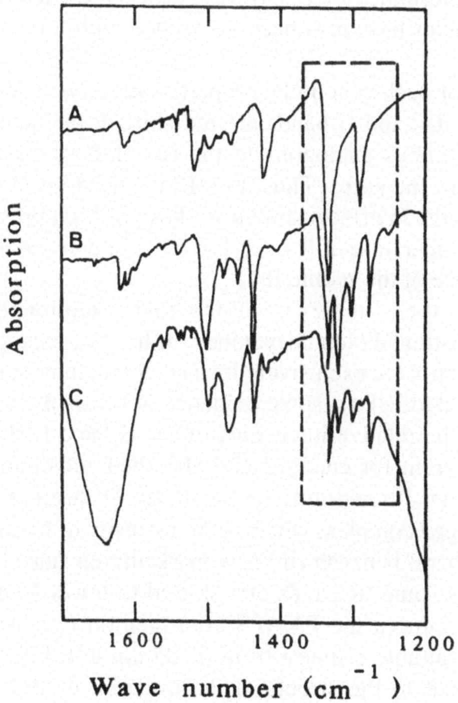

Figure 6. IR spectra of (A) FePc, (B) H_2Pc, and (C) mixture of FePc and H_2Pc encaged in Y zeolites.

The valence of the TMI in entrapped Pc has been the subject of XPS investigations by Romanovskii and Gabrielov [66,78]. Fe, Ni, and Co occur in the divalent state, while Os and Ru are trivalent. In the latter case, the positive charge of the complex must be compensated by the anionic zeolite lattice. Rh on the contrary is easily reduced to Rh(II) or even to Rh(I) during Pc synthesis. This implies that negatively charged [Rh(I)(Pc)] complexes may be entrapped in the anionic zeolite medium [65].

The close resemblance between zeolite-entrapped and dissolved Pc's has been proven by a multitude of techniques, such as Mössbauer spectroscopy [79], cyclic voltammetry [80], and EPR spectroscopy [81,82].

While the pore system of the VPI-5 molecular sieve provides sufficient space for the Pc ring, the faujasite supercage may be somewhat too small. The size of a Pc ring is about 1.5 nm, whereas the free diameter of a supercage is only 1.3 nm. Therefore, based on molecular graphics analysis, a saddle-type deformation of the aromatic system was proposed. In this model, the four benzene rings of the chelate occupy the four twelve-membered ring openings of the supercage [63,83]. Some indirect (and weak) evidence for this saddle-type deformation comes from the broadening of the N_{1s} lines in XPS [66], and from a coordination number of 3.6 (instead of 4) and Fe-N distances of 1.84 nm (instead of 1.83 nm), determined by EXAFS [72]. Very small shifts of the IR C=N and C=C stretching frequencies have also been associated with this saddle-type deformation [72].

Several other effects of occlusion on Pc properties have been documented. In the DRS spectra of occluded Pc, B-, and Q-bands are red-shifted; the latter also become weaker [65,71,72]. This behavior is analogous to the red shift of the Q-band of crystalline [TMI(Pc)] upon pressure increase. Thus, [TMI(Pc)] must be in a spatially constrained environment, which results in effects similar to those of high pressures [75]. Moreover, the optimal synthesis temperature of [Co(Pc)] in the Y supercage is 542 K, which is 15 K higher than in the absence of the zeolite [84].

The confinement of the complexes to the inner zeolite volume is even more questionable for the substituted phthalocyanines. It has, for example, been demonstrated that nitro-substituted FePc's are exclusively located at the outer surface of zeolite Y [85]. For tetra-*t*-butyl substituted FePc's, some evidence in favor of encapsulation in NaY has been presented [86]. The relative intensities of the B-band (280 nm) and the Q-bands (550, 580 nm) are different for encaged and absorbed substituted Pc's. The Q-band, which arises from $\pi \rightarrow \pi^*$ transitions in the inner Pc aromatic system, is strongly suppressed for the encaged complex, whereas the intensity of the B-band, which is due to transitions in the peripheral benzene rings, is markedly enhanced. Furthermore, the Q- and B-bands, as well as some IR bands, are shifted to lower frequencies. Encaging of *t*-butyl substituted Pc reduces the EXAFS coordination number from 3.9 to 3.6 and increases the Fe-N interatomic distance from 0.185 nm to 0.190 nm. All these spectral differences are attributed to the structural deformation of the Pc plane and a slight departure of the Fe atom from the Pc plane. Catalytic results on the other hand disfavor the encapsulation hypothesis, since large substrates, such as stilbene, are epoxidized on the *t*-butyl substituted phthalocyanines. It is unlikely that this occurs in the supercages, which are already occupied by *t*-butyl substituted Pc's [86].

4.5.2 *Porphyrins*

Tetraphenylporphyrins are very successful homogeneous catalysts, but two major problems restrict their occlusion inside zeolites. Firstly, in situ synthesis is difficult, as the synthesis of porphyrins requires 8 monomers (i.e., 4 aldehydes and 4 pyrroles). Moreover, formation of linear polypyrrole is an important side reaction. A second, more important limitation is the size of tetraphenylporphyrins, which are markedly larger than Pc's. Nevertheless, Chan and Wilson report the in situ synthesis of tetraphenylporphyrin in the supercages of a Y zeolite [87].

Nakamura et al. proposed the in situ synthesis of the less bulky tetramethylporphyrin in zeolite Y [88]. Replacing the phenyl groups by methyl substituents should largely meet any steric problems. DRS spectra clearly show the Soret band of the porphyrin, but the Q-band was not observed. This was interpreted as a consequence of distortion of the porphyrin.

5. Conclusions

Several synthesis routes have been devised to synthesize zeolite immobilized TMI complexes. Almost all synthesis methods lead to full coordination by the external ligand, restricting the role of the zeolite lattice to that of an anion and/or a solvent. However, complexes with free coordination sites on the TMI are often necessary for control of reactivity and catalysis. Partial coordination can be achieved in three ways: (1) controlled decomposition of (pseudo)-octahedral complexes; (2) complexation by multidentate ligands with 3,4 or 5 coordinating atoms; and (3) in the case of monodentate ligands, by maintaining well-defined ligand:TMI ratios, in any case below 6. The first and third methods usually lead to mixtures of complexes with various coordination geometries. The second method has the advantage that stable and relatively immobile complexes are formed. However, coordination of the TMI may be incomplete and some donor atoms may not participate in the complexation. This shows that there is a need for devising selective synthesis routes giving one specific coordinatively unsaturated complex in the zeolite matrix.

In this respect, it may be worthwhile to extend the range of ligands studied to well chosen S- and P-containing molecules. This should lead to an extension of the area of heterogenization of homogeneous catalysts. The wide range of zeolite structure types and of chemical compositions allows for better chemical and geometrical matching of complexes and zeolites than shown in the limited number of attempts published so far.

In solution, complexation is described by stepwise stability constants. In the case of water and monodentate ligands we have:

$$[TMI(H_2O)_6]^{n+} + L \rightarrow [TMI(H_2O)_5(L)]^{n+} + H_2O \tag{26}$$

$$[TMI(H_2O)_5(L)]^{n+} + L \rightarrow [TMI(H_2O)_4(L)_2]^{n+} + H_2O \tag{27}$$

where $K_1 = \frac{[[TMI(H_2O)_5(L)]^{n+}]}{[[TMI(H_2O)_6]^{n+}][\,L\,]}$ and $K_2 = \frac{[[TMI(H_2O)_4(L)_2]^{n+}]}{[[TMI(H_2O)_5(L)]^{n+}][\,L\,]}$

Estimation of these sequential stability constants has not been possible in zeolites to date, except for the complexation of Cu^{2+} by en in X- and Y-type zeolites [28]. In this particular case, the mono-complex was slightly stabilized and the bis-complex slightly destabilized with respect to the corresponding aqueous solution species. Up to now, the question of the stabilizing or destabilizing effect of the zeolite matrix on the encapsulated complexes has not been quantitatively solved.

There are three reasons why the zeolite may affect the stability of the complexes:

(1) By encapsulating the complexes, the solvent is, at least partially, replaced by the zeolite. If the solvent competes for the TMI, as in the case of water, and if it is replaced by the structural oxygens with weaker coordinating power, then the complex is stabilized in the zeolite. If the solvent is non-coordinating, then only the environment of the complexes is changed and only minor solvent effects should be observed.

(2) Competing reactions may be more pronounced in the zeolite than in solution. This is especially the case for protonation reactions. Zeolites always contain some acidic sites and water molecules may be polarized by the exchangeable cations, giving more extensive protonation of basic ligands. This has a negative effect on their complexation ability.

(3) Bulky complexes may be deformed by encapsulation in the zeolites, thus changing their stabilities. Pc complexes are the most thoroughly studied examples. It is surprising, however, that, although distortion must be invoked for geometrical reasons, it is not clearly revealed in the spectroscopic data.

The data of the present review qualitatively confirm these three effects. However, they are small and in many cases negligible. This and the fact that the concentration of free ligands is necessary (and almost impossible to obtain in zeolite systems) makes determinations of stability constants in zeolites an onerous task.

Acknowledgments

D.E.D.V., P.P.K.G., and B.M.W. acknowledge the Belgian National Fund for Scientific Research (N.F.W.O.) for a research grant as research assistant. R.F.P. thanks the same institution for a post-doctoral fellowship. This work was supported by the Belgian Government in the frame of an Interuniversitary Attraction Pole (I.U.A.P.).

Glossary

^{3}CT	triplet charge transfer
bpy	2,2'-bipyridine
bpz	2,2'-bipyrazine
Cp	cyclopentadiene
dmb	4,4'-dimethyl-2,2'-bipyridine
dmgH	mono-anion of dimethylglyoxime
$dmgH_2$	dimethylglyoxime
DMSO	dimethylsulfoxide
DCB	dicyanobenzene
DRS	diffuse reflectance spectroscopy
en	ethylenediamine
EPR	electron paramagnetic resonance
EXAFS	extended X-ray absorption fine structure
MCM-41	a very large pore aluminosilicate molecular sieve with pore sizes in the range 20 - 100 Å.
Me	methyl, CH_3
MLCT	metal to ligand charge transfer
NIR	near infrared
O_l	an oxygen atom of the zeolite lattice
Pc	phthalocyanine
py	pyridine
SALEN	dianion of bis(salicylaldehyde)ethylenediimine
$SALENH_2$	bis(salicylaldehyde)ethylenediimine
smdpt	bis(salicylaldehyde)methyldipropyltriamine
terpy	2,2',2"-terpyridine
tetren	tetraethylenepentamine
THF	tetrahydrofuran
TMI	transition metal ion
UV	ultraviolet
Vis	visible
VPI-5	aluminophosphate molecular sieve with 18-membered rings
XANES	X-ray absorption near edge structure
XPS	X-ray photoelectron spectroscopy

References

1. J. Lunsford, *ACS Symp. Ser.,* **40**, 473 (1977).
2. W. Mortier and R. Schoonheydt, *Prog. Solid State Chem.,* **16**, 1 (1985); R. Schoonheydt, D. Roodhooft, and H. Leeman, *Zeolites*, **7**, 412 (1987).
3. (a) K. Klier, R. Kellerman, and P. Hutta, *J. Chem. Phys.,* **61**, 4224 (1974); (b) R. Kellerman and K. Klier, *ACS Symp. Ser.,* **40**, 120 (1977).
4. K. Balkus, C. Hargis, and S. Kowalak, *ACS Symp. Ser.,* **499**, 347 (1992).
5. S. Kowalak and K. Balkus, *Collect. Czech. Chem. Commun.,* **57**, 774 (1992).
6. A. Borvornwattananont and T. Bein, *J. Phys. Chem.,* **96**, 9447 (1992); A. Borvornwattananont, K. Moller, and T. Bein, *J. Phys. Chem.,* **96**, 6713 (1992).
7. W. De Wilde, R. Schoonheydt, and J. Uytterhoeven, *ACS Symp. Ser.* **40**, 132 (1977).
8. J. Lunsford and E. Vansant, *J. Chem. Soc., Faraday Trans. II,* **69**, 1028 (1973).
9. G. Bergeret, P. Gallezot, and B. Imelik, *J. Phys. Chem.,* **85**, 411 (1981); G. Bergeret, T. Tri and P. Gallezot, *J. Phys. Chem.,* **87**, 411 (1983).
10. D. Flentge, J. Lunsford, P. Jacobs, and J. Uytterhoeven, *J. Phys. Chem.,* **79**, 354 (1975).
11. E. Vansant and J. Lunsford, *J. Phys. Chem.,* **76**, 2860 (1972).
12. E. Vansant and J. Lunsford, *Adv. Chem. Ser.,* **121**, 441 (1973).
13. R. Howe and J. Lunsford, *J. Am. Chem. Soc.,* **97**, 5156 (1975).
14. C. Naccache and Y. Ben Taarit, *Chem. Phys. Lett.,* **11**, 11 (1971).
15. P. Dai and J. Lunsford, *Inorg. Chem.,* **19**, 262 (1980).
16. Y. Ukisu, A. Kazusaka, and M. Nomura, *J. Mol. Catal.,* **70**, 165 (1991).
17. Y. Yamada, *Bull. Chem. Soc. Japan,* **43**, 2661 (1970).
18. Y. Yamada, *Bull. Chem. Soc. Japan,* **45**, 64 (1972).
19. R. Schoonheydt, D. Van Wouwe, and H. Leeman, *Zeolites,* **2**, 109 (1982).
20. R. Schoonheydt, D. Van Wouwe, and M. Van Hove, *J. Colloid Interface Sci.,* **83**, 279 (1981).
21. R. Schoonheydt, R. Van Overloop, M. Van Hove, and J. Verlinden, *Clays and Clay Minerals,* **32**, 74 (1984).
22. R. Schoonheydt, D. Van Wouwe, and H. Leeman, *J. Chem. Soc., Faraday Trans. I,* **76**, 2519 (1980).
23. R. Schoonheydt, D. Van Wouwe, M. Van Hove, E. Vansant, and J. Lunsford, *J. Chem. Soc., Chem. Comm.,* 33 (1980).
24. I. Bresinska and R. Drago, *Stud. Surf. Sci. Catal.,* **68**, 101 (1991).
25. R. Taylor, R. Drago, and J. George, *J. Am. Chem. Soc.,* **111**, 6610 (1989).
26. R. Drago, I. Bresinska, J. George, K. Balkus, and R. Taylor, *J. Am. Chem. Soc.,* **110**, 304 (1988).
27. R. Taylor, R. Drago, and J. Hage, *Inorg. Chem.,* **31**, 253 (1992).
28. R. Schoonheydt, P. Peigneur, and J. Uytterhoeven, *J. Chem. Soc., Faraday Trans. I,* **74**, 2550 (1978); P. Peigneur, Ph.D. Thesis, K. U. Leuven (1976);

P. Peigneur, J. Lunsford, W. De Wilde, and R. Schoonheydt, *J. Phys. Chem,*. **81**, 1179 (1977).
29. R. Schoonheydt and J. Pelgrims, *J. Chem. Soc., Dalton Trans.*, 914 (1981).
30. R. Howe and J. Lunsford, *J. Phys. Chem.*, **79**, 1836 (1975).
31. P. Dutta and R. Zaykoski, *J. Phys. Chem.*, **93**, 2603 (1989).
32. R. Howe and J. Lunsford, in *Proc. 6th Int. Congress on Catalysis*, G. Bond, P. Wells, and F. Tompkins (Eds), The Chemical Society, London, **1**, p 540 (1976)
33. P. Dutta and C. Bowers, *Langmuir*, **7**, 937 (1991).
34. W. Lubitz, C. Winscom, H. Diegruber, and R. Möseler, *Z. Naturforsch.*, **42a**, 970 (1987).
35. H. Diegruber and P. Plath, *Stud. Surf. Sci. Catal.*, **12**, 23 (1982).
36. J. Strutz, H. Diegruber, N. Jaeger, and R. Möseler, *Zeolites*, **3**, 102 (1983).
37. H. Diegruber, Ph.D. Thesis, Universität Bremen (1984).
38. W. De Wilde and J. Lunsford, *Inorg. Chem. Acta.*, **34**, L229 (1979).
39. W. De Wilde, G. Peeters, and J. Lunsford, *J. Phys. Chem.*, **84**, 2306 (1980).
40. W. Quayle and J. Lunsford, *Inorg. Chem.*, **21**, 97 (1982).
41. W. Quayle, G. Peeters, G. De Roy, E. Vansant, and J. Lunsford, *Inorg. Chem.*, **21**, 2226 (1982).
42. K. Mizuno and J. Lunsford, *Inorg. Chem.*, **22**, 3484 (1983).
43. K. Mizuno, S. Imamura, and J. Lunsford, *Inorg. Chem.*, **23**, 3510 (1984).
44. S. Imamura and J. Lunsford, *Langmuir,* **1**, 326 (1985).
45. M. Borja and P. Dutta, *Nature*, **362**, 43 (1993).
46. K. Balkus, I. Bresinska, and S. Young, in *Proc. 9th. Int. Zeolite Conference*, R. von Ballmoos, J. Higgins, and M. Treacy (Eds), Butterworth-Heinemann, Stoneham MA, p 193 (1991).
47. L. Rankel and E. Valyocsik, *U. S. Patent 4,388,285* (1983).
48. K. Maruszewski, D. Strommen, K. Handrich, and J. Kincaid, *Inorg.Chem.*, **30**, 4579 (1991);
K. Maruszewski, D. Strommen, and J. Kincaid, *J. Am. Chem. Soc.*, **115**, 8345 (1993).
49. B. Durham, S. Wilson, D. Hodgson, and T. Meyer, *J. Am. Chem. Soc.*, **102**, 600 (1980).
50. P. Dutta and J. Incavo, *J. Phys. Chem.*, **91**, 4443 (1987) ;
J. Incavo and P. Dutta, *J. Phys. Chem.*, **94**, 3075 (1990);
W. Turbeville, D. Robins, and P. Dutta, *J. Phys. Chem.*, **96**, 5024 (1992).
51. N. Herron, *Inorg. Chem.*, **25**, 4714 (1986).
52. F. Bedioui, E. De Boysson, J. Devynck, and K.J. Balkus, *J. Chem. Soc., Faraday Trans.*, **87**, 3831 (1991).
53. D. De Vos, F. Thibault-Starzyk, and P. Jacobs, *Angew. Chemie, Int. Engl. Ed.*, **33**, 431 (1994).
54. C. Bowers and P. Dutta, *J. Catal.*, **122**, 271 (1990).
55. L. Gaillon, N. Sajot, F. Bedioui, J. Devynck, and K. Balkus, *J. Electroanal. Chem.*, **345**, 157 (1993).

56. K. Balkus, A. Welch, and B. Gnade, *Zeolites,* **10**, 722 (1990).
57. S. Kowalak, R. Weiss, and K. Balkus, *J. Chem. Soc., Chem. Commun.*, 57 (1991).
58. D. De Vos and P. Jacobs, in *Proc. 9th Internat. Zeolite Conf.*, R. Von Ballmoos, J. Higgins, and M. Treacy (Eds), Butterworth-Heinemann, Stoneham MA, p 615 (1991).
59. A. Zakharov and B. Romanovskii, *Vest. Mosk. Univ. Ser. Khim.*, **20**, 94 (1979).
60. S. Gudkov, E. Shpiro, B. Romanovskii, G. Antoshin, and K. Minachev, *Izv. Akad. Nauk SSSR Ser. Khim.*, **11**, 2248 (1980).
61. G. Meyer, D. Wöhrle, M. Mohl, and G. Schulz-Ekloff, *Zeolites*, **4**, 30 (1984).
62. N. Herron, G. Stucky, and C. Tolman, *J. Chem. Soc., Chem. Commun.*, 1521 (1986).
63. R. Parton, L. Uytterhoeven, and P. Jacobs, *Stud. Surf. Sci. Catal.*, **59**, 395 (1991) ; D. Huybrechts, R. Parton, and P. Jacobs, *ibid.*, **60**, 225 (1991) ; R. Parton, D. Huybrechts, P. Buskens, and P. Jacobs, *ibid.*, **65**, 47 (1991); R. Parton, D. De Vos, and P. Jacobs, in *Zeolite Microporous Solids: Synthesis, Structure and Reactivity*, E. Derouane et al. (Eds), Kluwer Academic Publishers, Amsterdam, p 552 (1992).
64. Z. Weide, Y. Xingkai, and W. Yue, *J. Mol. Catal. (China),* **5**, 168 (1991).
65. K. Balkus, A. Welch, and B. Gnade, *J. Incl. Phen. Molec. Recognition Chem.*, **10**, 141 (1991).
66. E. Shpiro, G. Antoshin, O. Tkachenko, S. Gudkov, B. Romanovskii, and K. Minachev, *Stud. Surf. Sci. Catal.,* **18**, 31 (1984).
67. B. Romanovskii, *Acta Phys. Chem.*, **31**, 215 (1985).
68. N. Herron, *J. Coord. Chem.*, **9**, 25 (1988).
69. E. Ignatzek, P. Plath, and U. Hündorf, *Z. Phys. Chem. Leipzig,* **268**, 859 (1987).
70. A. Zakharov and B. Romanovskii, *J. Incl. Phen.*, **3**, 389 (1985).
71. T. Kimura, A. Fukuoka, and M. Ichikawa, *Shokubai,* **30**, 444 (1988).
72. T. Kimura, A. Fukuoka, and M. Ichikawa, *Catal. Lett.*, **4**, 279 (1990).
73. A. Zakharov, B. Romanovskii, D. Luka, and V. Sokolov, *Metalloorg. Khim.*, **1**, 119 (1988).
74. A. Zakharov, *Mendeleev Commun.*, 80 (1991).
75. H. Diegruber and P. Plath, *Z. Phys. Chem. Leipzig,* **266**, 641 (1985).
76. G. Schulz-Ekloff, D. Wöhrle, V. Iliev, E. Ignatzek, and A. Andreev, *Stud. Surf. Sci. Catal.*, **46**, 315 (1989).
77. H. Diegruber, P. Plath, G. Schulz-Ekloff, and M. Mohl, *J. Mol. Catal.*, **24**, 115 (1984).
78. B. Romanovskii and A. Gabrielov, *J. Mol. Catal.*, **74**, 293 (1992).
79. M. Tanaka, Y. Sakai, T. Tominaga, A. Fukuoka, T. Kimura, and M. Ichikawa, *J. Radioanal. Nucl. Chem. Letters,* **137**, 287 (1989).
80. F. Bedioui, E. De Boysson, J. Devynck, and K. Balkus, *J. Electroanal. Chem.*, **315**, 313 (1991).
81. V. Iliev, L. Prahov, A. Andreev, E. Ignatzek, and G. Schulz-Ekloff, in *Proc. 6th Int. Symp. Heterog. Catal.*, Sofia, Part 2, 79 (1987).

82. T. Borisova, L. Izmailova, E. Kotov, and B. Romanovskii, *Zh. Fiz. Khim.*, **60**, 1195 (1986).
83. C. Tolman and N. Herron, *Catal. Today,* **3**, 235 (1988).
84. K. Balkus and J. Ferraris, *J. Phys. Chem.*, **94**, 8019 (1990).
85. R. Parton, C. Bezoukhanova, J. Grobet, P. Grobet, and P. Jacobs, *Stud. Surf. Sci. Catal.*, **83**, 371 (1994).
86. M. Ichikawa, T. Kimura, and A. Fukuoka, *Stud. Surf. Sci. Catal.*, **60**, 335 (1991).
87. Y. Chan and R. Wilson, *ACS Prepr. Div. Petrol. Chem.*, **33**, 453 (1988).
88. M. Nakamura, T. Tatsumi, and H. Tominaga, *Bull. Chem. Soc. Jpn.*, **63**, 3334 (1990).

Abstract. Researchers have devised several procedures for synthesizing transition metal ion (TMI) complexes in zeolite cavities and channels. Two main routes have been envisaged: synthesis with a preformed complex and complex assembly in the zeolite. An overview of the different preparation methods and of the TMI-complexes, encaged in zeolites, is given. These complexes have been mainly studied in zeolite Y. The zeolite is a non-coordinating anion, a solvent, or a ligand. It reduces the mobility of the complexes and provides chemisorption sites. Future developments will comprise the use of wide-pore zeolites and zeotypes (e.g., VPI-5, MCM-41, hexagonal faujasite) and the design of optically active surfaces. Furthermore, there is a need for quantification in general and for selectively synthesizing one type of complex.

Journal of Inclusion Phenomena and Molecular Recognition in Chemistry **21**: 215–237, 1995.

Zeolite Guest-Host Interactions: Implications in Formation, Catalysis, and Photochemistry

PRABIR K. DUTTA

1. Introduction

Zeolites are crystalline aluminosilicates with the composition $M_{x/n}\cdot(AlO_2)_x\cdot(SiO_2)_y\cdot wH_2O$ [1]. The framework is formed by -Si-O-Al(Si)- bonds and about 75 different topologies comprising of cages and channels with varying molecular dimensions are known [2]. The intracrystalline zeolitic space is occupied by H_2O molecules which can be removed upon heating. The M^{n+} cations, which neutralize the charge carried by the tetrahedral framework aluminum, are readily exchangeable. The close match in size between reactant molecules and the zeolite interior has been extensively exploited in hydrocarbon transformations, leading to many novel processes [3]. Reactions showing selectivity towards specific reactants, formation of specific products, as well as novel control of the transition state and molecular traffic control due to different intrazeolitic diffusivities of reactants and products, have all been observed [4].

We have exploited the close size match between the zeolite interior and structure-directing molecules to provide information on the assembly process of zeolites. In addition, entrapment of organometallic molecules in zeolite Y supercages has allowed us to examine reactant selectivity in olefin oxidation, stabilization of dioxygen complexes, and partitioning of molecules for directional photochemical transfer. We have also developed Raman spectroscopy, in both its spontaneous and resonance enhanced forms, to examine encapsulation. This technique has provided information about the interactions between zeolite guest and host at the molecular level. The examples discussed in this paper will illuminate how encapsulation can be exploited to understand various aspects of zeolite formation as well as zeolite guest-host reactivity.

2. Entrapment as a Monitor for Zeolite Assembly

It is well known that organic molecules direct the transformation of aluminosilicate gels to distinct framework structures [5]. In many of these syntheses, the organic molecule becomes trapped within the zeolite framework. The exact role played by these molecules in directing zeolite formation is an area of current research [6]. Various studies have

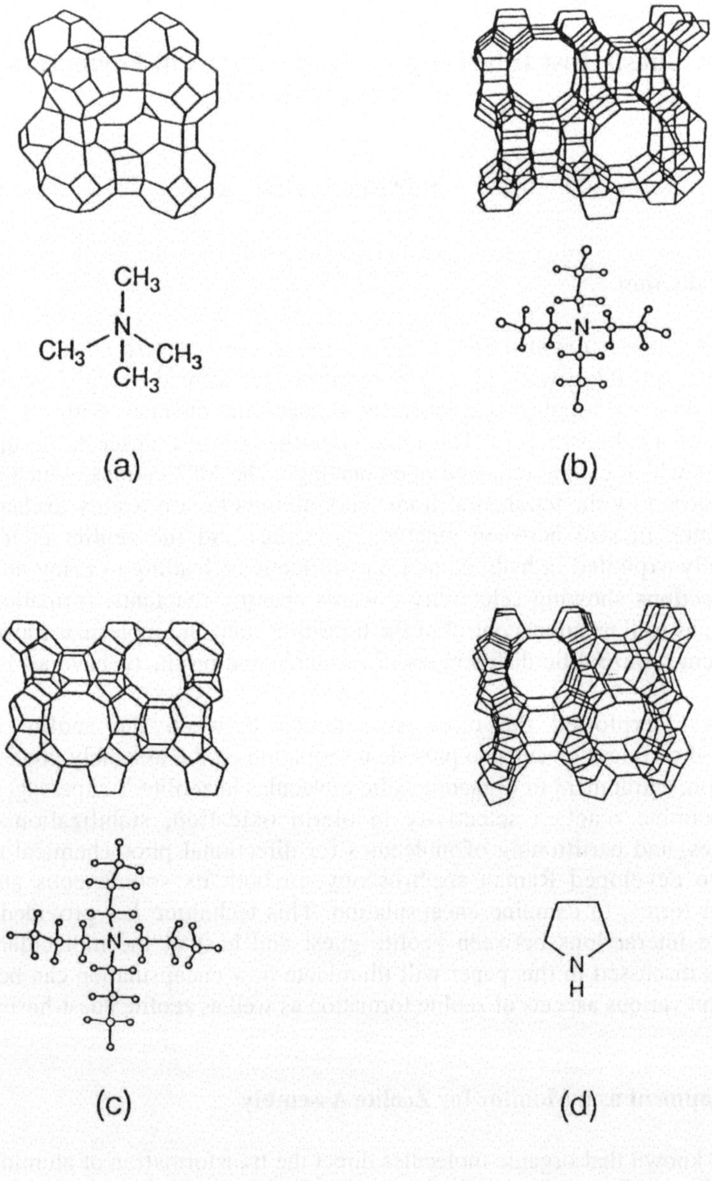

Figure 1. Structure of zeolitic frameworks and the corresponding structure directing agents: (a) TMA for ZK-4; (b) TEA for mordenite; (c) TPA for ZSM-5; and (d) pyrrolidine for ferrierite.

focused on the transformation of the inorganic aluminosilicate gel to zeolite, primarily by Raman [7-11] and solid state NMR spectroscopy [12-14]. Raman studies have shown that it is possible to detect four- and five-membered aluminosilicate rings in the gel [7-11], though how these units combine to form the long-range order inherent in these frameworks is not understood. Thus, what appears to be missing is information on the development of the porosity of these crystals. We have tried to address this shortcoming by focusing on the organic molecule as the organic-aluminosilicate system evolves towards a zeolitic framework. The strategy was to examine whether encapsulation of organic by the zeolite results in a strong enough perturbation to alter the characterization properties of the organic moiety. Information about zeolite assembly is then obtained by monitoring that structural perturbation. Raman spectroscopy is found to be particularly useful. Since the scattering cross-section of the organic moiety is considerably higher than the aluminosilicate framework, the Raman frequencies are a sensitive monitor of structural changes while the Raman scattering of water, the typical solvent in zeolite synthesis, has a small Raman scattering cross-section [15].

Figure 1 compares the structures of four structure-directing agents: tetramethylammonium ion (TMA), tetraethylammonium ion (TEA), tetrapropylammonium ion (TPA), and pyrrolidine (PL) and the zeolitic frameworks synthesized in the presence of these agents: zeolite A or ZK-4 (TMA) [16], mordenite (TEA) [17], ZSM-5 (TPA) [18], and ferrierite (PL) [19].

Figure 2 shows the Raman spectra of the structure-directing agents upon encapsulation into zeolites. Upon comparison with the spectra of the free structure-directing agents in solution, we have noted a number of spectral changes. These changes are highlighted as inserts in Figure 2 by showing the spectra of the unperturbed organic. Only in the case of TEA, which is responsible for the formation of mordenite, are no changes observed upon encapsulation, and thus no insert is shown.

By monitoring the Raman spectrum of the organic during zeolite growth, information about the entrapment and thereby the process of zeolite assembly and the development of porosity for three of these frameworks (ZK-4, ZSM-5, ferrierite) has been obtained. We will focus first on the results of the individual zeolite systems followed by a general discussion of zeolite assembly.

2.1 ZK-4 (ZEOLITE A FRAMEWORK)

The TMA cation results in synthesis of ZK-4, which has an architecture similar to that of zeolite A, but with Si/Al >1. As shown in Figure 1, there are two cages in ZK-4 which may contain the TMA ion, the α- or β-cages. The β-(sodalite)-cage has an internal dimension of ~6.4 Å, whereas the α-cage is ~8 Å. The dimension of the TMA cation is about ~6.4 Å. When positioned inside the α-cage [20], the Raman spectrum of TMA resembles that of the free-ion in solution. However, if it is held in the sodalite cage, the close match between the molecular size and the cage leads to a perturbation of the stretching frequencies: 752 cm^{-1} to 761 cm^{-1} for the C-N symmetric stretch and 950 to

955 cm^{-1} for the asymmetric C-N stretch [20,21]. This is evident from the Raman spectrum of TMA inside the zeolite sodalite where all of the TMA ions are entrapped in β-cages and is shown in Figure 3a.

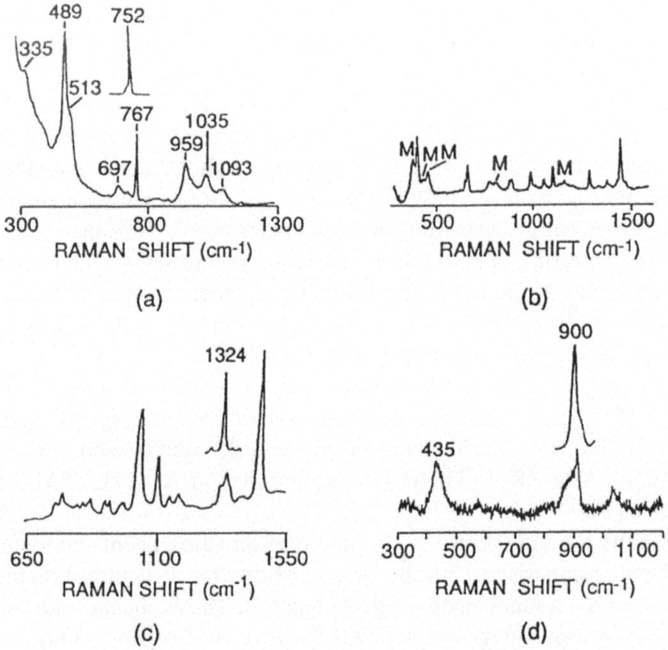

Figure 2. Raman spectra of structure-directing agents trapped in zeolite cages: (a) TMA in ZK-4; (b) TEA in mordenite; (c) TPA in ZSM-5; and (d) pyrrolidine in ferrierite. The insets on the figures show the spectrum of the structure-directing agent in solution.

The change in frequencies results from an increase in volume during the stretching vibrations which is met by resistance from the tightly surrounding cage framework and resembles the effect of increased pressure. Figure 3b represents this schematically. Therefore, these shifts in the Raman frequency of the TMA signal the formation of sodalite cages during the zeolite assembly process.

Figure 4 shows the Raman spectra of the aluminosilicate gel as a function of crystallization time for ZK-4. These spectra were obtained at room temperature from samples quenched at various times from the reaction mixture kept at 85° C [20].

The first appearance of the 767 cm^{-1} band (i.e., distinct sodalite cages) after 20 hours of synthesis coincided with the appearance of crystals as determined by X-ray diffraction experiments. This suggests that the formation of sodalite cages occurs with the assembly of the crystal itself. Thus, we conclude that the zeolite superstructure is not growing by assembly of the sodalite building units. Rather, the sodalite cages are formed concurrently with the zeolite superstructure of ZK-4.

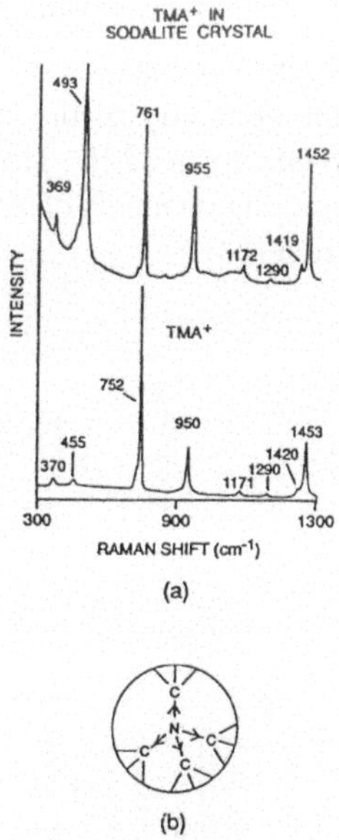

Figure 3. (a) Raman spectra of TMA in solution and in a sodalite crystal. (b) Symmetric stretching vibration of TMA entrapped in a sodalite cage (adapted from [20]).

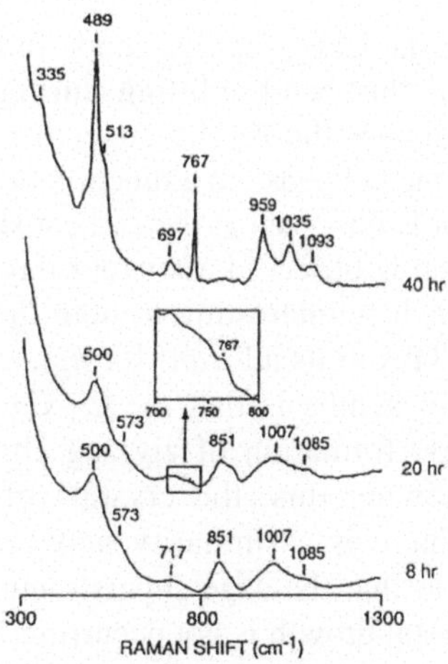

Figure 4. Raman spectra of the aluminosilicate gel during ZK-4 synthesis in the presence of TMA. The inset at 20 hour shows the first appearance of the entrapped TMA ion (adapted from [20]).

2.2 ZSM-5 (PENTASIL FRAMEWORK)

In the synthesis of ZSM-5, the vibrational changes of the TPA molecule highlighted in Figure 2 arise from a change in the geometry of the entrapped TPA cation. The TPA cation normally exists in its all *trans* form [22]. However, this geometry cannot be accommodated in the zig-zag intersecting channels of ZSM-5. A rotation around the C-C bond is required, as shown in Figure 5 [23].

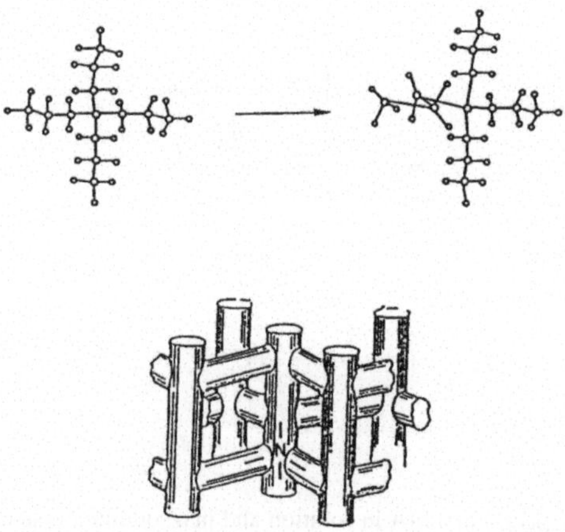

Figure 5. Conformational change of the TPA ion required to fit into the zig-zag channels of ZSM-5.

This results in splitting of the CH_2 wagging mode into two bands at 1324 and 1343 cm^{-1}. Thus, appearance of this band splitting during zeolite growth indicates the formation of the zig-zag chains of the ZSM-5 superstructure [22]. Figure 6 shows the Raman spectral evolution of the gel phase as a function of heating time.

Entrapment of the TPA ion is observed at the earliest stages of synthesis after one day of heating. Crystal formation is observed after four days of heating, as determined by X-ray diffraction. However, it is interesting to note that encapsulation at the earliest stages (< 2 days) leaves the TPA in its all *trans* form, i.e., no zig-zag chains are formed. It is only after three days of synthesis that the C_2-C_3 rotated conformer of TPA is observed, indicating the first formation of zig-zag chains and the superstructure of ZSM-5. Immediately following this, the crystals of ZSM-5 are observed in the diffraction experiments. There is a similarity between this system and that of the previously discussed ZK-4 in that the cage superstructure is formed at the end of the assembly process. Thus, crystal growth is not occurring by assembly of units consisting of trapped TPA ions at the intersection of channels, but rather the zig-zag chains are forming at the final stages of zeolite assembly.

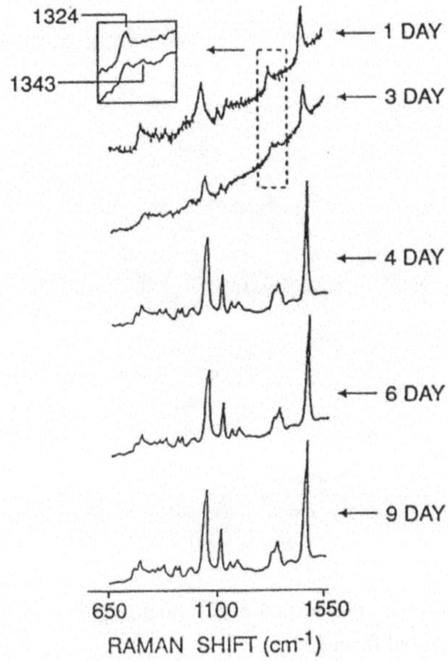

Figure 6. Raman spectra of the aluminosilcate gel during synthesis of ZSM-5 in the presence of TPA (adapted from [23]).

2.3 FERRIERITE FRAMEWORK

The above view of zeolite formation is also supported by the observations of the role of pyrrolidine in the synthesis of ferrierite [24]. Figure 7 shows the Raman spectral evolution of the gel as a function of synthesis time.

The earliest change observed, after 11 hours of heating, is the splitting of the 900 cm^{-1} band, which occurs as a result of protonation of the amine nitrogen and is assigned to the NH_2 wagging mode [25].

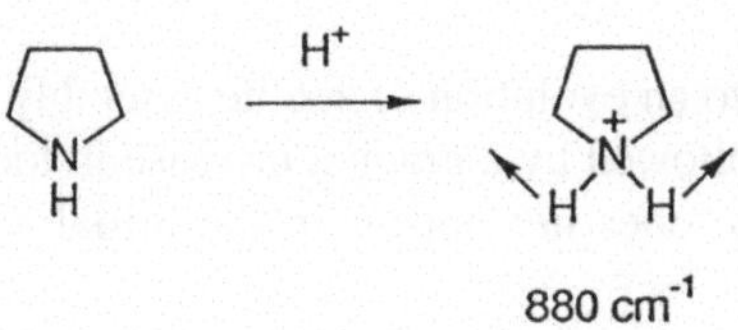

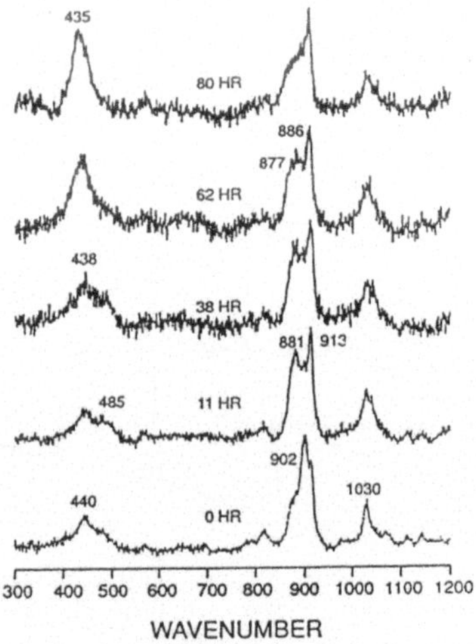

Figure 7. Raman spectra of aluminosilicate gel during synthesis of ferrierite in the presence of pyrrolidine (adapted from [24]).

This indicates that, at the early stages, the pyrrolidinium ion is being formed and is present in the aluminosilicate gel, the NH_2^+ group neutralizing the charge on the aluminum. The evolution of this wagging mode as a function of synthesis is of importance. It splits into several bands and then, at the final stage of crystallization, appears as a broad band. Since the broad band at 880 cm^{-1} evolves from a single band through intermediate stages of split bands, we have assigned these as due to the pyrrolidinium cation being in slightly different environments, with different $NH_2^+\cdots O$ hydrogen bonds with the aluminosilicate framework [24]. We have interpreted these changes as progressive entrapment of the pyrrolidinium ion as the zeolite superstructure forms.

2.4 THE ASSEMBLY PROCESS

These observations point to an evolution of zeolite assembly in which the primary overall structure is formed first followed by a gradual increase in local ordering. Thus, complete entrapment of the organic does not appear in the initial step in nucleation, but rather occurs in the final stages.

Within the scope of the above studies and other studies in the literature, the role of the entrapped organic in zeolite synthesis occurs at several levels:

(a) The organic can modify the gel chemistry, e.g., it is well known that TMA stabilizes 4-membered aluminosilicate rings, whereas TPA does so for

5-membered rings [26,27]. We have shown that pyrrolidine stabilizes gels containing 5-membered rings [24].

(b) The organic needs to associate strongly with the aluminosilicate gel, as we have noted above. This is done through electrostatic, hydrophobic, and H-bonding interactions.

(c) Entrapment results at the final stages of zeolite assembly.

These observations are outlined schematically in Figure 8 as a model of zeolite synthesis.

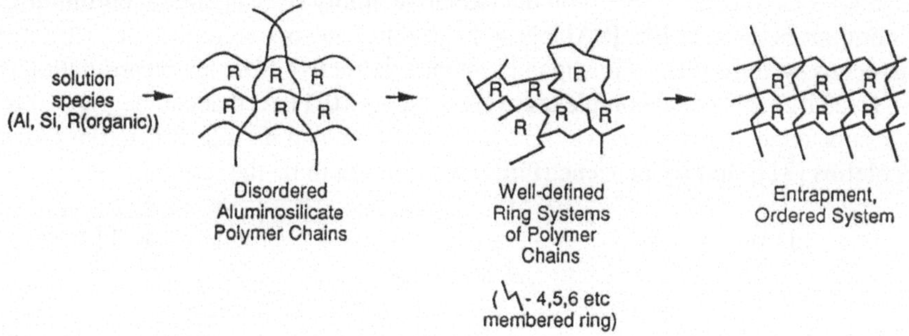

Figure 8. A schematic model of zeolite assembly.

3. Entrapment of Organometallic Complexes in Zeolite Y Cages

Organometallic complexes exhibit an unusually rich chemistry involving catalysis, ligand binding, and photochemistry in homogeneous solution, in biological systems, and on supports [28-29]. The cages of zeolites provide a unique host for these molecules, since they can be isolated from each other and the external world, yet can interact with reactants that penetrate through the ring system of the zeolite. Our primary emphasis has been on the large cages of zeolites A and Y as hosts. In section 2.1, we discussed the structure of zeolite A. A schematic of the zeolite Y supercage is shown in Figure 9. These super cages have dimensions of ~13 Å with openings of ~7 Å.

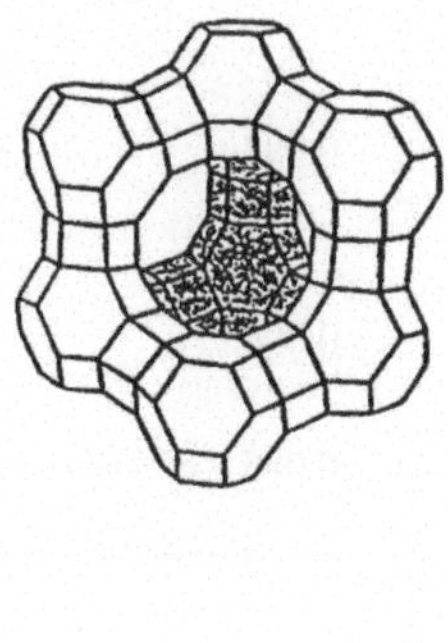

Figure 9. A zeolite Y supercage.

Complexes entrapped in these cages then have access to the external environment through the ~7 Å windows. Below we present our results on examining the interactions of the entrapped complexes in zeolite cages with reactants varying widely in size from aromatic olefins to dioxygen to photochemically generated electrons.

3.1 OLEFIN OXIDATION

Several studies have been reported on the oxidative ability of manganese, chromium, and nickel complexes of SALEN [SALEN = N,N'-ethylene-bis-salicylidene-aminato] in homo-geneous systems [30]. This ligand system is attractive from an encapsulation point of view, since it forms metal complexes of dimensions 10-11 Å, that can be trapped in the zeolite Y supercages. Our focus was on Mn^{III}SALEN-Y and its catalytic activity towards various olefins [31]. Iodosylbenzene (PhIO) was chosen to be the terminal oxidant:

$$\text{olefin} + \text{PhIO} \xrightarrow[\text{zeolite catalyst}]{} \text{products} + \text{PhI}$$

Table I shows the reactivity for a range of olefins that we have examined. The olefins are listed in order of increasing size: cyclohexene, 6.7 x 7.4 Å; styrene, 6.7 x 9.5 Å; *trans*-stilbene, 6.7 x 13.9 Å; and *cis*-stilbene, 8.0 x 10.3 Å.

In the case of cyclohexene, the amount of olefin that reacted is comparable to the PhI formed. The products were cyclohexene oxide and cyclohexenol, as identified by GC.

Table I. Olefin oxidation by MnSALEN-Y.

Olefin	olefin reacted (mmol)	PhI formed* (mmol)	Ratio (olefin/PhI)	Product mmol (%)
Cyclohexene	0.11	0.11	1	0.015 (14) cyclohexene oxide; 0.011 (10) cyclohexenol
Styrene	0.057	0.092	0.6	0.011 (19) styrene oxide; 0.015(26) $PhCH_2CHO$
Trans-stilbene	0.048	0.090	0.5	0.01(20) *trans*-stilbene oxide
Cis-stilbene	0.019	0.085	0.02	*cis*-stilbene oxide

* 0.11 mole of PhIO used.

For styrene and the stilbenes, the PhIO consumed is considerably greater than the olefin reacted. We have proposed that the solvent molecules (acetonitrile) are undergoing oxidation [31]. The competition with solvent oxidation increases as the olefins get larger and the diffusivity within the zeolite system becomes more difficult. The lower diffusivities of the larger olefins, however, leads to a reversal of the relative reactivities, as compared to homogeneous solutions, thus exhibiting size selectivity. With the zeolitic catalyst, cyclohexene is twice as reactive as styrene and three times as reactive as *trans*-stilbene. In the homogeneous system, the opposite trend, with *trans*-stilbene being one and a half times more reactive than cyclohexene, was found [30]. In the homogeneous system, too, oxidation of the solvent is a problem but it is not as severe as with the zeolite. This is because in homogeneous solution the reaction is complete within 30 minutes, but in the zeolite it takes over 16 hours. The MnSALEN-zeolite Y complex exhibited catalytic activity for over 60 hours; the long term stability presumably arising due to encapsulation.

What this study showed is that catalysts from homogeneous systems can indeed be transposed into the cages of zeolites and still promote the olefin oxidation reactions observed in solution. The zeolitic system also shows reversal of reactivity as compared to homogeneous solution because of restrictions in diffusivity. However, the problem is that the lower diffusivities lead to considerably lower rates of reactions. This may not have been a problem in itself, but becomes so because the catalyst is active enough to oxidize the solvent. Since the solvent molecules are considerably smaller than the substrates and are present at much higher concentrations, this competition detracts from the overall usefulness of the system.

3.2 DIOXYGEN BINDING

It is clear from the system discussed above that encapsulation in zeolite cavities allows for the isolation of metal complexes. This strategy has been used in the literature to stabilize complexes which in solution will dimerize and thereby deactivate [32-34]. Excellent examples are the O_2 dioxygen complexes of transition metals. It is well known that many low-spin Co^{II}-ligand complexes will bind O_2, but then rapidly dimerize to form peroxo complexes. By building the complex in a zeolite cage with a sufficiently large ligand, only a monomeric form can be accommodated. Herron has explored this with a CoSALEN complex using pyridine as the axial base [32]. Lunsford et al. have synthesized bipyridine and terpyridine complexes of Ru that also bind dioxygen [33].

We chose to synthesize dioxygen complexes using tetraethylenepentamine (tetren) as ligand (shown below) [35]. There were two advantages we saw with this ligand. Firstly, it is a pentadentate ligand, thus eliminating the need for an axial base. Secondly, it is soluble in water, thus allowing for ready transport into the zeolite pores. We examined both zeolite Y and A as supports, the cage diameters being ~13 and ~11.4 Å, respectively.

$NH(CH_2CH_2NHCH_2CH_2NH_2)_2$

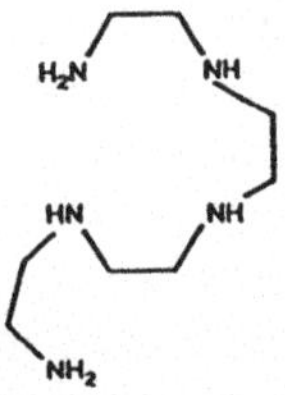

Resonance Raman spectra of the Co(tetren)-dioxygen complex in the cages of zeolites Y and A are shown in Figure 10 [35]. In the case of zeolite Y, the bands at 498, 643, and 808 cm^{-1} are readily assigned to Co-N, Co- O, and the O-O stretch of the peroxo complex [36]. In the case of zeolite A, new bands are observed at 580, 1034, and 1138 cm^{-1}. The bands at 580 and 1138 cm^{-1} are characteristic of the Co-O and O-O stretching modes of superoxo complexes [37].

Thus, the chemistry can be summarized as:

Zeolite Y: 2Co(tetren) + O_2 $\rightarrow$ (tetren) Co-O-O-Co(tetren)

Zeolite A: Co(tetren) + O_2 $\rightarrow$ (tetren) Co-O_2.

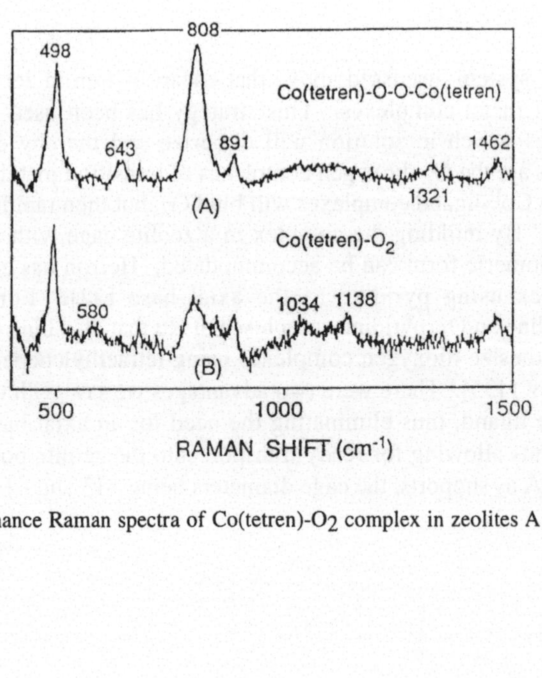

Figure 10. Resonance Raman spectra of Co(tetren)-O_2 complex in zeolites A and Y (adapted from [35]).

There is some peroxo complex also in the zeolite A sample. Modeling of the size of the Co(tetren) shows it to be between 7-8 Å. This means that the complex can exist as a dimer in zeolite Y supercages (~13 Å) but can exist only as a monomer in a zeolite A α-cage (~11.4 Å). The observation of a peroxo complex with the zeolite A sample probably arises from dimers formed on the outside surface of the zeolite crystallites, an observation not atypical for synthesis in aqueous solution [37].

EPR spectra of the complexes formed in zeolites Y and A support the Raman characterization and are shown in Figure 11. The complex in zeolite Y exhibits a very weak signal, probably arising from iron impurities in the zeolite. The μ-peroxo dimer is not expected to produce an EPR signal since the electrons are paired on

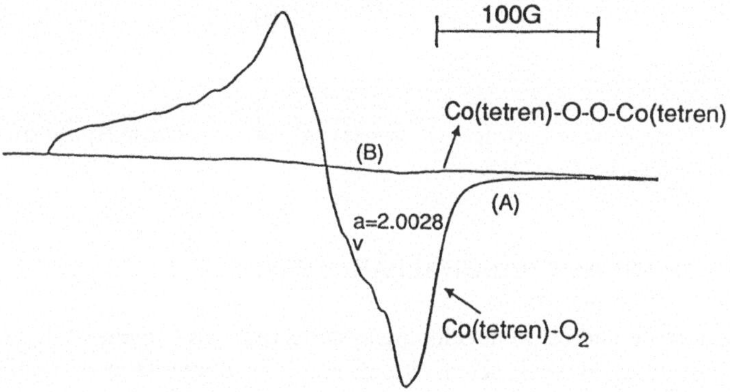

Figure 11. EPR spectra of Co(tetren)-O_2 complex in (A) zeolite A and (B) zeolite Y (adapted from [35]).

the Co-O-O-Co unit. The EPR spectrum from the complex in zeolite A is typical of a Co^{III}-O_2^- superoxo complex, with the hyperfine splitting expected due to the Co(I=$^7/_2$) nucleus [38].

It is possible to remove the O_2 from the Co(tetren)-O_2 complex in zeolite A by exposing it to N_2. The complex can be re-formed after re-exposure to O_2. This cycle is captured by the EPR spectra shown in Figure 12. The binding constant was estimated to be 2.2 x 10^3 M^{-1}, indicating that Co(tetren) binds O_2 very strongly.

Thus, this study showed that zeolite encapsulation leads to the stabilization of complexes not easily obtained in aqueous solution. Since O_2 is a much smaller reactant than the previously discussed olefins, its diffusivity into the zeolite was not a problem. The drawback in this case (from an application point of view for separating O_2 from N_2) was that, the binding of O_2 to Co(tetren) in zeolite A was very tight and reversed only with difficulty. This notwithstanding, it is clear that the close fit between interesting molecular systems and the zeolite cages allows for assembly of novel chemical systems.

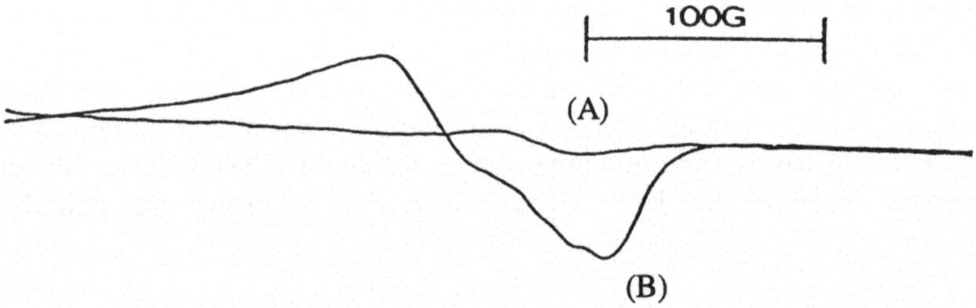

Figure 12. EPR spectra of Co(tetren)-O_2 complex (A) upon exposure to N_2 and (B) re-exposure to O_2 (adapted from [35]).

3.3 PHOTOCHEMICAL CHARGE SEPARATION

There is considerable interest in developing systems that can convert sunlight to chemical energy. In particular, if such schemes can result in conversion of readily available chemicals to fuels, it makes this chemistry particularly important [39].

$$2H_2O \rightarrow 2H_2 + O_2 \qquad \Delta G = 68.3 \text{ kcal/mole}$$

$$2H_2O + CO_2 \rightarrow CH_4 + 2O_2 \qquad \Delta G = 195.5 \text{ kcal/mole}$$

In both these examples, the uphill free energy can, in principle, be provided by solar energy. The processes also involve transfer of electrons, i.e., they are redox reactions. Attempts to use light energy for these reactions can exploit photochemical electron-transfer [40], as in:

$$D \xrightarrow{h\nu} D^*$$

$$D^* + A \rightarrow D^+ + A^-$$

$$4D^+ + 4OH^- \rightarrow O_2 + 4D + 2H_2O$$

$$2A^- + 2H^+ \rightarrow H_2 + 2A$$

where D is the molecular species that absorbs the light and has a long-lived excited state (D^*). In this excited form, it can donate an electron to an acceptor molecule, A. If D and A are chosen appropriately, then D^+ and A^- can bring about redox chemistry, such as shown above, in splitting of H_2O into H_2 and O_2. Since D and A are regenerated, the process can be made cyclic. However, the difficulty which needs to be circumvented in order for this strategy to work is to separate D^+ and A^- after they are formed, since there is a strong driving force for the back-electron-transfer reaction. The best example of such a system is found in photosynthesis, in which a membrane is used as a host for a complex arrangement of acceptors and donors, designed optimally for charge separation [41].

We have investigated the possibility of using the cages of zeolites as a partitioning medium for entrapping D and A in order to enhance photochemical charge separation. Our choice of D and A were $Ru(bpy)_3^{2+}$ and bipyridinium cations (viologens) for several reasons.

$$Ru(bpy)_3^{2+} \rightarrow Ru(bpy)_3^{2+*}$$

$$Ru(bpy)_3^{2+*} + MV^{2+} \rightarrow Ru(bpy)_3^{3+} + MV^{+\bullet}$$

Firstly, this system has been extensively examined in aqueous solutions [42]. Secondly, the D^+ and A^- forms have the requirements to convert H_2O to O_2 and H_2, respectively [43]. Thirdly, these molecules satisfy the architectural requirements for assembly within the zeolitic system, as explained below.

Figure 13a shows the results of photolysis with visible light (420-650 nm) of an aqueous solution of 0.1 M $Ru(bpy)_3^{2+}$ and 0.01 M methylviologen (MV^{2+} = $CH_3N^+C_5H_4\ C_5H_4N^+CH_3$) as measured by the electronic spectra. The band at 450 nm due to the MLCT band of $Ru(bpy)_3^{2+}$ remains unperturbed and no evidence of any electron transfer is detected. Figure 13b shows the same system with triethanolamine added. Within seconds of photolysis, bands at 390 and 610 nm due to the radical cation of methylviologen ($MV^{+\bullet}$) are observed. The chemistry is as follows:

$$Ru(bpy)_3^{2+} \xrightarrow[h\nu]{450nm} Ru(bpy)_3^{2+*}$$

$$Ru(bpy)_3^{2+*} + MV^{2+} \rightleftarrows Ru(bpy)_3^{3+} + MV^{+\bullet}$$

$$Ru(bpy)_3^{3+} + N(CH_2OH)_3 \rightarrow Ru(bpy)_3^{2+} + \text{degradation products}$$

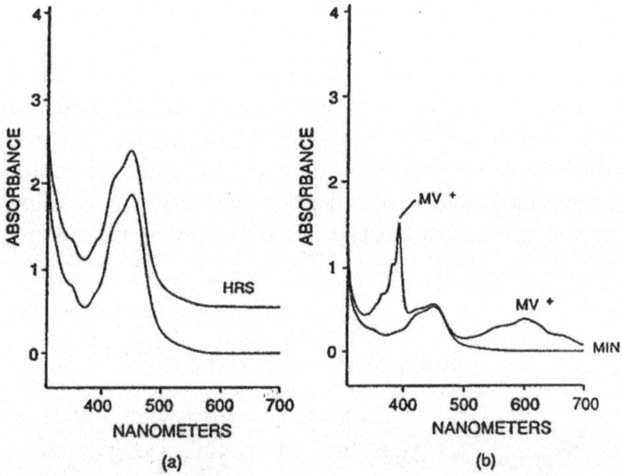

Figure 13. Visible light photolysis of $Ru(bpy)_3^{2+}$ and MV^{2+} in (a) an aqueous solution and b) in the presence of triethanolamine.

In the presence of the sacrificial electron donor, triethanolamine, a fraction of the $Ru(bpy)_3^{3+}$ reacts with it before the back-electron-transfer reaction can occur, thus stabilizing the $MV^{+\cdot}$ [44]. The absence of any $MV^{+\cdot}$ in Figure 13a is due to the rapid back-electron-transfer reaction.

The zeolite offers a novel way to arrange the $Ru(bpy)_3^{2+}$ and MV^{2+}. $Ru(bpy)_3^{2+}$ is a molecule of ~13 Å dimension and, if synthesized inside a zeolite supercage, is entrapped since it cannot escape through the ~7 Å openings [45-47]. However, it can still communicate with molecules in neighboring cages through the 7 Å windows. This complex can readily be synthesized in situ within the zeolite supercage by heating bipyridine with $Ru(NH_3)_6^{3+}$-exchanged zeolite Y, as shown below.

$$\text{Na-Y} + Ru(NH_3)_6^{3+} \xrightarrow{\text{ION-EXCHANGE}} \text{Na},Ru(NH_3)_6^{3+}\text{-Y}$$

$$\text{Na},Ru(NH_3)_6^{3+}\text{-Y} + 3\,\text{bpy} \xrightarrow{\text{HEAT}} \text{Na},Ru(bpy)_3^{2+}\text{-Y}$$

95% YIELD

$$Na,Ru(bpy)_3^{2+}\text{-}Y \xrightarrow[\text{EXCHANGE}]{\text{SOXHLET NaCl}} \text{PURIFIED SAMPLE}$$

The initial loading level of $Ru(NH_3)_6^{3+}$ determines how closely the Ru units are packed. This can vary from very dilute loadings, i.e., 1 Ru per 1000 supercages, to a maximum of 1 Ru per cage [48]. In the studies discussed in this paper, the packing is about 1 $Ru(bpy)_3^{2+}$ per 10-15 supercages.

The MV^{2+} cations can readily be ion-exchanged into the Na-$Ru(bpy)_3^{2+}$-Y from aqueous solution to a maximum loading of 2 MV^{2+} per supercage [49]. The dimensions of this molecule are ~13 x 6 Å, so it travels readily through the ~7 Å openings [50]. However, MV^{2+} is not likely to be in the supercages occupied by $Ru(bpy)_3^{2+}$, since this complex completely fills the supercage. Thus, a $Ru(bpy)_3^{2+}$,MV^{2+} - zeolite sample has the $Ru(bpy)_3^{2+}$ partitioned in a supercage and surrounded by MV^{2+} in the neighboring cages separated by the ~7 Å windows.

In order to establish that electron-transfer can occur upon photoexcitation in this architecture, we examined the time-resolved resonance Raman spectrum of the sample [51]. The 20 ns, 366 nm photon excites the $Ru(bpy)_3^{2+}$ and also serves as the excitation for the Raman spectrum. As Figure 14 shows, within the 20 ns pulse, we observe the spectrum characteristic of the $MV^{+\bullet}$ radical with bands at 1360, 1535, and 1661 cm^{-1} [52]. This is evidence for oxidative quenching through the 7 Å windows.

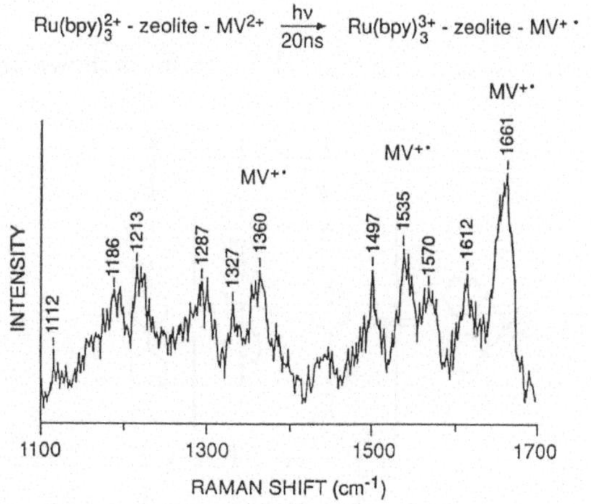

Figure 14. Time-resolved resonance Raman spectrum of $Ru(bpy)_3^{2+}$,MV^{2+}-Y (adapted from [51]).

Time-resolved diffuse reflectance spectra show that the $MV^{+\bullet}$ radical-cation formed, decays back with a half-life of 10 μsec and follows second-order kinetics as expected for the bimolecular back-electron-transfer reaction [51].

However, continuous photolysis (450-620 nm) showed that the radical cation ($MV^{+\bullet}$) can be stabilized, even though its rate of formation is very slow. Figure 15 shows the diffuse reflectance spectra as a function of photolysis time, and the bands due to $MV^{+\bullet}$ at 390 and 610 nm are found to grow. Upon turning off the light, the $MV^{+\bullet}$ decays back in tens of minutes [51].

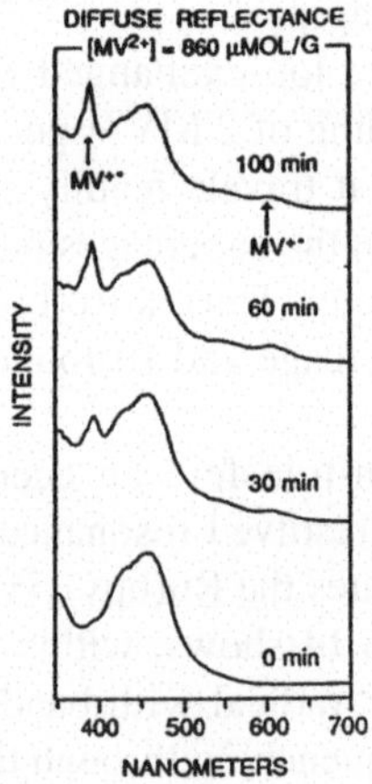

Figure 15. Diffuse reflectance spectra of $Ru(bpy)_3^{2+}$,MV^{2+}-Y as a function of photolysis time (420-650 nm).

A mechanistic scheme consistent with these observations is shown below.

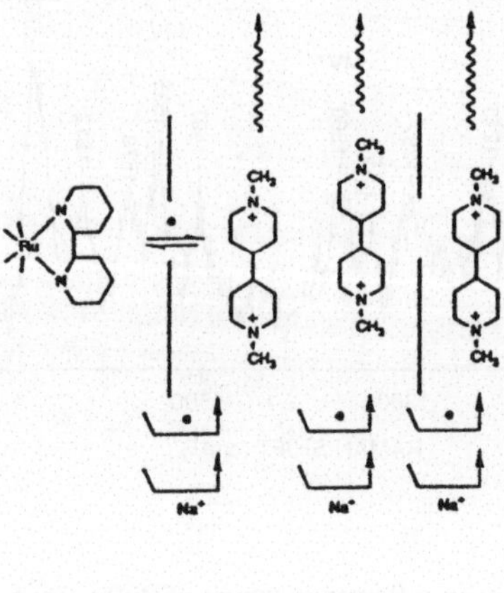

It is emphasized that upon oxidative quenching, the back-electron-transfer is the dominant reaction. However, there is another low efficiency pathway, in which the optimal packing of the viologen molecules leads to electron-hopping and charge separation. This process is facilitated in the zeolite by the migration of co-cations (i.e., Na^+) along with electron movement. Once the charge is separated in this fashion, the back-electron transfer is considerably slowed down.

The problem in this design is that most of the time the oxidative quenching is followed by the back-electron-transfer reaction. Clearly, a more efficient route is required. In order to do so, we have focused our attention on a ternary system, thus providing a pathway for a subsequent electron transfer step after the primary oxidative quenching [53]. This strategy is illustrated below schematically in Figure 16.

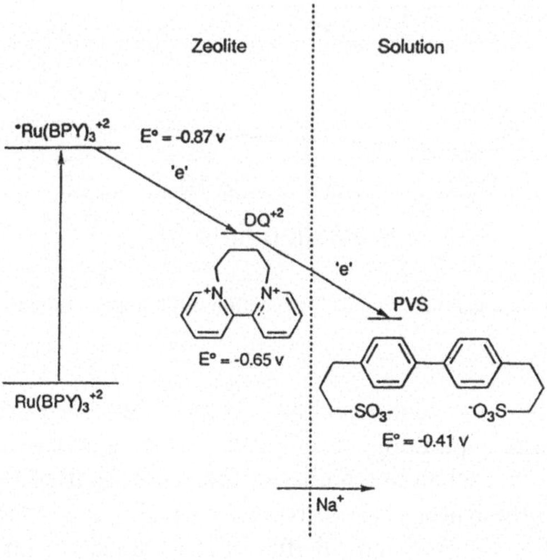

Figure 16. A strategy for efficient photochemical charge-transfer using zeolitic architecture.

In order to provide a directional push for electron-transfer, we chose two viologens with different reduction potentials. The viologen with the higher redox potential (DQ^{2+}, E_o = -0.65 V) [54] was ion-exchanged into the zeolite and surrounded the entrapped $Ru(bpy)_3^{2+}$. A neutral viologen (PVS) with a lower reduction potential (E_o = 0.41 V) [54] was introduced into the external aqueous medium. By choosing it to be a neutral species, there was no possibility of the PVS replacing the DQ^{2+} in the zeolite. The extent of photochemical charge separation was readily analyzed by monitoring the PVS radical formed in solution. Figure 17 compares the yield of radical as a function of time for the Na,$Ru(bpy)_3^{2+}$-Y and $Ru(bpy)_3^{2+}$,DQ^{2+}-Y with PVS in solution. There is a marked increase in the yield of the radical in the ternary system. If the intrazeolitic DQ^{2+} is replaced with MV^{2+}, whose reduction potential (-0.44 mV) is comparable to that of PVS, then the yield of radical drops again to that of the binary system, indicating that the strategy of vectorial electron-transfer is necessary.

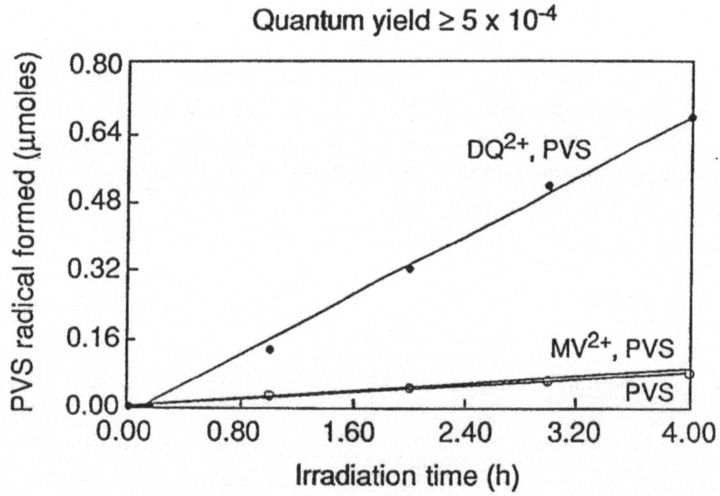

Figure 17. Yield of the PVS radical in solution as a function of photolysis time in the ternary system.

There are several aspects of the zeolite architecture that have been exploited in order to achieve this efficient charge separation. Firstly, the supercages isolate the $Ru(bpy)_3^{2+}$ molecule. Secondly, the ion-exchanging ability of the zeolite is used to hold the charged viologens in place around the $Ru(bpy)_3^{2+}$ sensitizer. Thirdly, the negative charge of the framework helps in repelling the PVS radical after interfacial electron-transfer, promoting charge separation. Fourthly, along with the injection of the electron into PVS in solution, a charge neutralizing Na^+ is also injected into solution, thus maintaining neutrality and preventing an opposing junction potential.

In summary, the purpose of this review, based primarily on the work done by our group has been to show that the close fit in size between zeolite cages and molecules can be exploited to derive information about zeolite assembly, as well as, assembling superstructures that allow for novel chemical and photochemical applications.

Acknowledgments

I would like to thank all my students and colleagues, whose names appear in the manuscripts. In addition, thanks are also due to the National Science Foundation and Department of Energy for financial support.

References

1. D. W. Breck, *Zeolite Molecular Sieves*, Krieger Publishing, Malabar, Florida (1984).
2. J. M. Newsam, *Science*, **231**, 1093 (1986).
3. J. W. Ward, *Applied Industrial Catalysis*, **3**, 271 (1984).
4. S. Csicsery, *Zeolites*, **4**, 202 (1984).
5. B. M. Lok, T.R. Cannan, and C.A. Messina, *Zeolites*, **3**, 282 (1983);
   R.M. Barrer, *Hydrothermal Chemistry of Zeolites*, Academic Press, London (1982).
6. M. E. Davis and R. F. Lobo, *Chem. Mater.*, **4**, 756 (1992).
7. P. K. Dutta and D. C. Shieh, *J. Phys. Chem.*, **90**, 2331 (1986).
8. P. K. Dutta, D. C. Shieh, and M. Puri, *J. Phys. Chem.*, **91**, 2332 (1987).
9. P. K. Dutta, D. C. Shieh, and M. Puri, *Zeolites*, **8**, 306 (1988).
10. P. K. Dutta, M. Puri, and D.C. Shieh, *Mater. Res. Soc. Symp. Proc.*, **111**, 101 (1988).
11. P. K. Dutta, K. M. Rao, and J. Y. Park, *J. Phys. Chem.*, **95**, 6654 (1991).
12. A. V. McCormick and A. T. Bell, *Catal. Rev. Ser. Eng.*, **31**, 97 (1988).
13. P. Bodart, J. B. Nagy, Z. Gabelica, and E. G. Derouane, *J. Chim. Phys.*, **83**, 777 (1986).
14. C. D. Chang and A. T. Bell, *Catal. Lett.*, **8**, 305 (1991).
15. D. A. Long, *Raman Spectroscopy*, McGraw-Hill, NY (1977);
    D. Lin-Vien, N. B. Colthup, W. G. Fateley and J. G. Grasselli, *Infrared and Raman Characteristic Frequencies of Organic Compounds*, Academic Press, NY (1991).
16. H. Kacirek and H. Lechert, *ACS Symp. Ser.*, **40**, 244 (1977).
17. P. A. Jacobs and J. A. Martens, *Synthesis of High-Silica Aluminosilicate Zeolites*, Elsevier, Amsterdam (1987);
    S. Ueda, H. Murata, and M. Koizumi, *Am. Mineral.*, **65**, 1012 (1980);
    E. N. Givens, C. J. Plank, and E. J. Rosinski, *U. S . Patent* 4,052,472, (1972).
18. G. T. Kokotailo, D. H. Olson, S. L. Lawton, and W. M. Meier, *Nature (London)*, **272**, 437 (1978).
19. A. Araya and B. M. Lowe, *J. Chem. Res.*, **192**, (1985).
20. P. K. Dutta, B. DelBarco, and D. C. Shieh, *Chem. Phys. Lett.*, **127**, 200 (1986).
21. B. D. McNicol, G. T. Pott, K. R. Loos, and N. Mulder, *Adv. Chem. Ser.*, **121**, 152 (1973).
22. A. Zalkin, *Acta Crystallogr.*, **10**, 557 (1957).
23. P. K. Dutta and M. Puri, *J. Phys. Chem.*, **91**, 4329 (1987).
24. P. K. Dutta, K. M. Rao, and J. Y. Park, *Langmuir*, **8**, 722 (1992).
25. D. Garfinkel, *J. Am. Chem. Soc.*, **80,** 3827 (1958).
26. D. Hoebbel and W. Weiker, Z. *Anorg. Allg. Chem.*, **384**, 43 (1971).
27. E. J. J. Groenen, A. G. T. G. Korbeek, M. Mackay, and O. Sudmeijer, *Zeolites*, **6**, 403 (1986).

28. J. M. Thomas and K. I. Zamarev, *Perspectives in Catalysis*, Blackwell, Oxford (1992).
29. J. K. Kochi, *Organometallic Mechanisms and Catalysis*, Academic Press, NY (1978).
30. K. Srinivasan, P. Michaud and J. K. Kochi, *J. Amer. Chem. Soc.*, **108**, 2309 (1986); B. Muenier, E. Guilnet, M. DeCarvalho, and R. Poliblanc, *J. Am. Chem. Soc.*, **106**, 6668 (1984).
31. C. Bowers and P.K. Dutta, *J. Catal.*, **122**, 271 (1990).
32. N. Herron, *Inorg. Chem.*, **25**, 4714 (1986).
33. S. Imamura and J. H. Lunsford, *Langmuir*, **1**, 326 (1985).
34. G. Meyer, D. Worhle, M. Mohl, and G. Schulz-Ekloff, *Zeolites*, **4**, 30 (1984).
35. P. K. Dutta and C. Bowers, *Langmuir*, **7**, 937 (1991).
36. T. Shibahara and M. Mari, *Bull. Chem. Soc. Jpn.*, **51**, 1374 (1978).
37. K. Nakamoto, Y. Nanaka, T. Ishiguro, M. W. Urbana, M. Suzuki, M. Kozuka, Y. Nishida, and S. Kida, *J. Am. Chem. Soc.*, **104**, 3386 (1982).
38. R. J. Taylor, R. S. Drago, and J. E. George, *J. Am. Chem. Soc.*, **111**, 6610 (1989).
39. V. N. Parmev and K. I. Zamarev, in *Photocatalysis-Fundamentals and Applications*, N. Serpone and E. Pelizzetti (Eds.), John Wiley, N.Y., p 656 (1989).
40. K. Kalyansundaram, *Photochemistry in Microheterogeneous Systems*, Academic Press, NY (1987).
41. *Energy Conversion by Plants and Bacteria*, (Ed) Govindjee, Academic Press, NY (1982).
42. J. Olensted III and T. J. Meyer, *J. Phys. Chem.*, **91**, 1649 (1987);
K. Mandal and M. Z. Hoffman, *J. Phys. Chem.*, **88**, 5632 (1984);
M. Z. Hoffman, *J. Phys. Chem.*, **92**, 3458 (1988);
C. Chiorleoli, M. T. Indelli, M. A. R. Scandola, and F. Scandola, *J. Phys. Chem.*, **92**, 156 (1988).
43. K. Kalyanasundaram, M. Gratzel, and E. Pelizzeti, *Coord. Chem. Rev.*, **69**, 57 (1986);
J. M. Lehn, J. P. Sauvage, and R. Ziessel, *Nouv. J. Chim.*, **3**, 423 (1979);
P. K. Ghosh, B. S. Brunschwig, M. Chou, C. Creutz, and N. Sutin, *J. Am. Chem. Soc.*, **106**, 4772 (1984).
44. G. Neshvad and M. Z. Hoffman, *J. Phys. Chem.*, **93**, 2445 (1989).
45. W. H. Quyale and J. H. Lunsford, *Inorg. Chem.*, **21**, 97 (1982).
46. J. A. Incavo and P. K. Dutta, *J. Phys. Chem.*, **94**, 3075 (1990).
47. K. Maruszewski, D. P. Strommen, K. Handrich, and J. R. Kincaid, *Inorg. Chem.*, **30**, 4579 (1991).
48. W. Turbeville, D. S. Robins, and P. K. Dutta, *J. Phys. Chem.*, **96**, 5024 (1992).
49. E. S. Brigham, P. T. Snowden, Y. I. Kim, and T. E. Mallouk, *J. Phys. Chem.*, **97**, 8650 (1993).
50. L. A. Summers, *The Bipyridinium Herbicides*, Academic Press, NY (1980).
51. P. K. Dutta and W. Turbeville, *J. Phys. Chem.*, **96**, 9410 (1992).

52. M. Forster, R. B. Girling, and R. E. Hester, *J. Raman Spectrosc.*, **12**, 36 (1982).
53. M. Borja and P. K. Dutta, *Nature*, **362**, 43 (1993).
54. I. Willner, A. Ayalon, and M. Rabinovitz, *Nouv. J. Chem.*, **14**, 685 (1990).

Abstract. This review primarily focuses on contributions of our research group in the area of zeolite guest-host chemistry. The perturbation of the Raman spectra of the organic structure-directing-agents during synthesis of zeolites ZK-4, ZSM-5, and ferrierite show clearly the entrapment of the organic by the aluminosilicate framework. The microordering around the organic coincides with the development of long-range-order typical of crystal formation. Using three examples involving zeolite entrapped organometallic complexes of manganese(III), cobalt(II), and ruthenium(II), we have demonstrated a wide range of chemistries using reactants varying from olefins to dioxygen to photogenerated electrons.

Journal of Inclusion Phenomena and Molecular Recognition in Chemistry **21**: 239–282, 1995.

Photochemistry of Organic Molecules Within Zeolites: Role of Cations

V. RAMAMURTHY and NICHOLAS J. TURRO

1. Introduction

Photochemistry in organized media has attracted considerable attention during the last decade [1]. A number of materials have been utilized to "hold" molecules in a fixed orientation so as to initiate reactions. Often, such media do not remain "passive" and they themselves interact with the guest molecules and influence their photobehavior. When this occurs the material is considered to be "active" [2]. In this chapter, we are concerned with zeolites, which depending on the guest and the reaction remain either active or passive. Zeolites have been known for over two centuries and have been actively investigated by chemists with interest in their catalytic behavior for over three decades. However, photochemists' interest in such systems is only recent and can be traced to a few classic papers published in the 1980's [3]. In recent years, increasing interest in utilizing zeolites to influence the photobehavior of molecules has become evident. This chapter provides a brief summary of the work carried out in our laboratories in the area of photochemistry and photophysics of organic molecules included within X and Y zeolites. Work carried out in other zeolites is not part of this review. Although our work has been placed in a proper context with reference to the publications of others, no attempt has been made in this article to provide an exhaustive, in depth, and critical coverage of the literature [4].

The two synthetic forms of faujasite are referred to as zeolite X and Y and have the following, typical unit cell compositions [5-7]:

X type $M_{86}(AlO_2)_{86}(SiO_2)_{106}\cdot 264\ H_2O$

Y type $M_{56}(AlO_2)_{56}(SiO_2)_{136}\cdot 253\ H_2O$

where M is a monovalent cation.

Charge-compensating cations present in the internal structure of zeolites are known to occupy three different positions in zeolites X and Y. As illustrated in Figure 1, the first type (site I), with 16 per unit cell (both X and Y), is located on the hexagonal prism faces between the sodalite units. The second type (site II), with 32 in number per unit cell (both X and Y), is located in the open hexagonal faces. The third type (site III), with 38 per unit cell in the case of X type and only eight per unit cell in the case of Y type, is located on the walls of the larger supercage cavity (one unit cell consists of eight supercages). Only cations of sites II and III are expected to be readily accessible to the

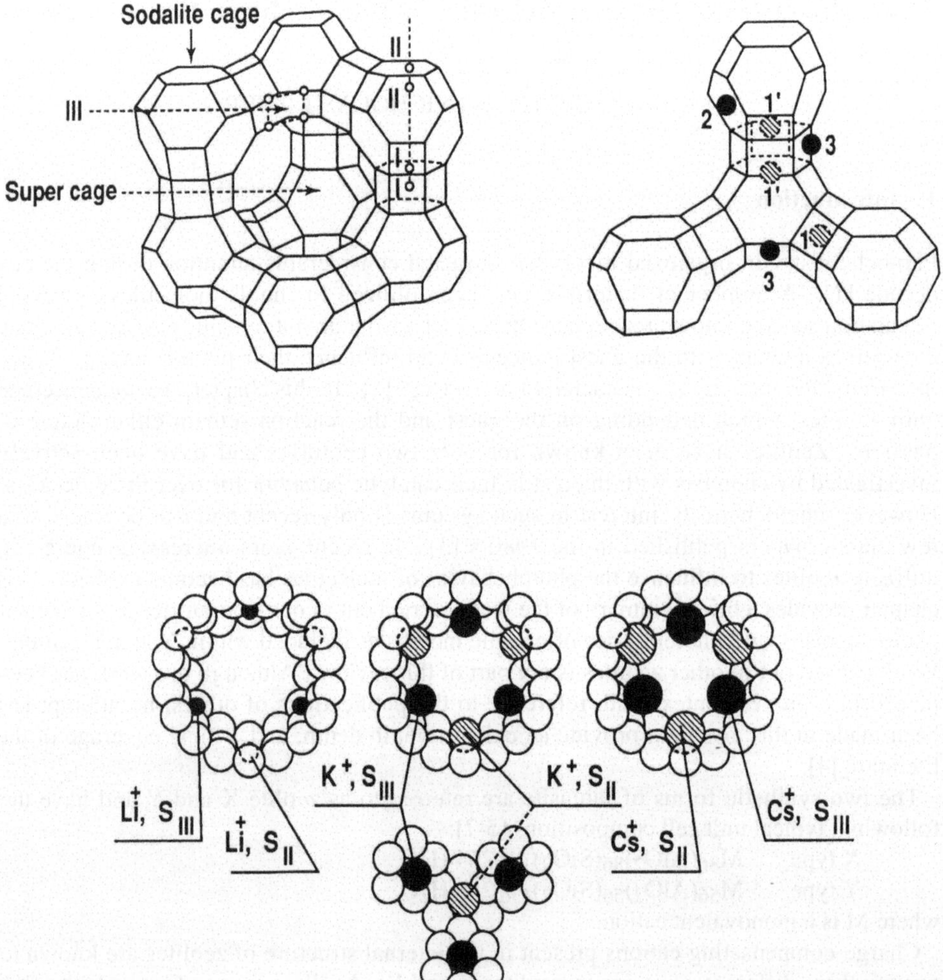

Figure 1. Supercage structure and cation location (I, II, III) within X and Y type zeolites. Bottom portion shows the reduction in available space (relative) within the supercage as the cation size increases.

absorbed organic. The free volume available for the organic within the supercage depends on the number (X vs. Y) and nature of the cations (M^+). Supercages are large as evident from their known capacity to include 28 molecules of water, 5.4 molecules of benzene, or 2.1 molecules of perfluorodimethylcyclohexane per supercage. The largest pore opening is a 12-ring with dimensions of about 7.4 Å. As the calculated supercage volumes given in Table 1 show, the free volume decreases as the cation size increases from Li to Cs [8]. The supercages form a three-dimensional network with each supercage connected tetrahedrally to four other supercages through the 12-membered ring openings.

Table 1. Cation dependence of supercage free volume in M-Y zeolites.

Cation	Ionic Radius of the	Vacant Space[b] within the Supercage (Å^3)	
(M^+)	Cation (Å)[a]	Y-Zeolite	X-Zeolite
Li	0.6	834	873
Na	0.95	827	852
K	1.33	807	800
Rb	1.48	796	770
Cs	1.69	781	732

a. R. J. Ward, *J. Catalysis,* **10**, 34 (1968).

b. Calculations of polyhedral volumes were performed using a modification of the POLYVOL Program [D. Swanson and R. C. Peterson, *The Canadian Mineralogist*, **18**(2), 153 (1980); D. K. Swanson and R. C. Peterson, *POLYVOL Program Documentation*, Virginia Polytechnic Institute, Blacksburg, VA] assuming the radius of the TO_2 unit to be 2.08 Å (equivalent to that of quartz).

2. Zeolite Cavities Viewed as Nanoscopic Reaction Vessels

Recently, a simple 'model' has been adopted to understand the influence of organized media on the photobehavior of guests included in them [2,9]. As per this model, a reaction cavity is considered "*passive*" and predictions concerning the behavior of the excited guest molecules can be made solely on the bases of cavity size, shape, and flexibility when there are no significant interactions between the guest molecule and a host at various points along its walls. The cavity is considered "*active*" and predictions concerning the excited-state properties require consideration of different (and, perhaps, time-variable) interactions of the guest within the cavity when significant interaction between the walls and the guest exists. Interactions may vary from weak van der Waals forces to strong hydrogen bonds or electrostatic forces between charged centers. Presence of cations and functional groups on the surfaces of zeolites, which can interact with guest molecules, prompts the zeolites to be considered to possess "active" reaction cavities. The silica-alumina surface constituting the walls of the zeolite matrix as well as cations can participate and influence the course of photochemical and photophysical events. Thus, the zeolite cavities possess characteristics which allow them to be classified as active cavities. However, depending on the guest molecules, the interaction between the guest organic molecules and the zeolite matrix may be so weak that one should view the zeolites as possessing "passive" reaction cavities. Examples are provided below wherein both active and passive characteristics of zeolite reaction cavities are highlighted.

3. Guest-Cation Interaction: Light-Atom Effect

Electronic interaction between the cation and the guest molecules stabilizes the guests at sites close to the cation. The strength of binding interaction between the cation and the guest molecule, as per ^{2}H NMR:phenanthrene in zeolite X of phenanthrene within M^+-X zeolites, is directly dependent on the charge density/electrostatic potential of the cation [10]. The higher the charge density (i.e., charge per unit volume of the cation), the stronger the binding (ΔH binding: Na^+, 14.9; K^+, 11.0; Cs^+, 7.9 kcal mole^{-1}). Photophysical properties of the guest molecules included within zeolites are also influenced by the cation; the nature of the influence depends on the cation. The phenomenon observed with smaller and lighter cations is termed the "light-atom effect" and is discussed in this section.

In aromatics (included within zeolites), the lowest energy symmetry forbidden electronic absorption band gains intensity with light cations [11]. Diffuse reflectance spectra recorded for pyrene, at a low loading level (Figure 2), within a number of cation exchanged X and Y zeolites suggest that the interaction between the cation and pyrene is strong and that the strength of interaction depends on the polarizability of the cation. In Figure 2, the intensity of the 0–0 transition of the S_0 to S_1 band is dependent on the charge density of the cation present within supercages. Further work is required to identify the cause of the above changes brought forth by cations. Tentatively, we believe that the symmetry reduction of the aromatic molecule occurs due to strong interaction between the cation and the aromatic along the π face resulting in slight bending of the aromatic plane.

Lifetimes of the excited singlet and triplet states of aromatics are influenced by the cations. A consistently shorter lifetime was measured in Li-X than in Na-X for the excited singlet state of a number of aromatic molecules (Table 2) [12]. It is quite likely, on the basis of what has recently been shown in the case of silica surfaces [13], that reduction in lifetime results from enhanced radiationless processes when the aromatic molecules interact strongly with the cations.

In the case of ketones, the "light-atom effect" manifests itself in altering the nature of the lowest triplet state [14]. There is a small difference in diffuse reflectance spectral maxima between Li-X– and Cs-X–valerophenone complexes. Excitation (250–300 nm) of valerophenone (VLP) included in zeolites at 77 K gives rise to phosphorescence in the region 375 to 550 nm. The emission and excitation spectra of VLP in Li-X and Cs-X are displayed in Figures 3 and 4. The triplet lifetimes for Na-, K-, and Rb-X are all

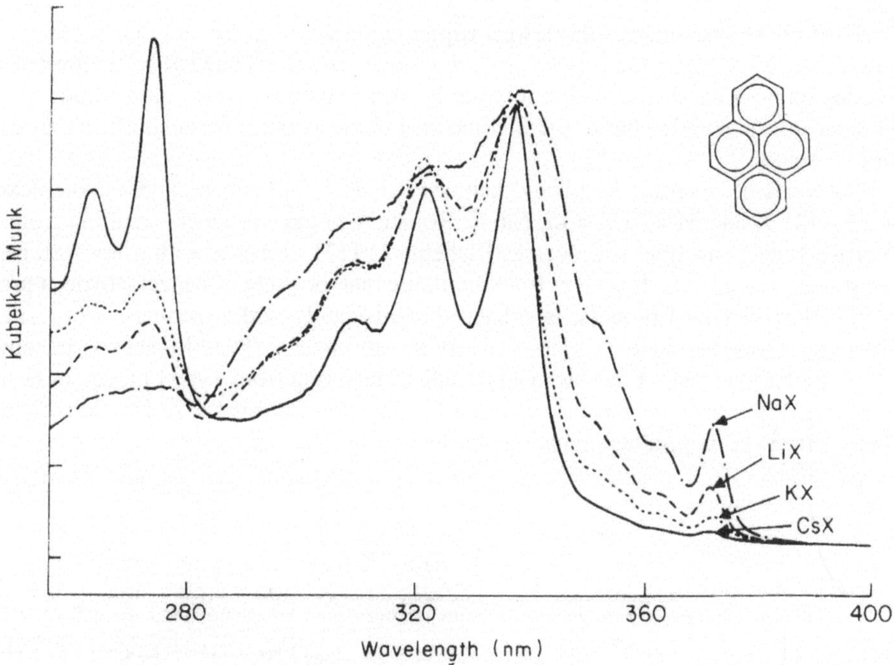

Figure 2. Diffuse reflectance spectra of pyrene included in cation-exchanged X zeolites. Note the intensity dependence of λ_{max} at ~370 nm on the cation.

Table 2. A comparison of the singlet lifetimes (ns) of aromatics in various cation-exchanged X zeolites

Zeolite	Naphthalene	Acenaphthene	Phenanthrene	Pyrene	Chrysene
Li-X	32.6	5.0	29.8	92.4	19.2
Na-X	49.1	11.6	36.6	115.4	27.1
K-X	19.4	13.1	27.8	161.4	39.2

between 0.4-1.2 μs indicating little effect of the cation, at least within the limitation of the rather large variations in triplet lifetime from one sample to the other. However, there are significant changes when the cation is changed to either Li or Cs. For the Li-X samples, the triplet decay is much longer (4–5 μs) and does not give satisfactory fits to a single exponential decay. It is quite likely that the long lifetime observed in Li-X is a consequence of the change in the character of the lowest triplet from nπ* to ππ*. Such a proposal is supported by the phosphorescence spectra. Examination of Figure 3 reveals that the emission observed in Li-X is distinctly different from that in Cs-X. While the emission in Li-X is characteristic of a ππ*, that in Cs-X is of a nπ* state [15]. Such a

reversal in the character of the lowest triplet is known to occur in polar solvents. The interior of Li-X being more polar [16] than other cation-exchanged X zeolites (see the section on polarity above) such a reversal is not unexpected. It should be borne in mind that part of the lengthening of the lifetime results from conformational effects, discussed briefly below.

Photolysis of several alkylphenyl ketones (Chart 1) in benzene and as complexes of zeolite M^+-X and M^+-Y (M = Li, Na, K, Rb, and Cs) gave products resulting from both Norrish type I and type II processes (Scheme 1) [17]. Results with a few ketones are illustrated in Figure 5. It is clear from the figure that the yield of benzaldehyde, a product of the Norrish type I process, is enhanced significantly within zeolites with respect to benzene. Cycloalkanones, which normally do not undergo Type I reactions in solution, upon irradiation in M^+-Y zeolites yield products resulting from Type I processes as major products (Figure 6). As illustrated with a few examples in Figure 7, the relative yield of Type I to Type II products depends on the cation; smaller cations such as Na have a larger influence.

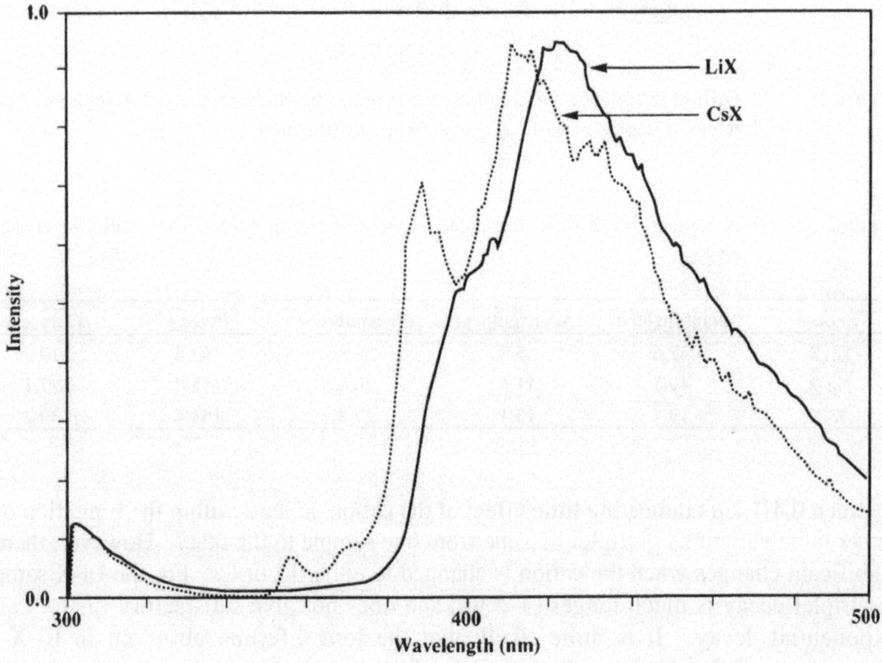

Figure 3. Phosphorescence emission spectra of valerophenone included in Li-X and Cs-X.

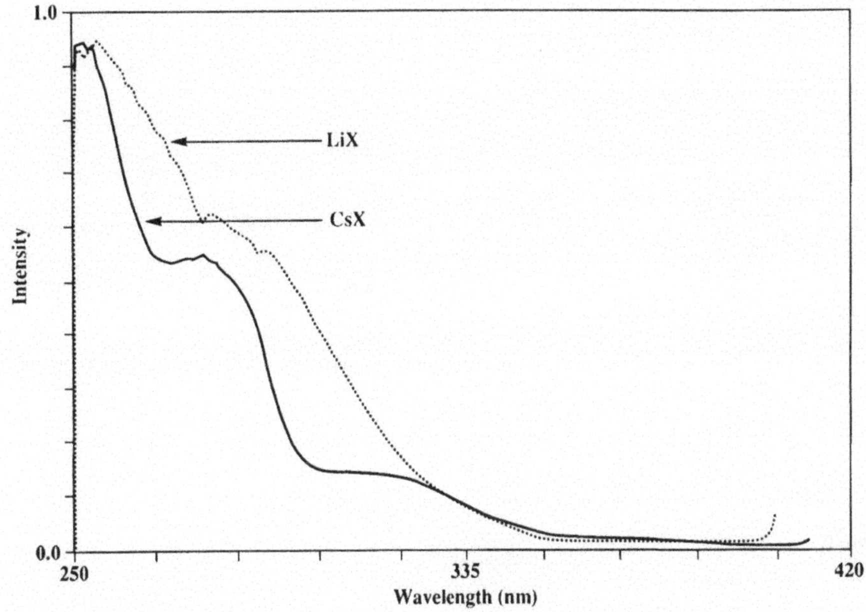

Figure 4. Phosphorescence excitation spectra of valerophenone included in Li-X and Cs-X.

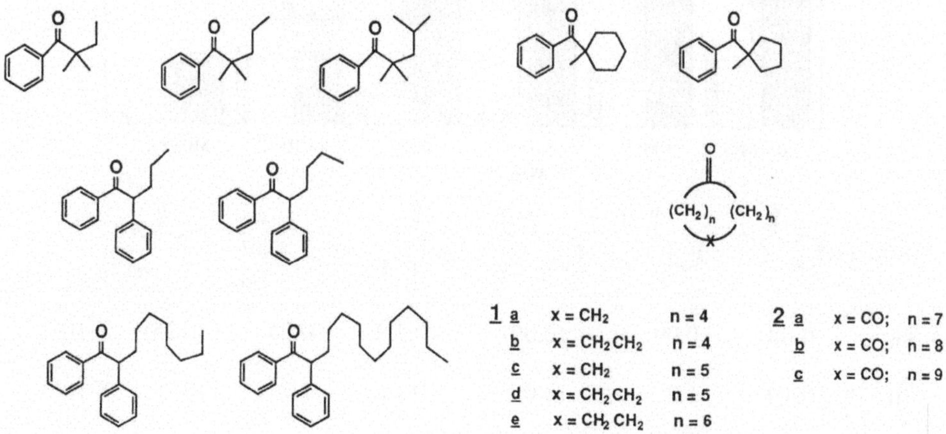

Chart 1.

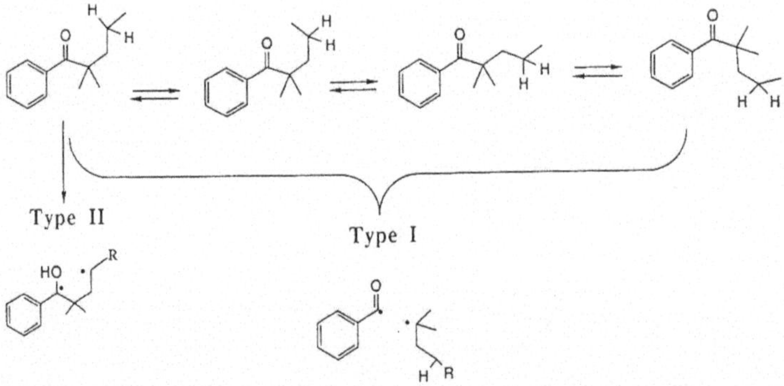

Scheme 1.

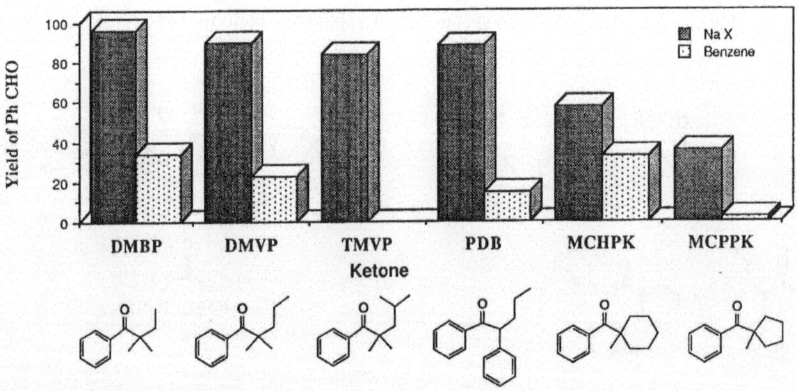

k_α	0.08	0.09	(0.09)	(0.21)	0.25	(0.25)	X 10^8
k_H	0.10	1.0	(5.0)	(1.0)	1.7	(0.25)	X 10^8

Figure 5. Type I, Type II reactivity controlled by cations within zeolite: studies with phenyl,alkyl ketones.

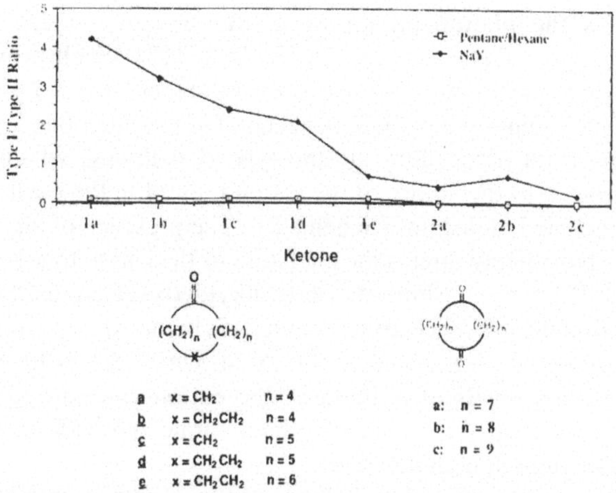

Figure 6. Type I, Type II reactivity controlled by cations within zeolite: studies with cyclic ketones.

It is proposed that the cations present within the supercage interact electronically with the carbonyl chromophore and thus, impede, sterically, the hydrogen abstraction (Scheme 1). Changes in preferred conformations are brought forth by cation binding. The effect is expected to be greater with Li^+ because it binds very strongly. Such a binding would affect the rate of hydrogen abstraction, but not the α-cleavage, and so would be expected to decrease the efficiency of the type II process alone. The mechanistic rationalization above is supported by triplet lifetime measurements for valerophenone and dimethylvalerophenone included within various cation-exchanged zeolites.

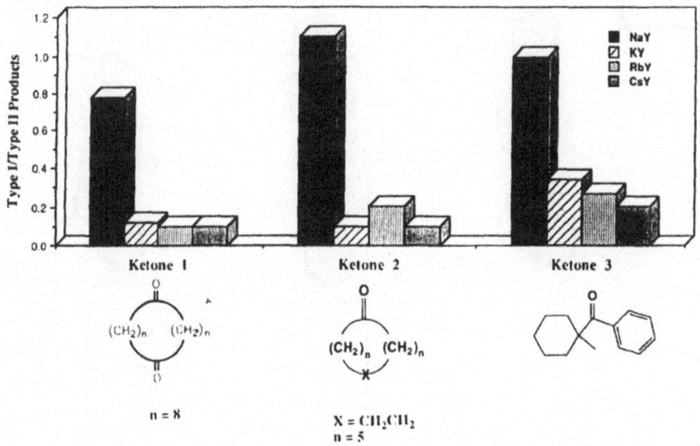

Figure 7. Dependence of type I to type II ratio on the cation. Note the higher influence by Na^+ with respect to other cations.

Another example is the photobehavior of α-alkylbenzoin ethers (X = O) and α-alkyldeoxybenzoins (X = CH_2; Scheme 2) (Table 3) [18] which provide important information. The zeolite cavity induces α-alkyl benzoin ethers to yield products derived via the type II pathway, a minor pathway in benzene. On the other hand, zeolites inhibit α-alkyldeoxybenzoins from proceeding via the type II pathway, which is favored in benzene. This is attributed to the ability of the cation present in the cavity to control the conformation of the included molecules (Scheme 2). The presence of an alkoxy chain in α-alkylbenzoin ethers most likely directs the chelation of the cation to a conformer that is favorable for the type II process. Similarly, in α-alkyldeoxybenzoins, the phenyl ring directs the conformational preference in the cavity (Scheme 3). Such a hypothesis is supported by the results on dealuminated zeolite-Y, in which the Si to Al ratio is very high (>550). At very low levels of aluminum, the cation concentration is also low. Therefore, conformational control is expected to be minimal and, indeed, type I products dominate the product mixture in both the cases.

Wada et al. have investigated the photobehavior of iron pentacarbonyl included in a variety of alkali-exchanged Y zeolites [19]. Photocatalytic isomerization of *cis*-but-2-enes was investigated. Although the details of the mechanism are yet to be understood, the activity decreased with the decrease in the electrostatic field of the cation: Li-Y, 27.3; Na-Y, 18.7; K-Y, 1.22; Rb-Y, 0.45; and Cs-Y, 0.71 x 10^{-6} mol min^{-1}.

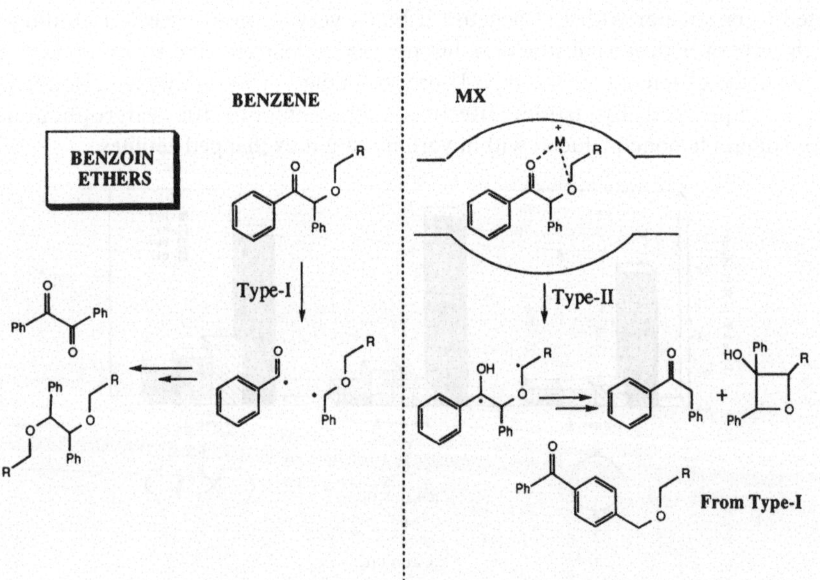

Scheme 2.

Table 3. Product distributions upon photolysis of benzoin methyl ether and α-propyldeoxybenzoin within zeolites[a].

Medium	Norrish Type I Products		Norrish Type II Products	
	Benzil/Pinacol ether	Rearrangement product	Deoxybenzoin	Cyclobutanol
(a) Benzoin methyl ether				
Benzene	26/67	1.0	1	7
Li-X	3	77	13	8
Na-X	4	72	10	14
K-X	7	48	14	18
Rb-X	5	46	18	22
Cs-X	8	34	17	31
(b) α-Propyldeoxybenzoin				
Benzene	5/24	--	54	17
Li-X	--	95	4	1
Na-X	--	88	5	7
K-X	--	48	31	21
Rb-X	--	32	22	45
Cs-X	--	21	27	42

a. Yields of products are based on GC analyses; see Schemes 2 and 3 for structure of products.

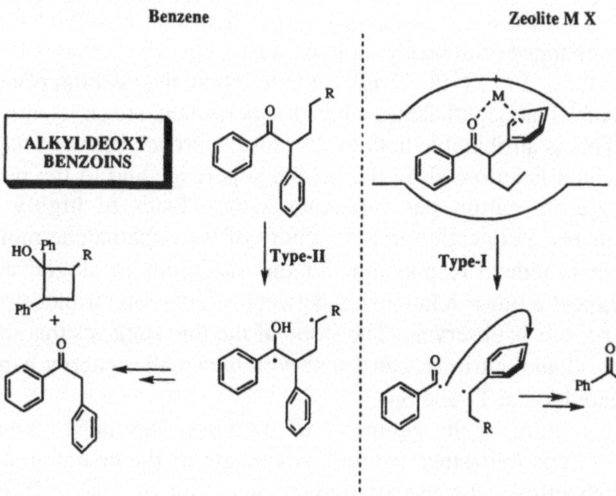

Scheme 3.

4. Guest-Cation Interaction: Heavy-Atom Effect

In the section above, we noted that light cations such as Li^+, Na^+, and K^+ interact strongly with guest molecules. A similar interaction, although more weakly, persists even with heavier cations such as Rb^+, Cs^+, and Tl^+. Weak interactions that exist between the cation and guest molecules lead to unique effects as illustrated in this section.

As shown in Figure 8, the emission spectrum of naphthalene is profoundly affected by inclusion in X and Y zeolites [20, 21, 22]. For low-mass cations such as Li^+ and Na^+, the emission spectra show the typical naphthalene blue fluorescence. However, as the mass of the cation increases (e.g., from Rb^+ to Cs^+ to Tl^+), there is a dramatic decrease in fluorescence intensity and a simultaneous appearance of a new vibronically structured low-energy emission band that is readily identified as naphthalene phosphorescence. Table 4 lists excited singlet (at 298 K) and triplet lifetimes (at 77 K) of naphthalene included within various cation-exchanged zeolites. It is clear that both these lifetimes are cation dependent. On the basis of the following observations, we conclude that the heavy-cation effect is responsible for the enhanced phosphorescence and decreased singlet and triplet lifetimes for naphthalene within K-, Rb-, Cs-, and Tl-cation-exchanged faujasites. It is well known that the effect of external heavy-atom perturbation scales with the square of the perturber's spin-orbit coupling constant, ξ^2 and that a log-log plot of τ_T^{-1} vs. ξ^2 should be linear with a predicted maximum slope of unity [23]. As shown in Figure 9, the expected dependence is observed. For comparison, we have also provided in Figure 9 the linear relationship observed in two systems, namely 1-halonaphthalenes [23] and 1,5-naphtho-22-crown-6 [24], where the external and internal heavy-atom effects, respectively, are presumed to operate. The magnitude of the heavy-atom effect observed in zeolites is significantly larger than that observed for the 1,5-naphtho-22-crown-6 exchanged with heavy-atom cations where the cation is rigidly held over the naphthalene π-face. In fact, the zeolite samples show heavy-atom effects nearly as large as for a series of 1-halonaphthalenes where the perturbers are covalently attached to the chromophore. This is attributable both to the close approach between naphthalene and the heavy-atom which is enforced by the zeolite supercage and to the presence of more than one heavy-atom cation per supercage which leads to highly effective concentrations of the heavy-atom cation in the vicinity of the naphthalene molecule. If the heavy-cation effect is indeed responsible for the variations in singlet and triplet lifetimes, one would expect a linear relationship between singlet- and triplet-decays with cation variation. Indeed, this is observed. The slope of the line suggests that the singlet-decay is enhanced more than the triplet, consistent with the smaller energy gap between the S_1 and T_1 as compared to the T_1 and S_0.

In order to obtain a picture of the geometry of the cation-aromatic (naphthalene) interaction in X- and Y-type faujasites, we took advantage of the heavy-atom induced phosphorescence which allows the use of optical detection of magnetic resonance (ODMR) in zero applied magnetic field [25]. The sublevel specific dynamics for absorbed naphthalene show a distinct increase in relative radiative character and total rate constant of the out-of-plane x-sublevel with increasing mass of the cation perturber. Enhancement from only triplet sublevel T_x is expected for a heavy-atom approach along the x-axis [26]. ODMR kinetic results suggests that the naphthalene is absorbed through its π-cloud at a cation site (Figure 10).

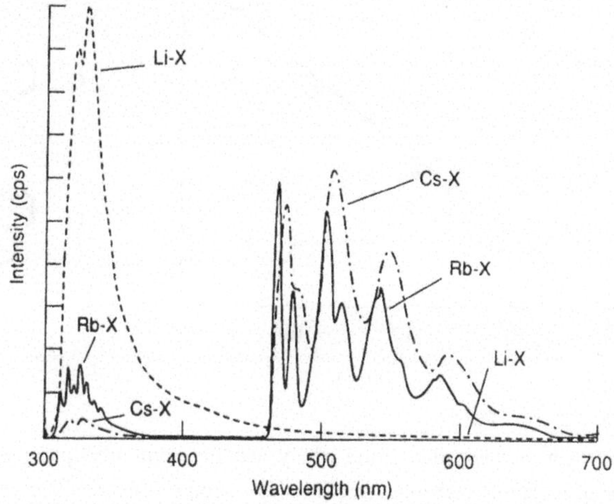

Figure 8. Emission spectra of naphthalene included in various cation–exchanged X zeolites. The ratio of fluorescence to phosphorescence depends on the cation.

Table 4. Photophysical parameters for naphthalene included within zeolites.

Zeolite Host	Triplet Lifetime[a] (s)	Singlet Lifetime[b] (ns)	P/F[c]
Li-X	-	33.0	1.0×10^{-4}
Na-X	-	35.4	7.3×10^{-2}
K-X	1.72	19.4	0.16
Rb-X	0.72	2.22	8.1
Cs-X	0.20	0.23 (87%), 1.87 (13%)	45
Tl-X	0.0012	—	only P
Li-Y	-	31.8	1.2×10^{-3}
Na-Y	-	25.1	1.0×10^{-3}
K-Y	-	13.8	0.1
Rb-Y	-	3.8	9.0
Cs-Y	-	0.7	60

a. The lifetime measured at 77 K.

b. The lifetime measured at 298 K.

c. Phosphorescence to fluorescence intensity ratio estimated at 77 K; the number is independent of the wavelength of excitation.

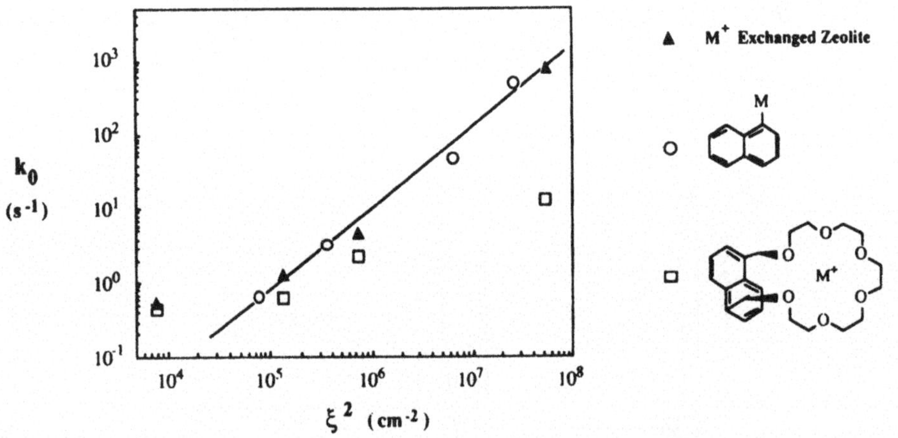

Figure 9. Correlation between the triplet decay and the spin–orbit parameter for naphthalene included in cation-exchanged zeolites. For comparison, results on other related systems are also shown.

The above heavy-cation effect within zeolites is found to be general. Enhanced phosphorescence is observed for a wide-range of different organic guests, such as anthracene, acenaphthylene, phenanthrene, chrysene, fluoranthene, pyrene, and 1,2,3,6,7,8-hexahydropyrene when included in Tl^+-exchanged faujasites (Figure 11). The only set of examples of guests for which phosphorescence is not observed are fused aromatics, which are too large in diameter to fit through the ~7.4 Å windows of the X- and Y-type zeolites (e.g., coronene and triphenylene). Also, this phenomenon is not restricted to faujasites alone. Enhanced phosphorescence has been observed for guest molecules within heavy-cation-exchanged zeolites ZSM-5, L, mordenite, omega, and beta.

It is easy to appreciate the potential of the unusual environment of the zeolite when one realizes that even olefins (Chart 2), systems that under normal conditions do not show phosphorescence, emit from their triplet states when included in Tl^+-exchanged zeolites.

Excitation of *trans*-stilbene included in Tl-X and in Tl-ZSM-5 emits phosphorescence and fluorescence both at room temperature and at 77 K [27]. The triplet emission spectra at 77 K for a number of substituted *trans*-stilbenes included in Tl-X are provided in Figure 12. The triplet emission maxima for stilbenes measured in the present study agree well with the literature reports [28]. The ability to record phosphorescence from stilbenes even at 298 K is significant as only very weak phosphorescence from *trans*-stilbene and several substituted *trans*-stilbenes has been recorded at 77 K in organic glass containing ethyl iodide as the heavy-atom perturber.

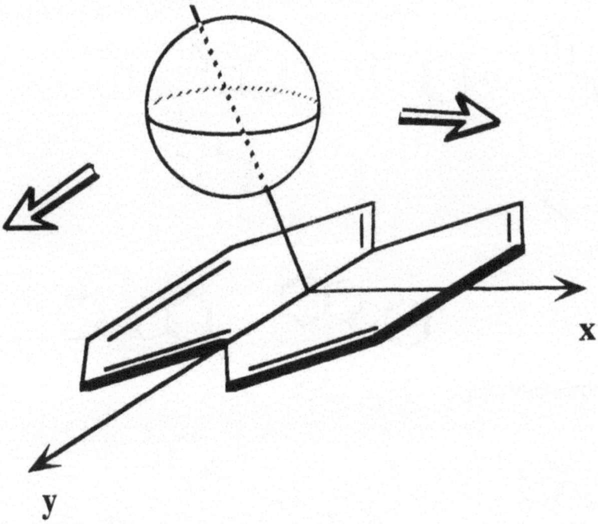

Figure 10. Geometry of cation-binding to naphthalene as inferred from ODMR experiments.

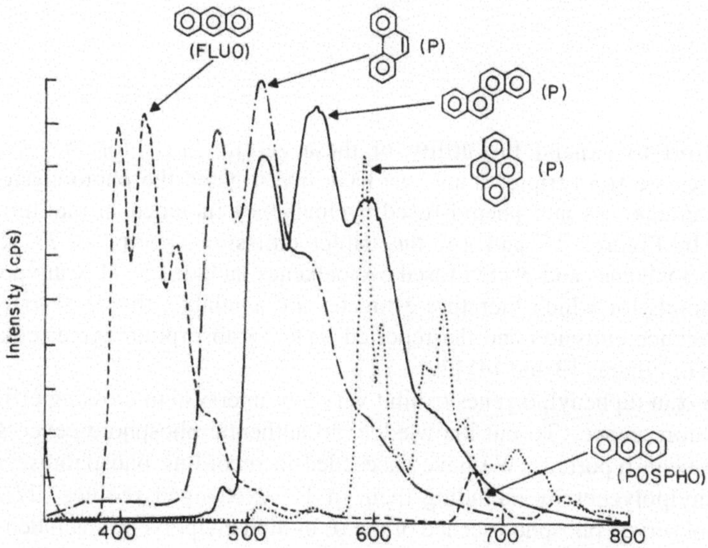

Figure 11. Emission spectra of aromatics included within Tl-X at RT. Emission consists mostly of phosphorescence.

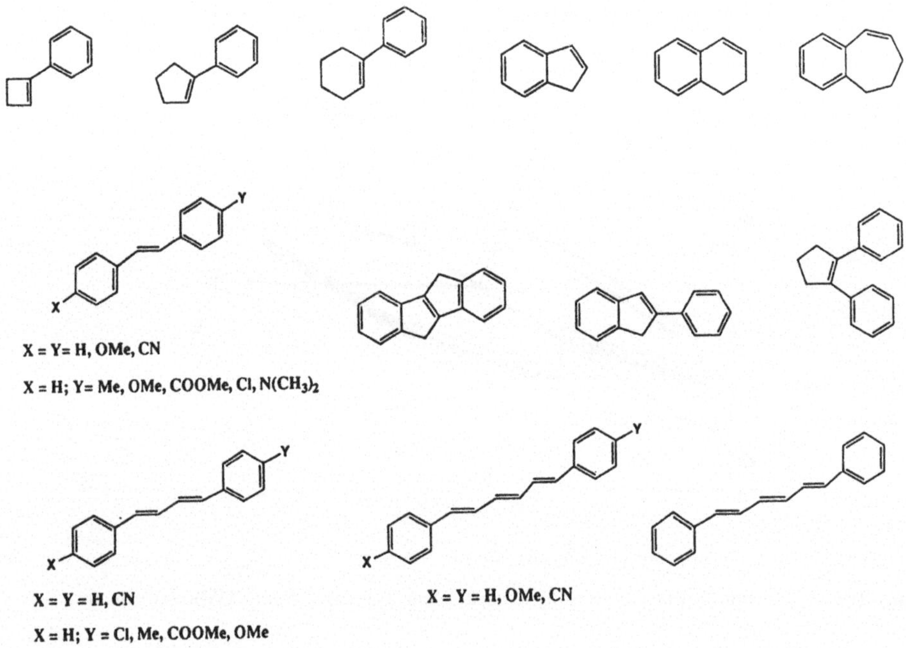

Chart 2.

In an effort to expand the utility of these zeolite hosts for the observation of phosphorescence from triplet states, we have investigated the photophysics of several 1-phenylcycloalkenes and phenyl-fused cycloalkenes included in thallium-exchanged zeolites. In Figures 13 and 14, the triplet emission spectra at 77 K of several 1-phenylcycloalkenes and phenyl fused cycloalkenes included in Tl-X are provided [27]. For compounds for which literature estimates are available, the 0—0 transition of the phosphorescence emission and the reported S_0 to T_1 absorption agree remarkably well (see inserts in Figures 13 and 14) [29].

All-*trans* α,ω-diphenylpolyenes exhibit very low intersystem crossing efficiencies and efficient fluorescence. To our knowledge, no authentic phosphorescence spectra from these have been reported. We have succeeded in recording phosphorescence of these α,ω-diphenylpolyenes by including them in Tl^+-exchanged zeolites [27]. Figure 15 shows the observed phosphorescence of the α, ω-diphenylpolyenes included in Tl-X. At 77 K, a well-resolved structured emission for each of the polyenes with a prominent vibronic spacing of 1200-1400 cm^{-1} (as expected for triplet phosphorescence) is observed. The singlet-triplet energy gaps ($\Delta_{T_1 \rightarrow S_0}$) obtained from the observed zero-zero lines are in excellent agreement with literature predictions [30].

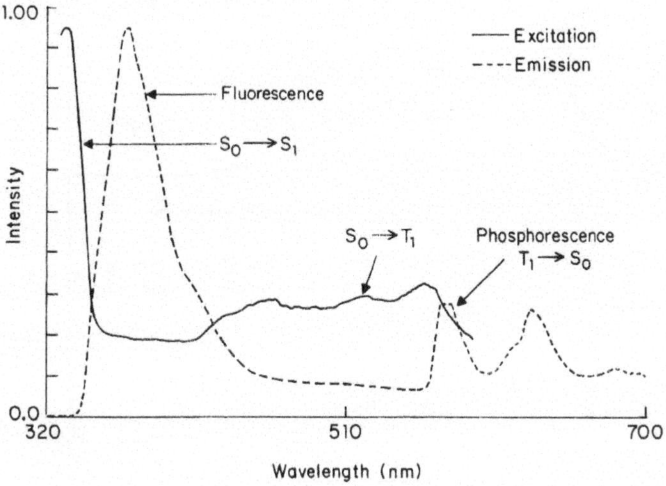

Figure 12. Emission and excitation spectra at 298 K of *trans*-stilbene included in Tl-X.

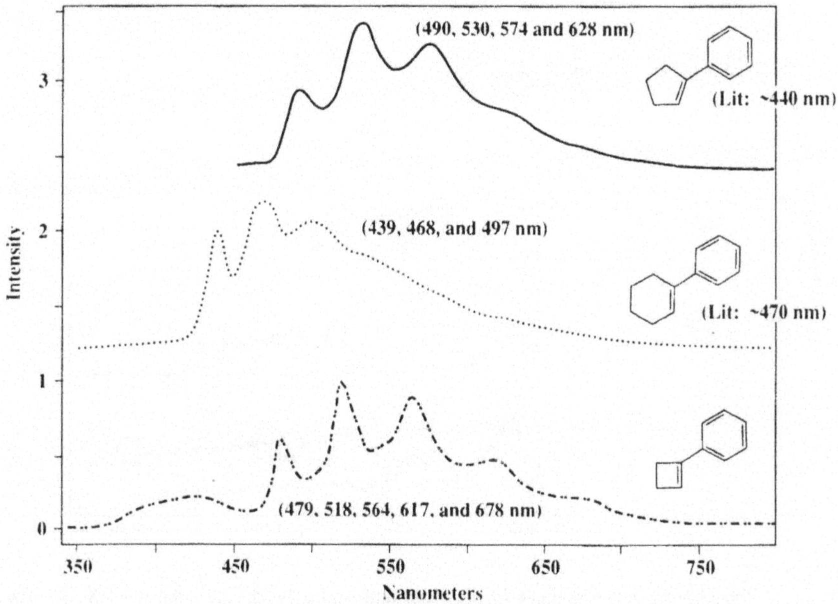

Figure 13. Phosphorescence spectra at 77 K of 1-phenylcycloalkenes included in Tl-X.

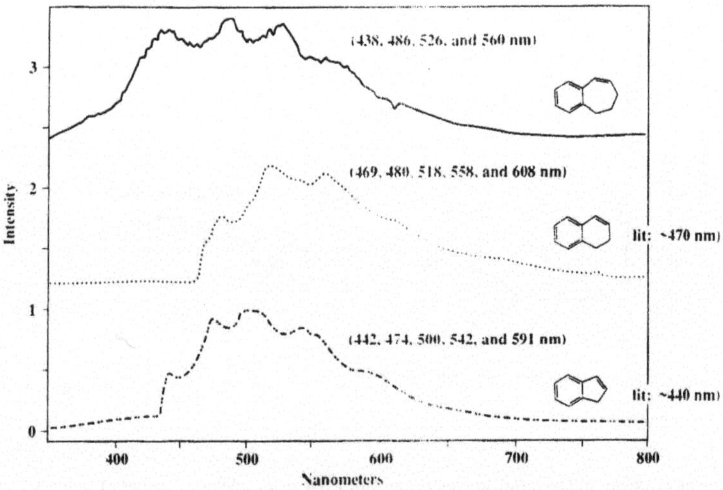

Figure 14. Phosphorescence spectra at 77 K of phenyl-fused cycloalkenes included in Tl-X.

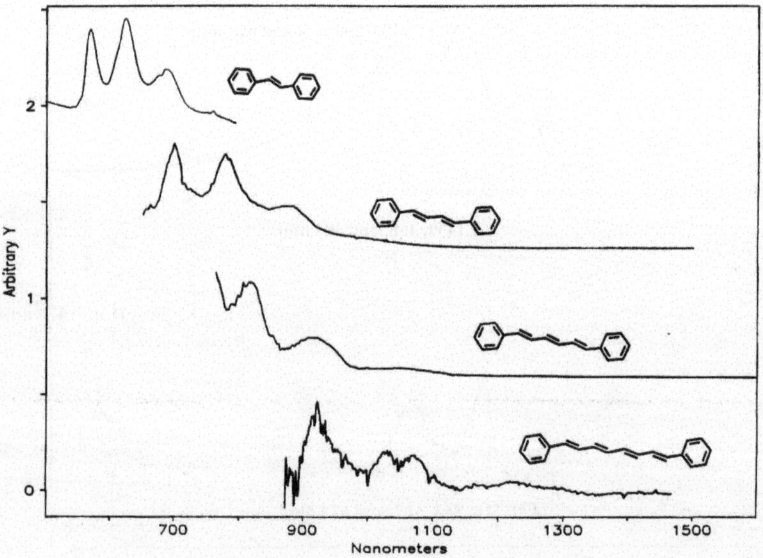

Figure 15. Phosphorescence spectra at 77 K of α,ω-diphenyl polyenes included in Tl-X. Excitation wavelength: stilbene, 295 nm; diphenylbutadiene, 340 nm; diphenylhexatriene, 350 nm; diphenyloctatetraene, 375 nm.

In addition to the above systems, enhanced cation-dependent phosphorescence from xanthone and *para*-dimethylaminobenzonitrile [31] included within ZSM-5 and Y zeolites has been observed recently.

As discussed above, cations can be utilized to control the efficiency of the triplet generation from an organic molecule included in the cavities of zeolites. Two examples provided below illustrates how such an effect can also be utilized to control product distribution in a photoreaction.

The photobehavior of acenaphthylene is unique in that it has been extensively studied in various constrained media and has been subjected to one of the largest heavy-atom effects on its dimerization [32]. The irradiation of acenaphthylene in solution yields the *cis*- and the *trans*-dimers; the singlet gives predominantly *cis*-dimer, whereas the triplet gives both *cis*- and *trans*-dimers in comparable amounts (Scheme 4). Photolyses of dry solid inclusion complexes of acenaphthylene in various cation (Li, Na, K, Rb)-exchanged Y zeolites gave the *cis*- and *trans*-dimers [33]. *Cis*- to *trans*-dimer ratio, relative efficiency of dimerization, relative triplet yields, and triplet lifetimes of acenaphthylene are dependent on the cation as summarized in Table 5.

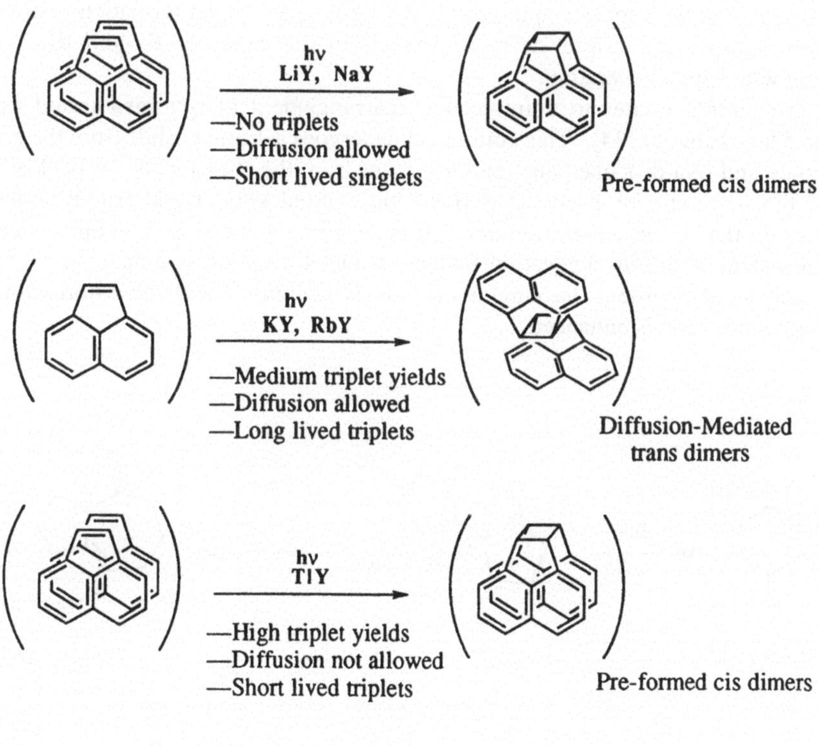

Scheme 4.

Table 5. Cation dependent photodimerization of acenaphthylene included in M^+-Y type zeolites (<S>=0.5)

Zeolite	*Cis-/Trans*-dimer Ratio	Relative Efficiency of Dimerization	Relative Triplet Yield	Triplet Lifetime (μs)
Li-Y	25	0.2	-	--
Na-Y	25	0.2	-	--
K-Y	2.3	0.4	0.2	9.6
Rb-Y	1.5	1.0	0.5	5.7
Cs-Y	4.2	0.8	0.7	2.1

The absence of triplet formation in Li-Y and Na-Y is consistent with the solution behavior in which the intersystem crossing yield from S_1 to T_1 is reported to be near zero. This, as well as the exclusive formation of *cis* dimer, support the conclusion that the dimerization in the supercages of Li-Y and Na-Y is from the excited singlet-state. The high triplet yield in K- and Rb-Y is thought to be a consequence of the "heavy-atom effect" caused by the cations present within the supercage. The trends observed in the variation of the triplet yield and the triplet lifetime with the increasing mass of the cation are consistent with the expected spin-orbit-coupling-induced triplet formation. Formation of the *trans*-dimer (the triplet-derived product) in the cages of K- and Rb-Y is in agreement with triplet generation.

Another example relates to unimolecular rearrangement of β,γ-unsaturated ketone (Scheme 5 and Table 6) [34]. This compound undergoes a 1,3-acyl shift from the excited singlet state and oxa-di-π-methane rearrangement from the triplet state. It is clear from Table 6 that the ratio of products derived via excited singlet and triplet states are dependent on the cation when irradiation of the ketone included in Y zeolite is carried out. Consistent with our suggestion that the triplet formation is enhanced in heavy-cation-exchanged zeolites, the singlet lifetime is decreased and the fluorescence to phosphorescence ratio is enhanced.

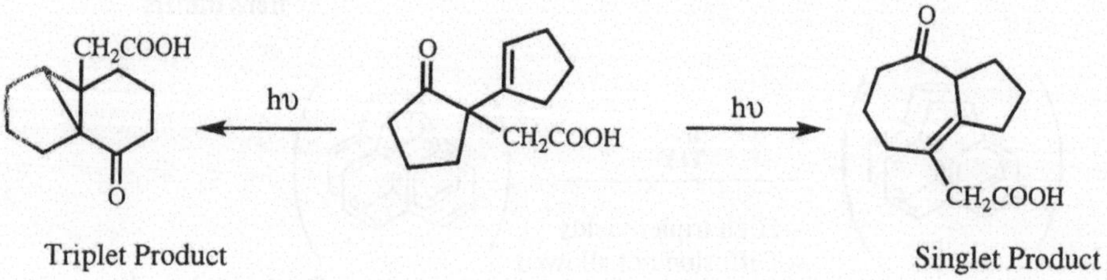

Scheme 5.

Table 6. Photophysical properties[a,b] and photochemical behavior of β,γ-unsaturated ketone (Scheme 5).

Zeolite	Singlet to Triplet Product Ratio	F/P[c]	S_1 lifetime at 77 K (ns)
Li-Y	>99:1	10:1	9.2
Na-Y	>99:1	10:1	8.9
K-Y	58:42	5:1	8.2
Rb-Y	47:53	2:1	6.9
Cs-Y	82:18	2:1	5.8
Tl-Y	38:62	0.2:1	4.4

a. Two emissions one with short lifetime (ns range; λ_{max}: 378 nm) and the other with long lifetime (ms range; λ_{max}: 430 nm) were recorded at 77 K. Both emissions had the same excitation spectra at 77 K. At room temperature the longer wavelength emission was absent. Short wavelength emission was attributed to the S_1 state and the long wavelength one to the T_1 state.

b. The singlet lifetimes were measured at 77 K on an Edinburgh single photon counting spectrometer.

c. Fluorescence to phosphorescence ratio at 77 K.

5. Guest-Cation Interaction: Acid-Base Interaction

Zeolites possess a number of acidic and basic sites. When the exchangeable cation is proton, Brønsted acid sites are also present. The strength of Lewis basic sites (12-ring oxygens in the windows of the supercages) in X and Y zeolites are known to be dependent on the cation. For example, in the case of Y-type zeolites, it has been estimated that the negative charge on the 12-ring oxygen atoms increases with the decrease in cation acidity (Li-Y, -0.345; Na-Y, -0.351; and K-Y, -0.381) [35]. Among a number of techniques, fluorescence has been popular in monitoring the surface acidity of zeolites [36]. Fluorescence probes such as 8-quinolinol, 8-hydroxyquinoline, quinoline, α- and β-naphthols, and acridine are used to measure the Brønsted surface acid strengths. Surface acidity is estimated from the emission intensities of neutral and acidic or basic forms. Lewis as well as Brønsted acid sites on the surfaces of zeolites have been monitored with pyridine. In all of these cases, interaction between the acid sites and the probe results in a new emitting species such as protonated probe, charge transfer complex, or species resulting from electron transfer from the probe to the Lewis site.

Three examples provided below illustrate the importance of considering the possibility of acid-base interactions when organic molecules are included within zeolites:

Benzophenone included within cation- (H^+, Li^+, Na^+, K^+, Rb^+, and Cs^+) exchanged ZSM-5 yields phosphorescence characteristic of both protonated and hydrogen-bonded forms [37]. The ratio of these emissions depends on the temperature of zeolite activation and the nature of the cation. Smaller cations such as Na^+ and H^+ favor the protonated form. It is suggested that the cation changes the surface acidity and larger cations such as Rb^+ and Cs^+ do not favor the formation of free proton.

Irradiation of *trans*-azobenzene included within H^+-Y and H^+-β zeolites, yields a cyclization product benzo[c]cinnoline and a disproportionation product benzidine (Scheme 6) [38]. These products are not obtained upon irradiation in a neutral medium and in Na-Y. Formation of these products within zeolites is suggested to result from the protonated form of azobenzene.

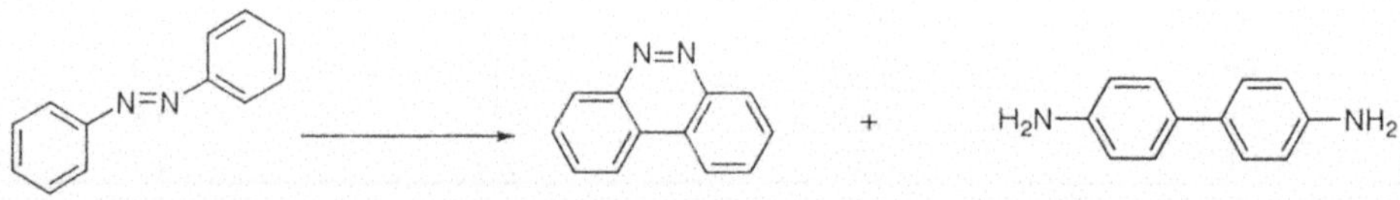

Scheme 6.

Cozens, Garcia, and Scaiano have shown recently that the photobehavior of 1,1-diphenyl-2-propanone within zeolites (Na-Y and in H-Y) depends on the availability of protons [39]. α-Cleavage is the primary process in both cases but the resulting products when 1,1-diphenyl-2-propanone is included in Na-Y and in H-Y are different (Scheme 7). Differing final products result from the fact that reacting species under the two conditions are different–in one case it is the neutral species whereas in the other it is the protonated form.

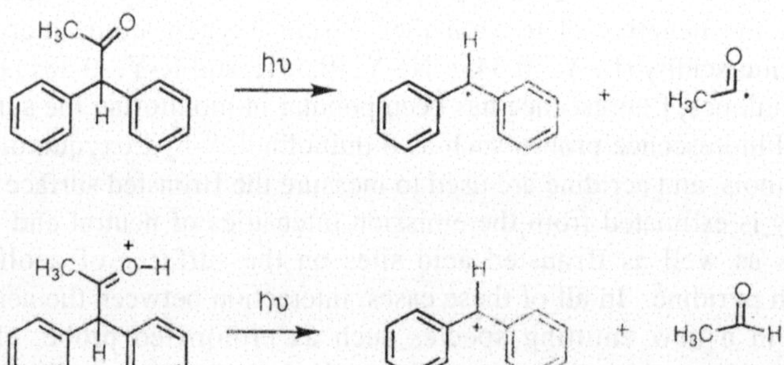

Scheme 7.

6. Cations as Reaction Cavity Free Volume Controllers: Lebensraum Effect

In the sections above, zeolites are considered to have active reaction cavities. However, several early studies pointed out that they do possess cavities which are passive in character with respect to certain reactions. We illustrate below with several examples why the cations, under certain circumstances should be considered nothing more than inert fillers which reduce the free volume available for the reactants within zeolites.

Irradiation of dibenzylketone in unrestricted environments such as solution media is characterized by the formation of benzyl radicals following α-cleavage and decarbonylation processes [40,41]. The termination process of these radicals is generally the uncorrelated (random) coupling to yield 1,2-diphenylethane. Photolysis of unsymmetrically substituted derivatives (represented as A(C=O)B), results in the formation of three radical coupling products, AA, AB and BB, in a ratio of 1:2:1 (Scheme 8). In contrast, when escape of the radicals is impeded by confining environments, the radical recombination probability is modified by favoring the formation of geminate products of the AB type. It has also been observed that recombination of the primary radical pair can sometimes occur before decarbonylation (or escape) giving rise to rearranged ketone photoproducts and recovered starting material (Schemes 8 and 9) [42]. Recombination of the primary radical pair generates 1-phenyl-*ortho*-methyl acetophenone (o-MAP) and 1-phenyl-*para*-methyl-acetophenone (p-MAP) by *ortho*- and *para*-coupling, respectively, in the case of DBK. Thus, in using DBKs as probes, one measures the yields of diphenyl ethanes and rearranged products and, in certain cases, the isotope enrichment factor (see below). The cage-effect, defined as {AB-(AA+BB)/(AA+AB+BB)}, is a measure of restriction (a factor dependent on topology) and is calculated from the yields of diphenyl ethanes. The isotope enrichment factor, (α), defined as {rate of disappearance of ^{12}C ketone/rate of disappearance of ^{13}C ketone}, is measured from quantum yields of disappearance of the ^{12}C and ^{13}C ketones.

The results obtained upon photolysis of evacuated samples of DBK under conditions of a relatively low loading (2% w/w) (which represents a nominal occupancy of ~15%) in the cation-exchanged X and Y zeolites are shown in Table 7. The first important observation comes from the fact that the product distribution depends significantly on the cation in the M-X zeolites while it remains relatively constant in the M-Y zeolite series.

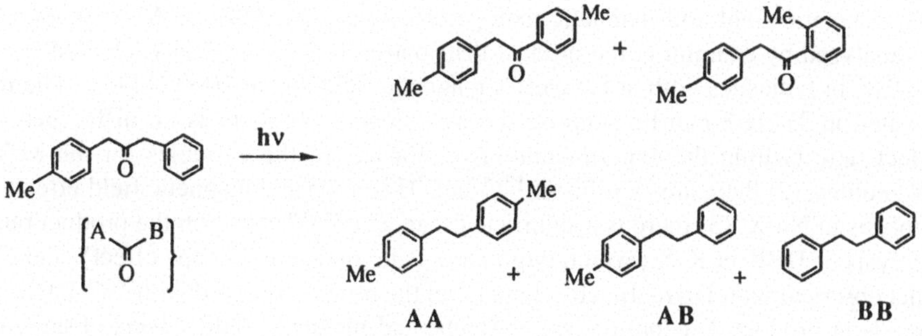

Scheme 8.

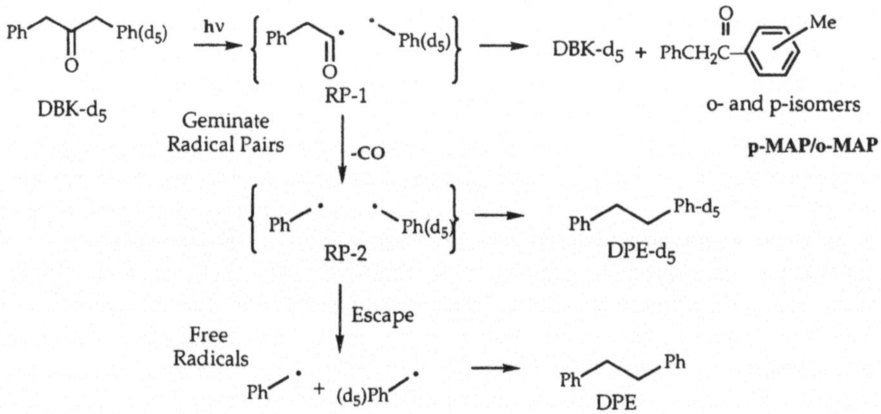

$$\% \text{ Cage effect} = \frac{\text{DPE-d}_5}{\text{DPE + DPE-d}_5 + \text{DPE-d}_{10}} \times 100$$

Scheme 9.

The results of the cage-effect obtained with isotopically labelled, DBK-d_5 at 2% w/w loading (~15 % occupancy), are also shown in Table 7 [43]. The yields of DPE-d_5, DPE-d_{10}, and DPE-d_0 determined by GCMS were used to calculate the cage-effect (Scheme 9). The most important observation relating to cage-effect is that the cage-effect increases from Li- to K- in the M-X zeolites, and that it remains constant in the M-Y zeolite series.

The magnetic field and magnetic isotope effects in the M-X and M-Y zeolites have been analyzed by examining the product distribution from isotopically labelled (^{13}C and ^{2}H) DBK in faujasites both at the earth's magnetic field and at 2000 G [44]. The results presented in Table 8 can be summarized as follows: (1) there is no difference in the product ratio (within the experimental error) for any of the variables for the M-Y and Li-X zeolites, (2) there are significant ^{13}C and ^{2}H isotope and magnetic field effects upon photolysis in Na-X, (3) there is a significant magnetic field effect on the product ratio for photolysis of DBK in K-X (even though there is no magnetic isotope effect), and (4) the results observed with DBK-d_{10} closely parallel the results observed with ^{13}C-DBK.

Trends in product distribution, cage-effects, and magnetic field effects observed with M-X and M-Y zeolites are consistent with the paradigm that the reaction cavity free volume controls the reaction profile and that is dependent on the size and number of cations present within the supercages [44]. In the case of Li-X and M-Y zeolites, substantial yields of decarbonylation products, relatively modest cage effects (Table 7) and small amounts of rearranged products (0-20%) emphasize the importance of the globally closed zeolite spaces. The lack of magnetic field and magnetic isotope effects in the products of primary radical recombination are also consistent with a relatively large space where the radical centers can separate and explore the global environment. In contrast, in K-X zeolites, large yields of rearranged products are obtained. Large magnetic field effects and no magnetic isotope effect in this zeolite are consistent with most coupling reactions occuring in a tight local environment where escape is difficult and enforced orbital overlap is important. These results suggest that spin-orbit coupling, which is isotope independent, may be the intersystem crossing mechanism in K-X zeolites. Finally, in agreement with our expectations, photolyses in Na-X give evidence of an environment that is intermediate between that observed in Li-X and M-Y on one hand, and K-X on the other. The local inter-radical distance and the ease of escape are expected to be intermediate between them. As the local space is also expected to be larger, intersystem crossing can be controlled by nuclear-electron hyperfine couplings (which are isotope dependent) as well as by strong external magnetic fields.

The product distribution obtained upon photolysis of benzoin alkyl ethers, α-alkyldeoxybenzoins, and α-alkyldibenzyl ketones is dependent on the cation as summarized in Tables 3 and 9 with one example from each class [18, 45, 46]

Table 7. Product distribution[a] and cage-effect from photolysis of DBK (2% w/w) in ion-exchanged faujasites (Schemes 8 and 9).

Zeolite	Percentage Yield			
	DPE	*o*-MAP	*p*-MAP	Cage Effect[b]
Li-X	80	3	16	5
Na-X	55	17	26	22
K-X	40	40	16	73
Li-Y	100	0	0	19
Na-Y	95	0	5	17
K-Y	94	2	4	19

a. Calculated error limit, 10%.

b. Cage-effect measured under the same conditions in a separate experiment by using DBK-d_5.

Photoprocesses undertaken by benzoin ethers and deoxybenzoins are shown in Scheme 2. In solution, the termination process of the benzyl radicals derived from α-alkyl-dibenzyl ketones consists only of the coupling between the two benzylic radicals and results in diphenylalkanes AA, AB, and BB in a statistical ratio of 1:2:1. Scheme 10 provides the structure of products and a mechanism for the formation of these products.

Within supercages, on the other hand, termination proceeds by both coupling and disproportionation (Table 9). A schematic diagram for the termination processes between the benzylic radicals is shown in Scheme 11. Preference for disproportionation within the supercage has been interpreted as follows: the association between benzylic radicals which would favor coupling, would be prohibited inside the cavity, especially in the presence of large cations, because of the reduction in free volume. Further, more drastic overall motion would be required to bring benzylic radicals together for head-to-head coupling than to move an alkyl group so that one of its methylene hydrogens would be in a position for abstraction by the benzylic carbon radical. It is logical to expect the radical pair to prefer the pathway of "least volume and motion" when the free space around it is small. Thus, as smaller cations are replaced with larger ones and as shorter alkyl chains are replaced with longer ones, one would indeed expect enhanced yields of olefins as observed in the reported study (Figure 16).

Table 8. Magnetic isotope and field effects on the product distribution from photolysis of DBK in M-X zeolites (Schemes 8 and 9)[a]

System	Percentage Yield		
	DPE	*o*-MAP	*p*-MAP
Li-X			
DBK-^{12}CO	81 [85][c]	3 [3]	16 [12]
DBK-^{13}CO (90%)[b]	81 [85]	3 [2]	16 [13]
DBK-d_{10} (95%)	79	2	19
Na-X			
DBK-^{12}CO	56 [65]	17 [13]	26 [22]
DBK-^{13}CO (90%)	27 [33]	37 [25]	36 [42]
DBK-d_{10} (95%)	67 [59]	10 [13]	23
K-X			
DBK-^{12}CO	40 [68]	40 [14]	16 [18]
DBK-^{13}CO (90%)	45 [62]	32 [13]	20 [25]
DBK-d_{10} (95%)	45	35	20

a. Results for M-Y zeolites are similar under all conditions (90–100% yield of DPE and very little isomers).

b. Isotope content of the sample.

c. Bracketed number is the product yield in the presence of 2000 G external magnetic field.

Table 9: Product distribution upon photolysis of α-hexyl dibenzyl ketone within zeolites[a].

Medium	Percentage of		
	Olefin	(AB)	Rearrangement Product
Li-X	39	17	37
Na-X	19	18	57
K-X	23	29	36
Rb-X	38	23	29
Cs-X	60	15	22

a. see Scheme 10 for structures.

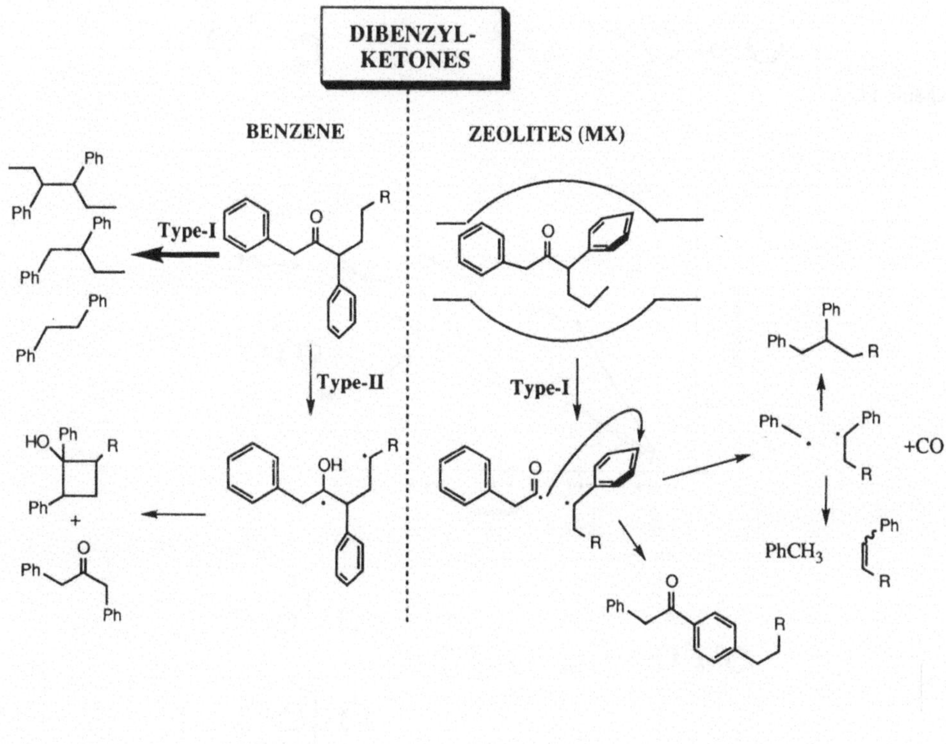

Scheme 10.

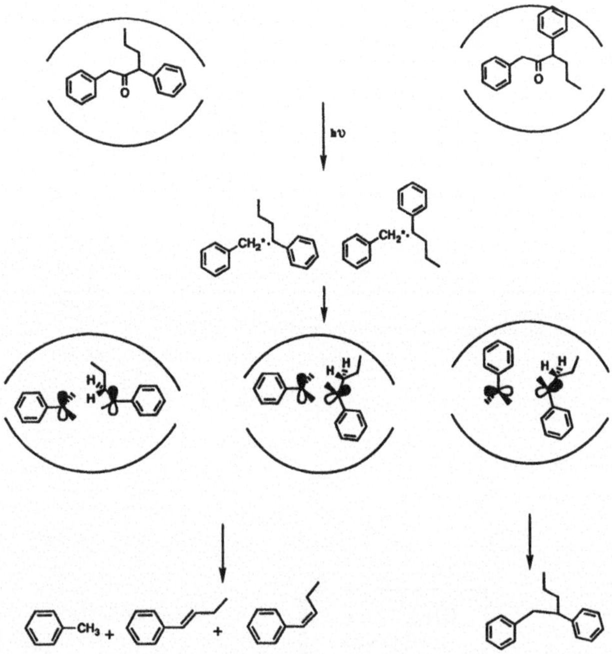

Scheme 11.

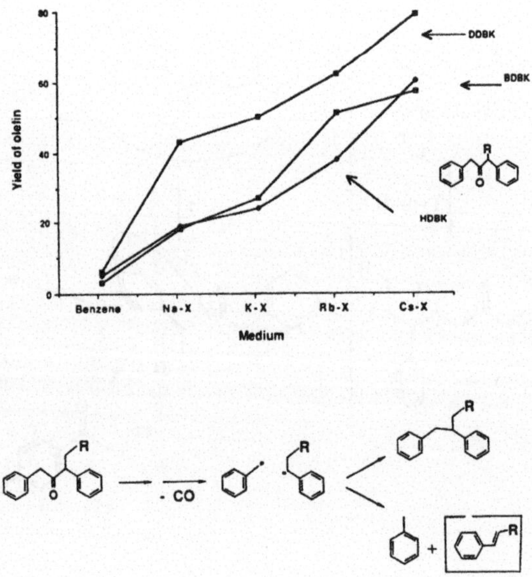

Figure 16. Photolysis of α-alkyl dibenzylketone: product dependence on cation.

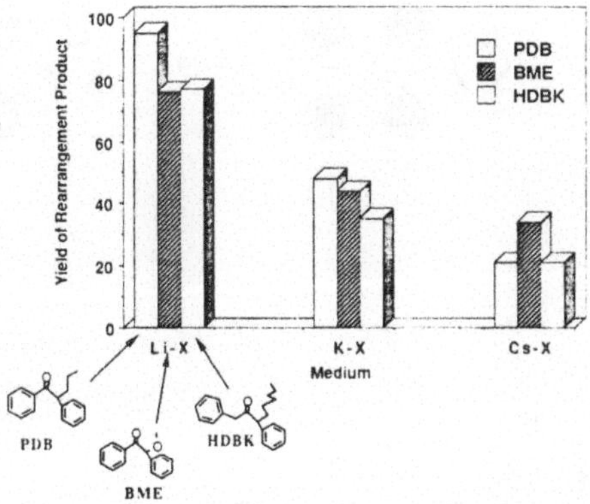

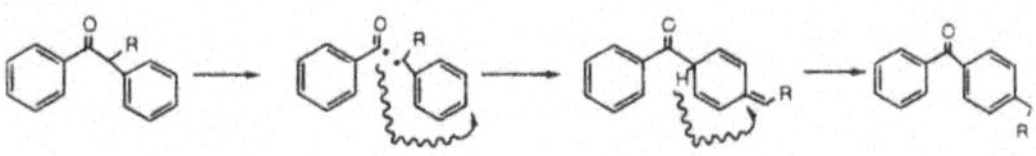

Figure 17. Photolysis of alkyldeoxybenzoin: cation dependent product distribution.

The above conclusion is also supported by the pathways undertaken by the primary triplet radical pair (Scheme 10) generated by the α-cleavage of the α-alkyl dibenzylketones and α-alkyl benzoin ethers and deoxybenzoins (Schemes 2 and 3). Perusal of Tables 3 and 9 reveals that while the rearrangement takes place in all cation-exchanged X and Y zeolites, the yield of the rearrangement product varies depending on the cation (Figure 17). The yield decreases as the cation present in the supercage is changed from Li^+ to Cs^+. Such a trend is attributed to the decrease in the free space within the supercage. As the available free space inside the supercage is decreased by the increase in the size of the cation, the translational and rotational motions required for the rearrangement process become increasingly hindered (Scheme 12). Under these conditions, competing paths, such as coupling to yield the starting ketone and decarbonylation, both of which require less motion, dominate. Examination of Table 3 reveals that in the case of α-methylbenzoin ether as well as in α-propyldeoxybenzoin the C/E ratio, i.e., the ratio of the yield of cyclobutanol, the cyclization (C) product, to that of deoxybenzoin, the elimination product (E) resulting from the 1,4-diradical derived via the Norrish type II γ-hydrogen abstraction process, depends on the cation present in the supercage. Also, the C/E ratio increases as smaller cations are replaced with the larger ones, i.e., C/E increases from Li^+ to Cs^+. The above dependence of the product

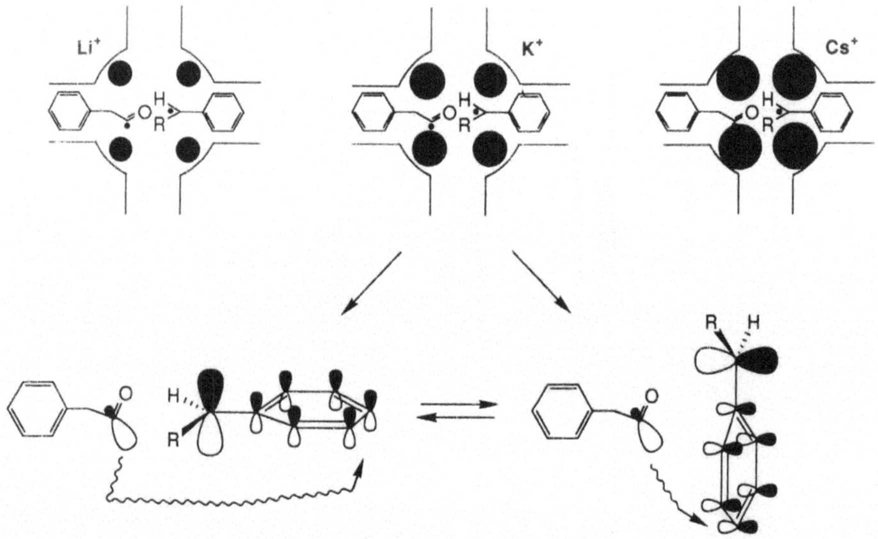

Scheme 12.

distribution on the cation can be understood on the basis of the well-understood mechanism of the type II reaction. The enhancement of cyclization within the supercages of X and Y zeolites in the presence of larger cations is believed to reflect the rotational restriction brought on the skewed-*transoid-cisoid* 1,4-diradical inter-conversion (Scheme 13). We propose that as the cation size increases, the 1,4-diradical is forced to adopt a compact geometry due to reduction in the available supercage free volume. Thus, the skewed 1,4-diradical first formed would be encouraged to relax to the *cisoid-* rather than to the *transoid-* conformer. Severe constraints would be imposed by the supercage on the *cisoid-transoid* interconversion and the barrier for the *cisoid* to *transoid* conversion would be accentuated. These factors are expected to enhance the yield of cyclobutanol.

One final example under this category deals with the photo-Fries rearrangement of phenyl benzoate [47]. As shown in Scheme 14 , irradiation of phenyl benzoate in polar and non-polar solvents yields *ortho-* and *para-* isomers almost in equal amounts. However, irradiation of phenyl benzoate included in X and Y zeolites yields the *ortho*-isomer preferentially (Table 10). A clear trend in the *ortho-para* ratio with respect to cation is evident both in X and Y zeolites. Increasing size of the cation reduces the chances of the benzoyl fragment migrating to the *para*-position. Consistent with the paradigm presented above reduction in the *ortho-para* ratio is much larger in X than in Y zeolites. Smaller cations with larger numbers (X zeolites) do the same job, reduce the free volume, as larger cations with fewer numbers (Y zeolites). Thus, from the examples above, it is clear that one has a handle on the size of the nanoreactors where the photoreaction is conducted.

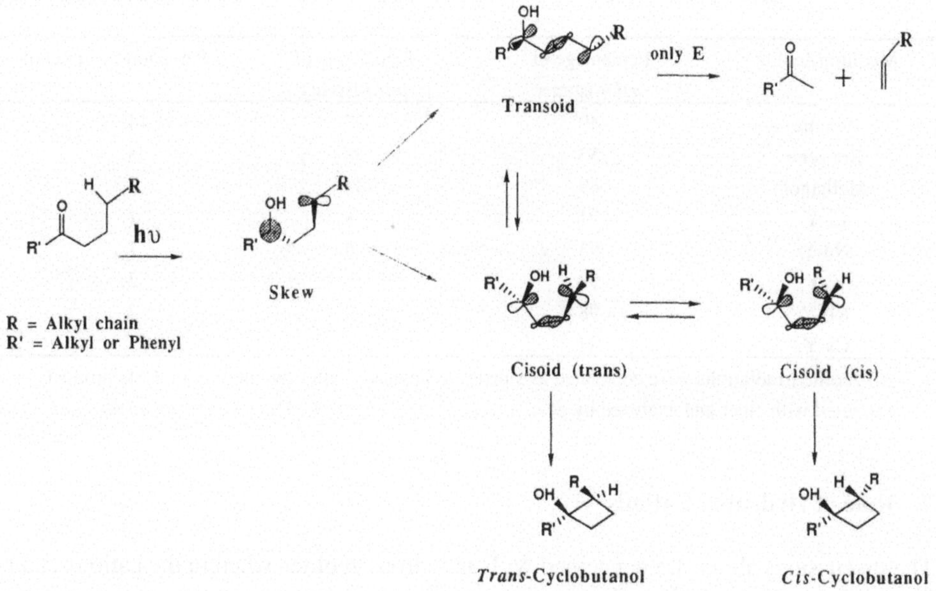

Scheme 13.

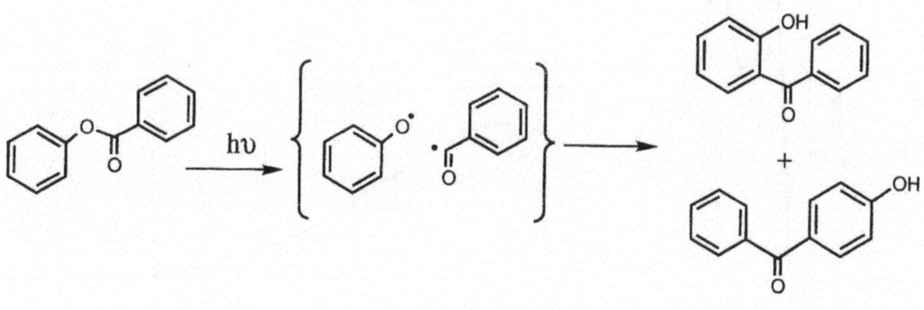

Scheme 14.

Table 10: Photolysis of phenylbenzoate in M-Y zeolites: effect of cation on the rearrangement process — *ortho-*, *para-*isomers (Scheme 14).[a]

Medium/Zeolite	Percentage of *ortho-*isomer	Percentage of *para-*isomer	Percentage of phenol
Hexane	49	37	14
Benzene	53	38	8
Methanol	49	43	6
Li-Y	91	8	1
Na-Y	93	6	1
K-Y	98	-	2
Rb-Y	98	-	2
Cs-Y	97	-	3

a. All zeolite irradiations were conducted as slurries in hexane. After two hour photolysis products were extracted with ether and analyzed by gc.

7. Role of Hydrated Cations

The discussions above are concerned with anhydrous zeolites wherein the cations can be considered to be "naked". These cations are free to interact with the guest molecules. However, it is well known that zeolites are highly hygroscopic and the absorbed water molecules are co-ordinated to the cations present within supercages. In this section, we discuss how the presence of water within zeolite supercages alters the photochemical and photophysical behavior of guest molecules.

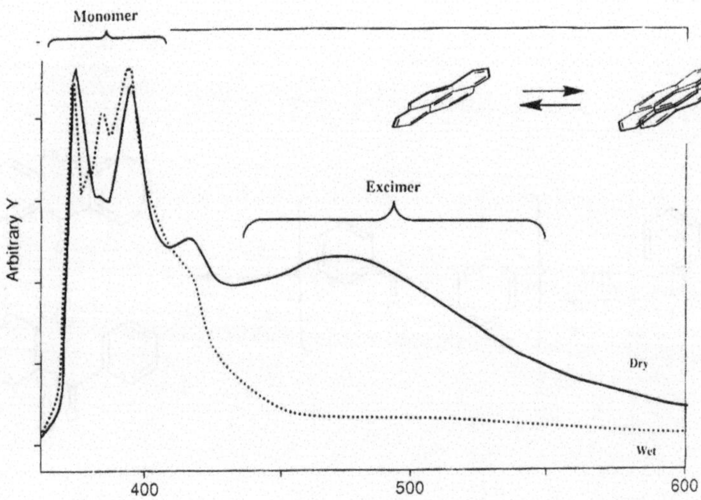

Figure 18. Influence of water on the emission spectra of pyrene included in Na-Y zeolites.

Excitation of pyrene included within anhydrous Na-Y results in two emissions as shown in Figure 18, one from the monomer and the other from the excimer [48]. The so-called excimer emission is definitely not an emission from typical dynamic excimer, since the excitation spectra for the monomer and the excimer emissions are different (Figure 19). An obvious choice of the latter emission being due to microcrystals was ruled out on the basis of following observations: (a) Both the wavelengths of emission and excitation are not identical to those of crystals reported in the literature; (b) The internal void space of Na-Y is too small to accommodate microcrystals; however, if micro-crystals are formed on the external surfaces they should be washable with excess hexane. Hexane wash did not change the ratio of intensities of emission from monomer and excimer. These observations suggested that the excimer results from pre-aggregated dimers (static excimers) present within the supercages of Na-Y. Such a conclusion is also supported by the following observations: (a) No growing-in of the excimer in the nanosecond time scale was noticed when the excimer emission decay was monitored by time resolved single photon counting techniques; (b) No negative pre-exponential term for excimer decay was obtained; (c) The ratio of the intensities of excimer to monomer emission increased slightly upon lowering the temperature; (d) Variation of loading level between 0.002 and 0.2 (average number of molecules per supercage) did not alter the pyrene monomer lifetime significantly.

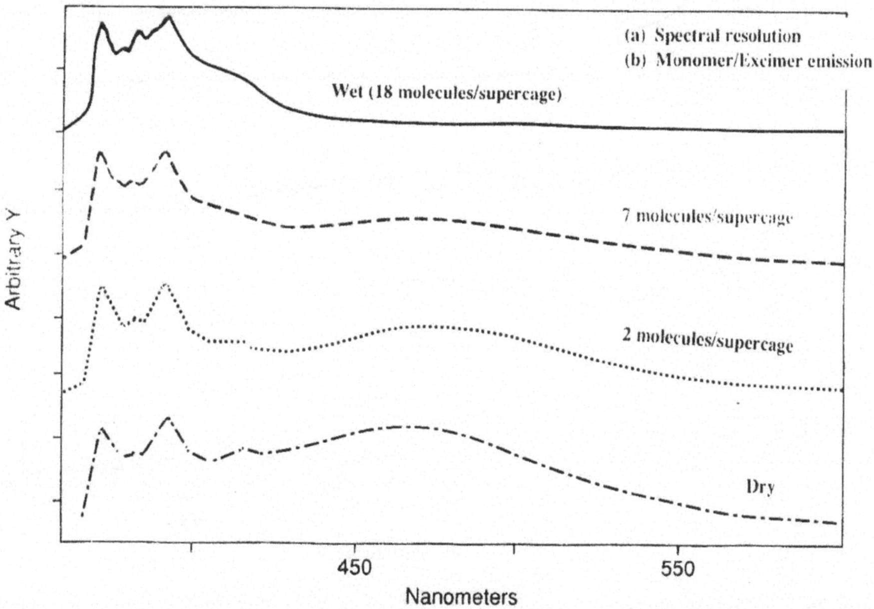

Figure 19. Excitation spectra for the monomer and excimer emissions of pyrene included within "dry" Na-Y: (a) ——, monomer, em l: 380 nm; (b) – – –, excimer, em l: 480 nm.

Co-absorption of water has a striking influence on the nature of aggregation of pyrene within zeolites. As seen in Figure 18, the intensity of excimer emission from pyrene present within fully hydrated Na-Y is fairly low. In fact, there is a correlation between the ratio of intensities of monomer to excimer emissions and the water content (Figure 20). Decrease in excimer emission with the increase in water content indicates that pre-aggregation is not favored within "wet" zeolites. Such may be the result of reduction in polarizing power of the cation; pre-aggregation is proposed to be a result of polarization of pyrene molecules by the cation (Figure 21).

The photophysical properties of phenanthrene bring out yet another unique characteristic of guest occupancy distribution with respect to water content within zeolites [49]. Emission spectra of phenanthrene included within anhydrous zeolite is non-unique (<S> varied between 0.002 to 0.1); fluorescence from the monomer is detected and no excimer emission is seen. However, when the zeolite containing higher loadings of phenanthrene (<S> above 0.1) absorbs water, in addition to the monomer emission, fluorescence from microcrystals is detected (Figure 22). A change in the occupancy status (monomer to microcrystals) is also reflected in diffuse reflectance spectra. At lower loading levels (<S> lower than 0.1), independent of the hydration status emission spectra corresponded only to that of the monomer and diffuse reflectance spectra of both wet and dry samples were nearly identical.

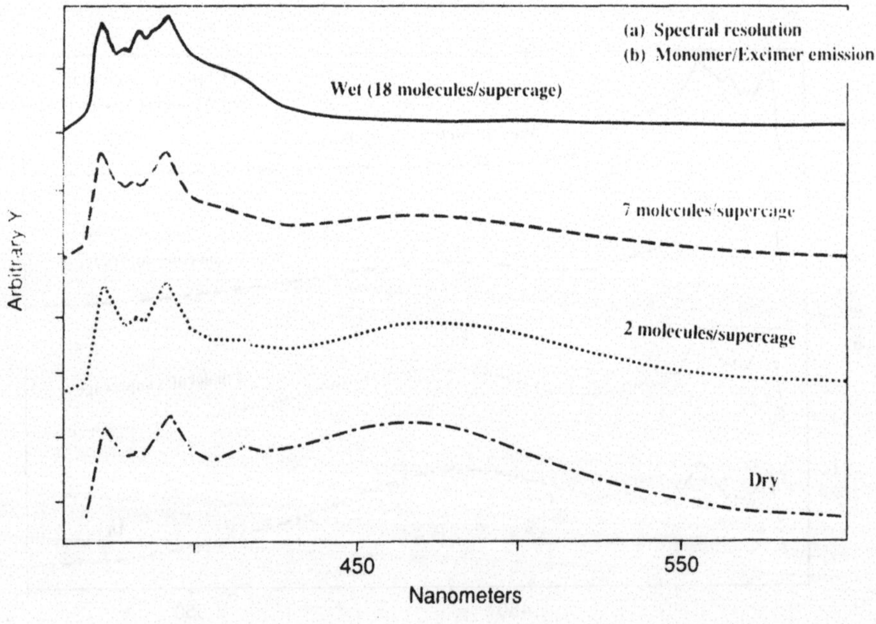

Figure 20. Emission spectra (excitation l: 340 nm) of pyrene within Na-Y at various loading levels of co-absorbed water: (a) – · – · –, "dry", no water; (b) · · · · · · <S>: 2; (c) – – – – –, <S>: 7; (d) ———, <S>: 18.

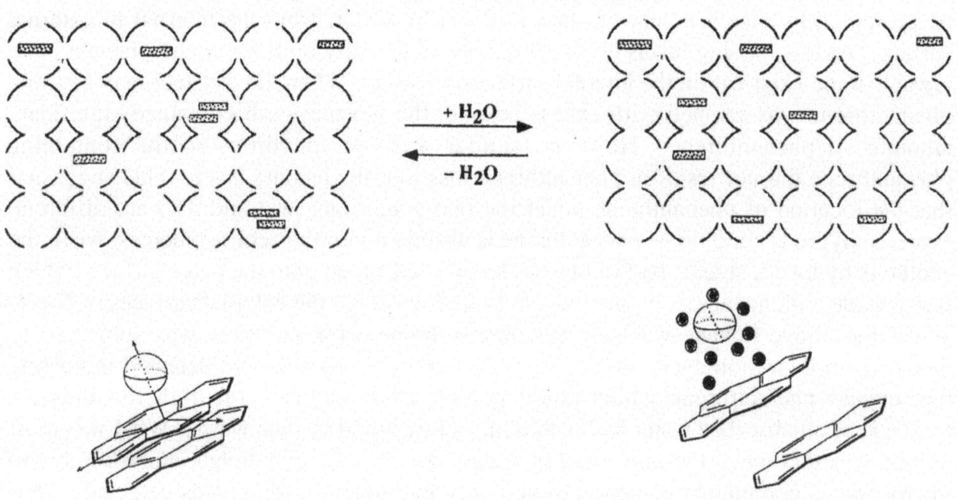

Figure 21. A schematic illustration of the influence of water on pyrene aggregation. Cation interaction leads to polarization of pyrene.

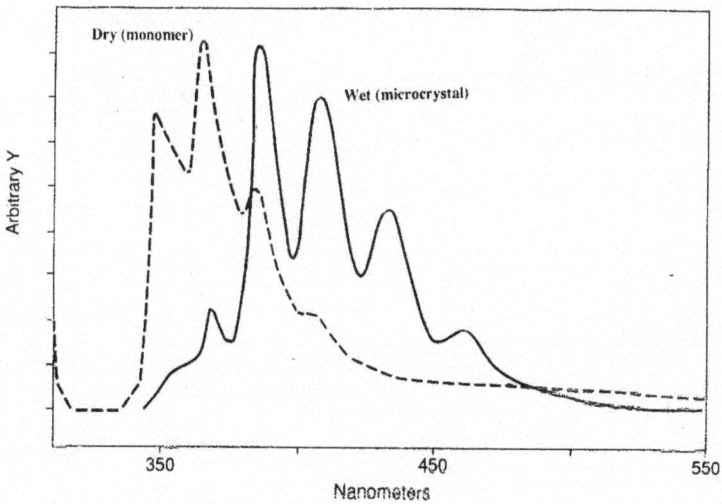

Figure 22. Emission spectra of phenanthrene, <S>: 0.4, included within "dry" and "wet" Na-Y: (a) - - - -, "dry" Na Y, ex l: 293 nm; (b) ———, "wet" Na-Y, ex l: 320 nm.

The association of phenanthrene to form micro-crystals at higher loadings is believed to be prompted by displacement of phenanthrene, by water, from the internal to external surface. At low loading levels of phenanthrene, apparently both water and phenanthrene are able to co-exist within the internal surfaces of Na-Y. When the wet zeolite containing phenanthrene was washed with excess hexane, the hexane wash contained significant amounts of phenanthrene. However, similar wash of anhydrous zeolite containing phenanthrene did not result in phenanthrene loss into the hexane layer. This suggested that the location of phenanthrene under the two conditions (wet and dry) are different. Under anhydrous conditions, phenanthrene is absorbed into the zeolite whereas, when the zeolite is hydrated, at least part of phenanthrene is adsorbed onto the outer surface (which is washable with hexane), i.e., microcrystals are located on the external surfaces of Na-Y. When the above hexane-washed, wet phenanthrene,Na-Y complex was subjected to dehydration-hydration cycle, no emission due to micro-crystals was detected indicating that excess phenanthrene which could not be accommodated (at high loadings) in presence of co-absorbed water has been removed by initial hexane wash. What was most unique was that when the unwashed hydrated sample was dehydrated, emission due to microcrystals completely disappeared and only monomer emission was detected. Thus dehydration-hydration cycles promote monomer-microcrystal cycles.

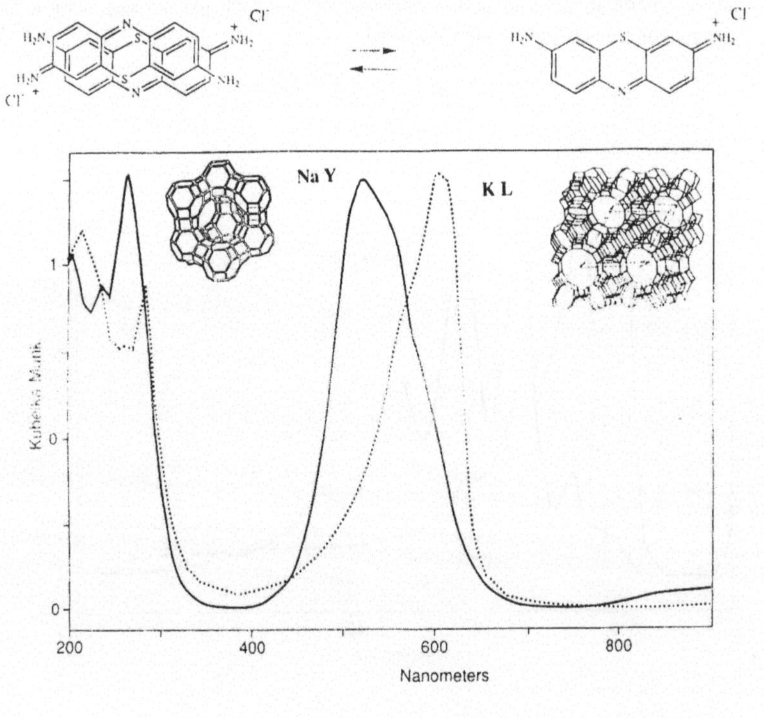

Figure 23. Diffuse reflectance spectra of thionin included within dry K-L and Na-Y.

In another interesting study within zeolites, the control of aggregation of dyes by co-absorbed water has also been observed [50]. Among the dyes studied were methylene blue, oxazine170, nile blue A, acridine orange, and cresyl violet. With thionin as an example, it was shown that complexation can be achieved from aqueous solutions by cation exchange. Thionin possesses a delocalized positive charge and is capable of partially displacing (ca 5%) Na^+ and K^+ from Y- and L-type zeolites. The dye-exchanged Y and L zeolites were shown to have different spectroscopic properties assigned to the state of aggregation of the dye (Figure 23). While monomeric thionin displays a strong absorption maximum at about 610 nm, dimeric thionin, also referred to as H-dimers, absorbs at about 540 nm. On visual inspection, thionin-exchanged K-L-zeolites appear pink while thionin exchanged Na-Y zeolites appear blue. These results suggest that the formation of thionin aggregation is not possible within the straight channel structure of L-zeolites. While the ~7.5Å diameter of L-zeolite channels may accommodate thionin molecules along their long axis, they seem unable to accommodate the face-to-face H-aggregated dimers. Complexation of thionin in the larger supercages of the Na-Y zeolites, on the other hand, results in the formation of hydrated H-dimer-complexed zeolites. It was also found that dehydration of the zeolite complex under high vacuum at moderate temperatures resulted in no change in the L-zeolite but displaced the

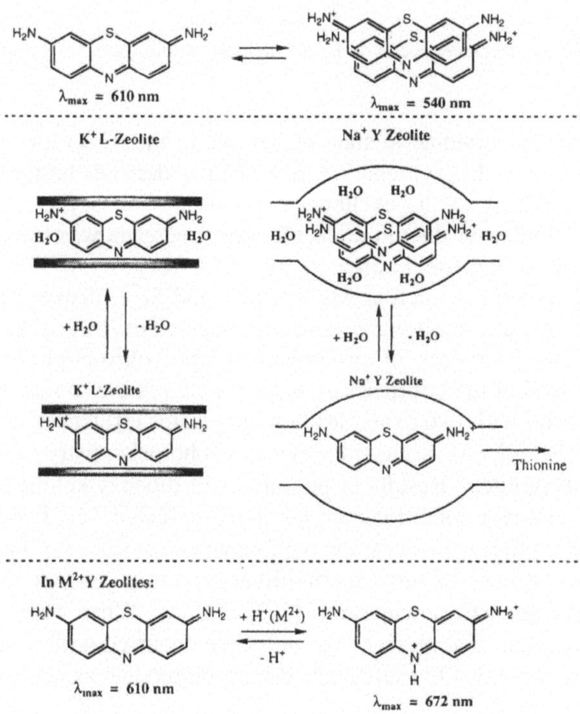

Scheme 15.

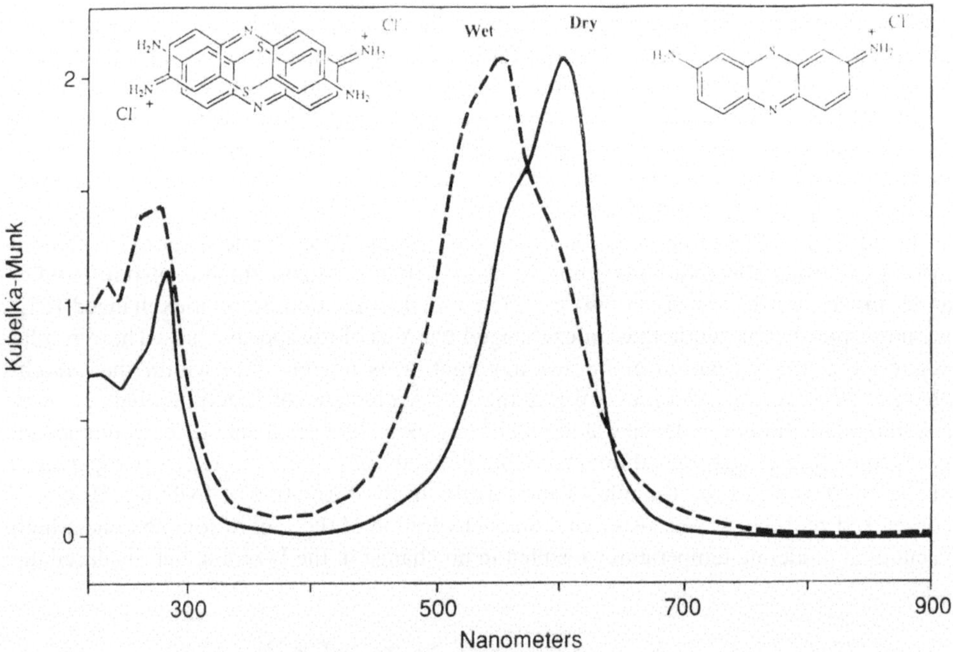

Figure 24. Diffuse reflectance spectra of thionin included within Na-Y: effect of included water.

equilibrium towards the monomeric state of the dye in the Y-zeolites (Scheme 15 and Figure 24). It was shown that aggregation in Y-zeolites depends on the concentration of the dye and on the nature of the exchangeable cation. No aggregate formation was observed when the number of thionin molecules per supercage was lower than 0.0005 or with Cs-Y zeolites, which are known to posses a very small free volume in the supercages. Divalent cations, such as Mg^{2+}, Ca^{2+}, and Sr^{2+}, displayed a behavior that was similar to that of Na-Y until water concentrations were lower than ca. 3%. Under those conditions, the formation of protonated thionin chromophores was observed suggesting that removal of the last traces of water resulted in the formation of free H^+.

We close this section with two examples in which water influences the photochemical product formation [51,52]. As discussed previously, photochemistry of dibenzyl ketones is influenced within zeolites. Results of photolyses of dibenzylketone included in M-X zeolites under dry and wet conditions are tabulated in Table 11. It is clear that when water is present within the supercages, the rearrangement process is completed inhibited. The cage effect in the case of *para*-methyldibenzylketone is reduced to 8% under hydrated conditions from 40% in anhydrous Na-X zeolites. When water-content is varied methodically, it is evident that much of the influence occurred with 4 or 5 molecules of water per supercage (note that the supercage can accommodate as many as 28 molecules of water).

Table 11. Photolysis of dibenzylketone in M-X zeolites: effect of hydration[a].

Zeolite	Condition	Percentage		
		DPE	*o*-MAP	*p*-MAP
Li-X	Dry	81	3	16
	Wet	100	0	0
Na-X	Dry	55	17	26
	Wet	100	0	0
K-X	Dry	40	40	16
	Wet	98	1	1

a. See scheme 9 for product structures.

A similar observation was made during photolysis of alkanophenones within Na-X zeolites. As illustrated in Figure 25, the ratio of cyclization to elimination products of the 1,4-biradical is influenced by the number of water molecules present within supercages. We believe that the first few molecules of water that enter the supercages coordinate to the cations and that these are the ones which have maximum influence on reactions.

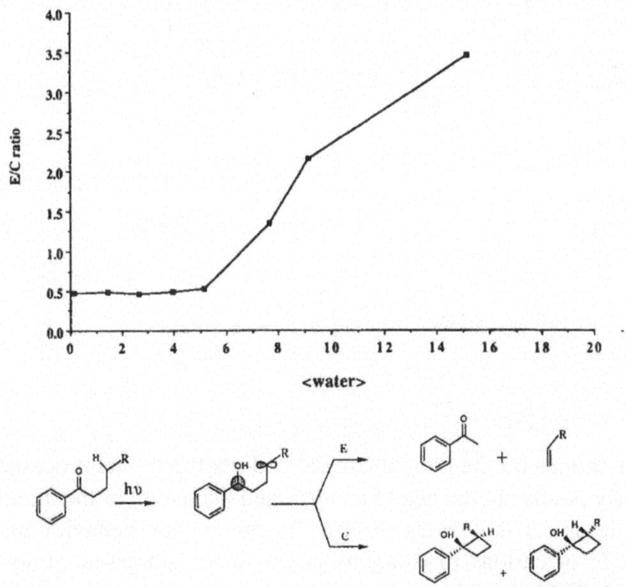

Figure 25. Influence of water on the elimination to cyclization product ratio from the 1,4-diradical resulting via γ-hydrogen abstraction in the case of valerophenone.

Molecules of water that come in afterwards merely fill the supercages. Further work is needed to understand this process. Water molecules that are present within the supercages interact with the hydroxyl of the 1,4-biradical and prevent internal return of the hydrogen and also influence the *transoid-cisoid* equilibrium of the diradical (Scheme 16).

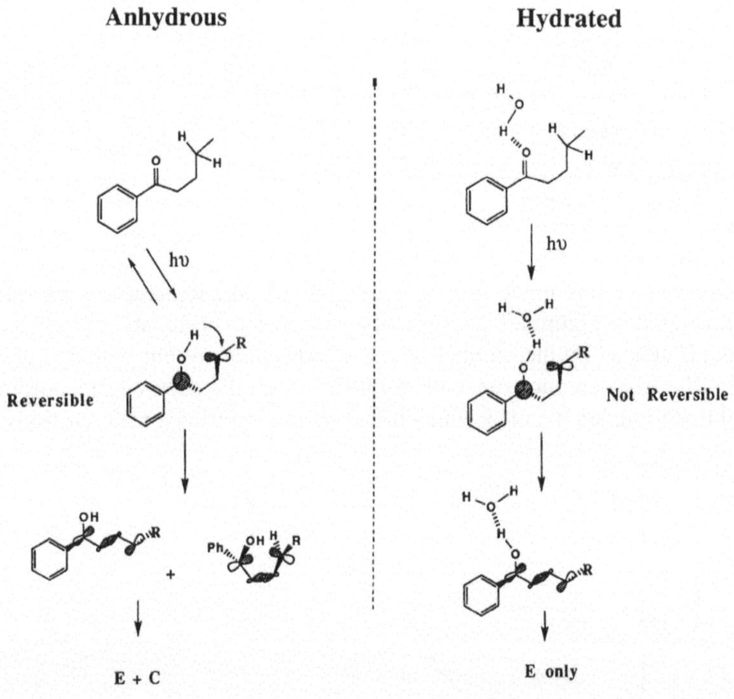

Scheme 16.

8. Summary

The influence of cations on the photochemical and photophysical processes described in this chapter clearly points out the need to understand the nature of the interaction between the cation and the guest molecules in order to predict the behavior of guests within zeolites. The role of cations in catalysis is yet to be addressed. Photophysical and photochemical studies of the type described in this article are expected to play an important role in the future towards understanding the role of cations in a number of thermal processes in which zeolites play an unique catalytic role.

Acknowledgment: VR thanks Tulane University for financial support and NJT thanks the National Science Foundation and the Air Force Office of Scientific Research for their generous support of this research.

References

1. K. Kalyanasundaram, *Photochemistry in Microheterogeneous Systems*, Academic Press, New York (1987);
T. Matsuura and M. Anpo (Eds), *Photochemistry on Solid Surfaces*, Elsevier, Amsterdam (1989);
V. Ramamurthy (Ed), *Photochemistry in Organized and Constrained Media*, VCH, New York (1991);
H.-J. Schneider, and H. Durr (Eds), *Frontiers in Supramolecular Organic Chemistry and Photochemistry*, VCH; Weinheim (1991);
V. Balzani and F. Scandola, *Supramolecular Photochemistry*, Ellis Horwood, Chichester, England (1991).
2. V. Ramamurthy, R. G. Weiss, and G. S. Hammond, *Adv. Photochem.*, **18**, 67 (1993).
3. N. J. Turro and P. Wan, *Tetrahedron Letters,* **25**, 3655 (1984);
H. L. Casal and J. C. Scaiano, *Can. J. Chem.*, **62**, 628 (1984);
S. L. Suib and A. Kostapapas, *J. Am. Chem. Soc.*, **106**, 7705 (1984).
4. For a summary see:
N. J. Turro, *Pure & Appl. Chem.,* **58**, 1219 (1986);
N. J. Turro, in *Molecular Dynamics in Restricted Geometries*, J. Klafter and J. M. Drake (Eds), John Wiley, New York, p 387 (1989);
V. Ramamurthy, in *Photochemistry in Organized and Constrained Media*, V. Ramamurthy (Ed), VCH, New York, p 429 (1991);
N. J. Turro and M. Garcia-Garibay, in *Photochemistry in Organized and Constrained Media,* V. Ramamurthy (Ed), VCH, New York, p 1 (1991);
V. Ramamurthy, D. F. Eaton, and J. V Caspar, *Acc. Chem. Res.*, **25**, 299 (1992);
V. Ramamurthy, *Chimia,* **46**, 359 (1992);
P. Kamat, *Chem. Rev.*, **93**, 267 (1993);
J. K. Thomas, *Chem. Rev.*, **93**, 301 (1993);
K. B. Yoon, *Chem. Rev.*, **93**, 321 (1993);
V. Ramamurthy and D. F. Eaton, *Chem. Materials,* **6**, 1128 (1994).
5. D. W. Breck, *Zeolite Molecular Sieves: Structure, Chemistry, and Use*, John Wiley and Sons, New York (1974);
A. Dyer, *An Introduction to Zeolite Molecular Sieves*, John Wiley and Sons, Bath (1988);
H. van Bekkum, E. M. Flanigen and J. C. Jansen (Eds), *Introduction to Zeolite Science and Practice*, Elsevier, Amsterdam (1991);

R. Szostak, *Molecular Sieves. Principles of Synthesis and Identification*, Van Nostrand, New York (1989).
6. D. W. Breck, *Zeolite Molecular Sieves: Structure, Chemistry, and Use*, John Wiley and Sons, New York, pp 529-592 (1974).
7. W. M. Meier and D. H. Olson, in *Atlas of Zeolite Structure Types*, Butterworths, Cambridge (Second Revised Edition), pp 96–97 (1987);
R. Szostak, *Handbook of Molecular Sieves*, Van Nostrand Reinhold, New York, pp 183–188 (1992).
8. Calculations of polyhedral volumes were performed using a modification of the POLYVOL Program [D. Swanson and R. C. Peterson, *The Canadian Mineralogist,* **18**, 153 (1980); D. K. Swanson and R. C. Peterson, "POLYVOL Program Documentation", Virginia Polytechnic Institute, Blacksburg, VA] assuming the radius of the TO_2 unit to be 2.08Å (equivalent to that of quartz).
9. R. G. Weiss, V. Ramamurthy, and G. S. Hammond, *Acc. Chem. Res.*, **26**, 530 (1993).
10. M. Hepp, V. Ramamurthy, D. R. Corbin, and C. Dybowski, *J. Phys. Chem.*, **96**, 2629 (1992).
11. V. Ramamurthy, *Mol. Cryst. Liq. Cryst.*, **240**, 53 (1994).
V. Ramamurthy, D. R. Sanderson, and D. F. Eaton, *J. Phys. Chem.*, **97**, 13380 (1993).
12. V. Ramamurthy and J. V Caspar, *Mol, Cryst. Liq. Cryst.*, **211**, 211 (1992).
13. Y. S. Liu, P. de Mayo, and W. R. Ware, *J. Phys. Chem.*, **97**, 5987, 5995 (1993).
14. V. Ramamurthy, D. R. Corbin, and L. J. Johnston, *J. Am. Chem. Soc.*, **114**, 3870 (1992).
15. E. J. Baum, J. K. S. Wan, and J. N. Pitts, *J. Am. Chem. Soc.*, **88**, 2652 (1966);
D. R. Kearns and W. Case, *J. Am. Chem. Soc.*, **88**, 5087 (1966);
A. A. Lamola, *J. Chem. Phys.*, **47**, 4810 (1967);
P. J. Wagner, A. E. Kemppainen, and H. Schott, *J. Am. Chem. Soc.*, **95**, 5604 (1973).
16. V. Ramamurthy, D. R. Sanderson, and D. F. Eaton, *Photochem. Photobiol.*, **56**, 297 (1992).
V. Ramamurthy and D. F. Eaton, in *Proceedings of the 9th International Zeolite Conference*, R. von Ballmoos, J. B. Higgins and M. M. J. Treacy (Eds), Butterworth-Heinemann, Boston, p 587 (1992).
17. V. Ramamurthy, D. R. Corbin, N. J. Turro, and Y. Sato, *Tetrahedron Lett.*, **30**, 5829 (1989);
V. Ramamurthy, X. G. Lei, N. J. Turro, T. R. Lewis, and J. R. Scheffer, *Tetrahedron Lett.,* **32**, 7675 (1991).
18. D. R. Corbin, D. F. Eaton, and V. Ramamurthy, *J. Am. Chem. Soc.*, **110**, 4848 (1988).
19. Y. Wada, Y. Yoshizawa, and A. Morikawa, *J. Chem. Soc., Chem. Commun.*, 319 (1990).

20. V. Ramamurthy, J. V Caspar, E. W. Kuo, D. R. Corbin, and D. F. Eaton, *J. Am. Chem. Soc.*, **114**, 3882 (1992).
21. J. V. Caspar, V. Ramamurthy, and D. R. Corbin, *Coordination Chem. Rev.*, **97**, 225 (1990).
22. V. Ramamurthy, J. V. Caspar, D. R. Corbin, and D. F. Eaton, *J. Photochem. Photobiol. A: Chemistry*, **50**, 157 (1989).
23. M. Kasha, *J. Chem. Phys.*, **20**, 71 (1952);
D. S. McClure, *J. Chem. Phys.*, **17**, 905 (1949).
24. J. M. Larson and L. R. Sousa, *J. Am. Chem. Soc.*, **100**, 1942 (1978);
S. Ghosh, M. Petrin, A. H. Maki, and L. R. Sousa, *J. Chem. Phys.*, **87**, 4315 (1987);
S. Ghosh, M. Petrin, A. H. Maki, and L. R. Sousa, *J. Chem. Phys.*, **88**, 2913 (1988).
25. V. Ramamurthy, J. V. Caspar, D. R. Corbin, B. D. Schlyer, and A. H. Maki, *J. Phys. Chem.*, **94**, 3391 (1990).
26. G. Weinzierl and J. Friedrich, *Chem. Phys. Lett.*, **80**, 55 (1981).
27. V. Ramamurthy, J. V Caspar, E. W. Kuo, D. R. Corbin, and D. F. Eaton, *J. Am. Chem. Soc.*, **114**, 3882 (1992);
V. Ramamurthy, J. V Caspar, and D. R. Corbin, *Tetrahedron Letters*, **31**, 1097 (1990).
28. J. Saltiel, G. E. Khalil, and K. Schanze, *Chem. Phys. Lett.*, **70**, 233 (1980);
D. F. Evans, *J. Chem. Soc.*, 1351 (1957);
R. H. Dyck and D. S. McClure, *J. Chem. Phys.*, **36**, 2326 (1962);
H. Gorner, *J. Phys. Chem.*, **93**, 1826 (1989).
29. T. N. Ni, R. A. Caldwell, and L. A. Melton, *J. Am. Chem. Soc.*, **111**, 457 (1989);
P. M. Crosby, J. M. Dyke, J. Metcalfe, A. J. Rest, K. Salisbury, and J. R. Sodeau, *J. Chem. Soc., Perkin II*, 182 (1977).
30. G. Heinrich, G. Holzer, H. Blume, and D. Schulte-Frohlinde, *Z. Naturforsch.*, **25b**, 496 (1970);
D. F. Evans and J. N. Tucker, *J. Chem. Soc., Faraday Trans. II*, **68**, 174 (1972).
31. V. Ramamurthy and D. R. Sanderson, unpublished results.
32. D. O. Cowan and R. L. Drisko, *Elements of Organic Photochemistry*, Plenum, New York, p 435 (1976).
33. V. Ramamurthy, D. R. Corbin, C. V. Kumar, and N. J. Turro, *Tetrahedron Lett.*, **31**, 47 (1990).
34. B. Borecka, A. D. Gudmundsttir, G. Olovsson, V. Ramamurthy, J. R. Scheffer, and J. Trotter, *J. Am. Chem. Soc.*, **116**, 10322 (1994).
35. D. Barthomeuf, *J. Phys. Chem.*, **88**, 42 (1984).
36. For a review: A. M. Eremenko, *Adsorbtsiya Adsorbenty*, **8**, 48 (1980).
37. S. Okamoto, H. Nishiguchi, and M. Anpo, *Chem. Lett.*, 1009 (1992).
38. A. Corma, H. Garcia, S. Iborra, V. Marti, M. A. Miranda, and J. Primo, *J. Am. Chem. Soc.*, **115**, 2177 (1993).
39. F. L. Cozens, H. Garcia, and J. C. Scaiano, *J. Am. Chem. Soc.*, **115**, 11134 (1993).

40. (a) N. J. Turro, *Proc. Natl. Acad. Sci.*, **80**, 609 (1983);
(b) G. F. Lehr and N. J. Turro, *Tetrahedron*, **37**, 3411 (1981);
(c) N. J. Turro and B. Kraeutler, *Acc. Chem. Res.*, **13**, 369 (1980).
41. (a) P. S. Engel, *J. Am. Chem. Soc.*, **92**, 6074 (1970);
(b) W. K. Robins and R. H. Eastman, *J. Am. Chem. Soc.* , **92**, 6076 (1970).
42. N. J. Turro and G. C. Weed, *J. Am. Chem. Soc.,* **105**, 1861 (1983).
43. Z. Zhang, Ph.D Thesis, Columbia University (1989)
44. N. J. Turro and Z. Zhang, *Tetrahedron Letters*, **30**, 3761 (1989)..
45. V. Ramamurthy, D. R. Corbin, and D. F. Eaton, *J. Org. Chem.* **55**, 5269 (1990).
46. D. R. Corbin, D. F. Eaton, and V. Ramamurthy, *J. Org. Chem.*, **53**, 5384 (1988); V. Ramamurthy, D. R. Corbin, D. F. Eaton, and N. J. Turro, *Tetrahedron Lett.*, **30**, 5833 (1989).
47. K. Pitchumani and V. Ramamurthy, unpublished results.
48. V. Ramamurthy and D. R. Sanderson, *J. Phys. Chem.*, **97**, 13380 (1993).
49. V. Ramamurthy, *Mol. Cryst. Liq. Cryst.*, **240**, 53 (1994).
50. V. Ramamurthy, D. R, Sanderson, and D. F. Eaton, *J. Am. Chem. Soc.*, **115**, 10438 (1993).
51. Z. Zhang and N. J. Turro, unpublished results.
52. V. Ramamurthy and D. R. Sanderson, unpublished results.

Abstract. In this article, we illustrate how one can utilize the cation embedded in a zeolite matrix to control the photophysical and photochemical behavior of guest molecules included in zeolite cages/cavities. Three aspects of cation-guest interaction are highlighted. Strong electronic interaction between the cation and the guest leads to mobilization of guests within zeolite supercages, alteration of the lowest electronic configuration ($n\pi^*$, $\pi\pi^*$) and restriction of conformation of molecules. Less obvious to a non-photochemist is the ability of cations, depending on their atomic weight, to induce spin conversion (singlet-triplet) in molecules. Photophysical studies carried out with aromatics and photodimerization of acenaphthylene highlight this point. The power of the heavy-atom cation effect in zeolites has been demonstrated by recording phosphorescence from several olefins whose phosphorescence has not previously been recorded. Cations by their sheer size can influence the mobility of molecules within zeolites. Restriction of motion of reactive intermediates results in product selectivity in reactions as illustrated with a few examples. Properties of cations and consequently their influence can change depending on whether they are hydrated or uncoordinated to water. Aggregation of dyes and aromatics and reactivities of carbonyl compounds are influenced by water present within zeolites. We have shown in this article that zeolites can be used as unique reaction vessels for photochemical reactions.

Journal of Inclusion Phenomena and Molecular Recognition in Chemistry **21**: 283–298, 1995.

Zeolites as Hosts for Novel Optical and Electronic Materials

NORMAN HERRON

1. Introduction

Small, molecular clusters of metals or semiconductors, having hybrid properties somewhere between those of the molecule and the bulk phase represent a fascinating new class of materials and, as such, have come under intense investigation [1,2]. In some senses, these materials represent the logical extension of the trend toward ever smaller electronic devices. As the size of such devices decreases, one reaches a regime where physics and chemistry meet and novel properties develop. At this size extreme, collective electronic properties become "distorted" by the fact that, at this point, a continuous wave function no longer describes the cooperative interaction of electrons and nuclei in the solid. The system, in fact, tends toward the classical "particle-in-a-box" quantum description.

In preparing materials for the study of phenomena associated with the resulting "quantum dots", a basic problem facing synthetic chemists is the control of surface reactions of such particles so as to arrest their growth at the small cluster stage. Thermodynamics usually dictates that the small cluster nuclei will tend to aggregate and develop into the bulk crystalline structure. Many approaches have been explored for the preparation of the desired small clusters - including the use of micelles [3], colloids [4], polymers [5], glasses [6], and surface-capping agents [7] in an effort to control the aggregation problem. In almost all cases, however, the cluster sizes and crystallinities are poorly defined and one would like to find an approach to this class of materials which produces a mono-dispersion of cluster sizes in a well-defined and characterizable array. The ability to produce such a regular array of single-sized clusters is key to defining the optical and electronic behaviors of these materials, unmuddied by size dispersion effects. These criteria would seem well met by an inclusion-type approach using a host lattice as the template within which the clusters could be constructed and confined.

One example of this type of approach involved CuCl crystallites of ~20-50 Å which were prepared within single crystals of NaCl or KCl by a melt procedure and were shown to exhibit exciton absorption features which were blue-shifted (a quantum confinement effect) from the bulk material [8]. Another approach, and the one which will be explored in this chapter, is the use of a *porous* crystalline host lattice, a zeolite molecular sieve, into which the active material (metal, semiconductor, etc.) is either directly imbibed or else created in situ . The former technique is exemplified by the sublimation of elemental mercury, selenium, tellurium, etc. into a variety of aluminosilicate zeolites. The second, "ship-in-a-bottle" approach [9] has been used for the generation of numerous clusters and superclusters (cluster arrays) in zeolites [10].

The well-defined and well-ordered cavities provided by a zeolite host are a potentially ideal environment for synthesizing single-sized clusters and cluster arrays. The windows of the zeolite pores provide ready access for transporting reagents to the cavities making a variety of synthesis schemes accessible. The additional potential for producing novel clusters of electronic materials and stabilizing unusual geometries and connectivities by using the surface passivating and templating effects of the host lattice is also most attractive [11].

1.1 WHY SMALL ELECTRONIC PARTICLES ARE INTERESTING

It is useful to understand why there is the current interest in very small particles of metals and semiconductors [1,2,12] and what limitations this interest places on the nature of useful materials. Bulk electronic, optical etc. properties of a material are often the result of the cooperative interaction of all of the molecules or atoms in the ensemble. Studying how these properties are altered as the number of constituent molecules/atoms in the ensemble becomes very small can reveal much about the fundamental source of the bulk property itself.

Theories of the mechanism of superconductivity and magnetism in highly constrained systems as well as interest in the metal-insulator boundary as a function of metallic cluster size has spurred research on small metallic clusters. The possibilities of generating novel magnetic recording systems based on nanoparticulate metals have also sharpened interest in these materials. Of course, there has been a long-standing interest within the catalyst community in small metallic clusters as catalytic sites and embedding these clusters inside zeolite pores introduces molecular sieving selectivity to any catalytic effects.

The concept of an all-optical or opto-electronic computer technology has attracted attention because of its potential for extreme speeds and parallel processing capabilities in areas such as image recognition. Such a technology requires several basic optical materials for the construction of devices which mimic their electronic counterparts. One such fundamental computing element is the optical transistor or bistable device which acts as a light switch or valve/amplifier. Basic requirements placed on materials for such a device are that they have a very rapid switching speed (ideally picosecond) and extreme photostability in order to perform trillions of switching operations/sec for years at a time. One realization of such a material could involve the use of third-order non-linear optical properties, χ^3, to effect a transient refractive index change. Illumination of such a material with intense laser light will cause a change of its refractive index leading to a switch from an opaque to transmissive state in an interferometer-type (e.g., Fabry-Perot) bistable device. While semiconductor materials themselves will perform this kind of switching at their band-edge wavelengths, the speed of the effect is slow - usually as a consequence of a long, free-carrier lifetime. This speed can be increased by providing more sites for efficient removal of these free-carriers - in other words more defect sites. One can view surface sites on a semiconductor particle as such defect sites and one simple way to increase their concentration is to increase the surface to volume ratio, i.e., to go to very small particles. The commercial color filters of Schott and Corning, based on CdS/Se nanoparticulates in a silica matrix, have verified the utility of this kind of material for non-linear-optical devices [6]. Researchers would like to explore a wide range of other semiconductors and matrices for these purposes and the zeolite host provides an almost ideal medium for such studies.

2. Metal Clusters in Zeolites

The incorporation of metallic clusters inside the pores of zeolites has a long history [13], mainly because of interest in the catalytic properties of the resultant materials. Typically, such materials have been prepared by ion-exchange of metal salts followed by hydrogen or CO reduction to produce metal clusters within the zeolite chambers. A classical example of such a process is in the preparation of Pt-loaded zeolites used as hydrogenation, dehydrocyclization, etc. catalysts [14,15]. Other encapsulation techniques involve (i) direct incorporation of the metal from the melt or the vapor and (ii) the absorption of an organometallic precursor which can then be thermally or optically decomposed to the metal inside the zeolite pores. Examples of each technique are described below.

A significant amount of the early work in this area concentrated on optimizing synthetic procedures in order to maintain a high dispersion and small cluster size for the metals during the incorporation protocol since this usually correlated directly with catalytic activity. In particular, the ability to maintain the metal clusters within the zeolite pore system and avoid their migration to the intercrystalline spaces and crystallite surfaces has been a major focus [16]. Later work began, however, to concentrate on the characterization of these small clusters and to explore their electronic properties as the loading density became sufficient to allow the clusters to interact electronically and eventually to form a contiguous metallic phase within the pore system. These electronic effects will be the main focus of this review.

2.1 ALKALI METALS

When vapors of the alkali metals Na, K, Rb, and Cs are exposed to dehydrated zeolites in their Na^+ or K^+ forms, the atoms of the alkali metals are spontaneously ionized inside the pores. The metal atoms essentially dissolve in the zeolite host solvent (*cf.* similar dissolution in liq. NH_3). The atoms of the incoming alkali metal form clusters with the resident cations of the zeolite to produce paramagnetic guest clusters within the zeolite host chambers. At low levels of Na incorporation, the resulting pink/red solids contain species such as Na_4^{3+}, Na_5^{4+} (which has been characterized crystallographically as a Na centered tetrahedron [17]) and Na_6^{5+} as identified by EPR [18]. As the loading of the alkali metal is increased, the pink zeolite eventually becomes blue/black - for example, up to four extra sorbed Na atoms per supercage of zeolite Na-Y can be achieved. At this loading level, the individual clusters begin to interact electronically and electron-hopping sets in. The resulting materials apparently allow electrons to hop between Na_4^{3+} clusters within the sodalite cages of zeolite Y provided the distance between such clusters is less than ~20 Å. Magnetic and transport properties of these physically constricted, highly electron deficient, dilute metallic states are of intense interest in studying the onset of an insulator-metal transition. The structural and electronic properties of these metallic superlattices has recently been reviewed [19]. Potassium clusters in zeolite LTA exhibit ferromagnetic properties in a limited range of K-loading with Curie temperatures <7 K. The magnetic properties show spin-glass behavior and are interpreted with a model of itinerant electron ferromagnetism [20]. For further details see Chapter 11 in this volume.

2.2 OTHER MAIN GROUP METALS

A limited number of studies of the incorporation of a continuous metallic phase within a zeolite pore system have been reported. Such studies typically involve the incorporation of the main group metal from the molten or vapor phase via the application of high pressure in order to overcome the surface tension effects. Liquid Ga at 10-20 kbar was forced into zeolite Na-X and the resultant conductivity of the entrained metal studied as a function of temperature and applied pressure [21]. Two types of metal-insulator transitions were detected corresponding to breaking connectivity between clusters inside the pore system. These same workers also studied In metal forced into zeolite Na-A at 20 kbar and 535 K. The inferred In_8 cluster which is produced appears to have optical and electronic properties intermediate between In atoms and bulk metallic In [22]. In both cases, release of the pressure causes immediate migration of the metal to the exterior of the pore system.

2.3 TRANSITION METALS

Classical work on the sorption of mercury vapor into zeolites [23] demonstrated that the sorption phenomenon could be effectively driven by a redox mechanism whereby the intrinsic cations of the zeolite are reduced by the incoming Hg vapor. This leads to anchoring of the growing Hg clusters at these original cation sites. In Hg-X and Ag-X zeolites, mercury uptake was "copious" and up to 25 metal atoms were absorbed per supercage. As this absorption proceeds, the zeolite changes in color from white through yellow, green, grey, and black suggestive of quantum size effects and cluster-cluster electronic interactions. In extreme cases, (especially Ag-X), the mercury filling of the pore system can be complete, producing a true nanofilamentary metallic system. Such systems have been investigated as composite superconductors with increased T_cs [24].

Silver ion-exchanged zeolite Y may be reduced by several methods [25] to produce metallic Ag clusters inside the pores. Yellow and red materials result and have been interpreted in terms of Ag_2 and Ag_3 charged clusters which can begin to interact as the loading density increases [26]. These materials can be highly luminescent and such luminescence is extremely sensitive to quenching by water.

Cadmium cluster species are produced in zeolite A by exposure of Cd ion-exchanged A, Cd_6-A, to Cd vapor at 550° C. The resultant red crystal has stoichiometry Cd_{11}-A and crystallographic analysis reveals a Cd_2 dimeric molecule with Cd-Cd distance of 2.97 Å in the midst of a low symmetry cluster of Cd_5^{4+} inside the A α-cage [27]. This identification of novel Cd clusters and reduced Cd^+ species demonstrates one of the powerful abilities of the zeolite framework to stabilize very unusual molecular species under otherwise normal conditions of temperature and pressure.

As was mentioned above, many metals in zeolite systems have been explored as catalysts and some of the best characterized are of the noble metals Pd and Pt. The metal-cluster size in the pore system is often controlled by careful manipulation of drying, calcination, and reduction conditions following ion-exchange with complex metal ions such as $Pd(NH_3)_4^{2+}$ [28]. A detailed study (using EXAFS) of Pd in zeolite X by Bein et al. [29] has revealed the Pd clusters as comprising 2-4 correlated atoms of Pd which have only weak Pd-oxygen coordination to the zeolite framework at 2.76 Å. The Pd clusters are proposed to lie within the sodalite units of the cage system as Pd_2 and Pd_4 units.

3. Semiconductors in Zeolites

The zeolite confinement approach has been developed to include many semiconductor guests in zeolite hosts. Again, the synthetic approaches mimic those adopted for inclusion of metallic species - e.g., direct absorption from the vapor phase or ion-exchange and subsequent reaction chemistry. The generation of compound semi-conductors inside the zeolite pore structure is another example of the "ship-in-a-bottle" synthesis technique where the individual components of the compound are introduced by any of a variety of methods and reacted together inside the pores to generate the semiconductor directly in situ as trapped, physically constrained particles. Our own work on Se in a variety of zeolites and CdS in zeolite Y represents some of the early steps towards exploring this new class of materials and is given as detailed examples below.

3.1 ELEMENTAL SEMICONDUCTORS

Some of the synthetically most accessible and thus most heavily studied examples of semiconductors in zeolites are the chalcogenides Se and Te. Simple sublimation of the pure chalcogenide into a dehydrated zeolite can lead to composites displaying band gap and optical absorption features which indicate size quantization. Our, and other, work on the Se materials is discussed in detail below. The Te-loaded zeolites have attracted additional interest because of their catalytic activity in hydrocarbon conversions including dehydrocyclization [30]. In the bulk phase, Te forms exclusively helical chain structures and no 8-membered rings - the analogs of those found, for example, for sulfur in zeolite A [31] - have been reported. In zeolites A and X, however, Te_8 rings have been suggested based on Raman spectroscopy (168 cm^{-1} band) of Te vapor absorbed into the zeolite crystals [32]. A similar Raman [33] and optical [34] study of Te in mordenite concluded that Te lay in the linear channels as relatively isolated and disordered chains similar to those proposed in the amorphous form of bulk Te. Finally, some unusual effects of dispersion of Te on its magnetic properties have been investigated by a careful study of Te dispersed (by vapor absorption) into zeolites X and A [35].

3.1.1 Se in Zeolite Hosts

Se in zeolite hosts has been extensively studied and numerous reports of the sundry properties which result have appeared. This is probably a direct result of the ease of synthesis of such materials via direct sublimation of Se into the pores of the dehydrated zeolite. Very substantial loadings of Se - for example 34 wt % Se in Y (~10 Se/13 Å supercage) [36] and 23 wt % Se in mordenite (~25 Se/15 Å channel length) [37] - have been reported and the pore systems can be essentially completely filled by a network of Se species. The exact nature of these size-restricted species has been the subject of many of the investigations. Raman spectra [33] of zeolite A and mordenite loaded with Se concluded that isolated rings and chains (respectively) were present, similar to those present in amorphous bulk Se. This use of the zeolite-entrained Se materials as structural models for amorphous Se has been extensively exploited in the work of Bogomolov et al. [38-41] who have studied both vibrational and electronic spectra in detail.

Our own work [36] has sought to further define in detail the structural and electronic consequences of the entrainment of Se in zeolites A, X, Y, mordenite, and $AlPO_4$-5. Using a combined MAS-NMR, electronic spectroscopy, and EXAFS approach, we have concluded that in all zeolites the Se-Se bond length is shortened relative to helical chains of t-Se (2.34 vs 2.37 Å). As might be expected, when only chains of Se are present (in the linear channel zeolites mordenite and $AlPO_4$-5), there is no interchain correlation indicating that the chains are isolated as "quantum wires" of the semiconductor and the band edge is shifted to significantly higher energy as a result of the chain isolation (1.98 $\rightarrow$ ~2.3 eV). The zeolites X and Y, with their 3-D interconnected channel system of large cages (13 Å), show the presence of both helical chains and ring structures with a wide variety of conformations. Zeolite A, having much smaller windows between the cages shows the presence of only the Se_8 entity (analogous to S_8 reported earlier [31]) with no extended chains. Again, the rings are essentially isolated from each other in separate cages and this isolation as "quantum dots" manifests itself as a blue-shift of the band edge from bulk α-Se (consisting of packed Se_8 rings) where inter-ring contacts modify the electronic behavior (2.53 $\rightarrow$ 2.85 eV).

3.2 COMPOUND SEMICONDUCTORS

AgI in zeolite mordenite has been shown to have unusual optical behavior in terms of its photosensitivity [42]. In a follow-up on this work, Ozin et al. [43] have generated silver halides in the sodalite zeolite family where reversible photo-, radiation-, thermal-, and pressure-sensitivity are observed and proposed for utility in a range of device applications, e.g., reversible image capture/data storage devices, [44] etc. Isolated Ag^+ ions in the sodalite cages can be reduced and then interact with one another forming an "expanded-metal" where the metal-metal distances are much greater than in metallic Ag and are modulated topologically by the interpenetrating sodalite framework.

WO_3 and MoO_3 are so called "intelligent" materials [45] because of their utility as components of electrochromic devices, rechargeable solid-state batteries, pH-microelectrochemical transistors, sensors, and solar cells. Absorption of the metal hexacarbonyl complex from the vapor state into zeolite Y followed by photooxidation yields encapsulated MO_3 clusters in the zeolite supercage system. Such loading can proceed to very high levels following a sequential absorption protocol such that up to 4 MO_3 units can be absorbed per supercage. Ordered supralattices of these clusters appear to form and they may reversibly lose and regain oxygen depending on annealing conditions, e.g., transforming from white $[WO_3]_n$ materials to blue $[WO_{3-x}]_n$ upon heating to >200° C in vacuum. The precise control over loading density translates into control over the optical and electronic properties which show loading dependent shifts similar to the behavior reported below for our own CdS system.

PbI_2 in X, Y, A, and L zeolites shows evidence of a strong size effect on the exciton absorption [46]. TiO^+ ion-exchanged in zeolites Y, L, and mordenite has been characterized by XPS and photochemistry and shows characteristic behavior of entrapped TiO_2 in the zeolite cages [47]. The entrapped clusters have a significant blue shift of the absorption band edge relative to anatase or rutile TiO_2. The material will photocatalyze the reduction of methylviologen (MV^{2+}) using 300 nm UV light and the reduced $MV^{+\cdot}$ radical cation was shown to have a remarkably long lifetime (hours) compared to equivalent non-zeolite systems. The indication is that the zeolite may be effectively moderating the back electron transfer between the TiO_2 hole and the radical cation.

PbS clusters in zeolite Y or mordenite have been briefly reported [10,48] being prepared from Pb-exchanged zeolite Y after exposure to H_2S. However, they have so far shown only featureless optical absorption spectra making it difficult to assess the true extent of quantum confinement effects. EXAFS analysis [49] suggests that isolated PbS_2^{2-} species exist and are strongly coordinated to the zeolite framework. The bright yellow PbS/Y material is very moisture sensitive, darkening in color upon exposure to atmospheric moisture, and eventually becoming brown/black. This behavior is interpreted as a migration of the physically entrapped PbS clusters out of the zeolite cages to the exterior zeolite surface where aggregation takes place forming bulk PbS. The same behavior is reported for CdS in zeolites and the observation seems to be general - zeolite encapsulated clusters may migrate under the influence of a coordinating solvent and form bulk phases on the zeolite exterior surface.

We have prepared one example of a III-V semiconductor, GaP, in zeolite Y [50]. Using a synthetic approach akin to that used for MOCVD deposition of GaP (i.e., absorption of $Ga(CH_3)_3$ vapor followed by calcination at ~400° C in an atmosphere of PH_3), a cluster of ~26-28 atoms (as revealed by electronic spectra, MAS-NMR, and EXAFS) was detected in the zeolite supercage. The absorption behavior of these clusters shows an interesting example of a red-shift of the band-edge relative to the bulk and may be a manifestation of a shift from an indirect band gap (as is seen in the bulk) to a direct band gap for the small clusters.

One of the most popular areas of investigation of compound semiconductors has been the II-IV materials, such as Zn/Cd:O/S/Se, etc. These materials have been exhaustively studied in the bulk phases and so an extensive literature exists for comparison with the behavior manifested by extremely small clusters entrapped in a zeolite host. In addition, the relative ease of synthesis - simply reaction of the metal cations with a source of chalcogenide anions - makes for simple sample preparation (see below). A pronounced blue shift of the optical absorption band edge of ZnO embedded inside zeolite A has been reported (from 370 nm (bulk) to ~280 nm) although characterization of the ZnO particles is poor [51]. CdS and CdSe in zeolites have been of particular focus and our own work, along with others, is reviewed below.

3.2.1 CdS in Zeolite Y

CdS in zeolites has been a very popular system for study because of the reasons given above and numerous reports on preparation and characterization of such systems have appeared [52-54]. Our own work on this system was some of the earliest [10,55,56] and provides a useful example of the whole approach towards synthesis and characterization of encapsulated electronic species and so is now described in some detail.

Cadmium ion-exchange of zeolite Y is carried out by slurrying 10 g of zeolite LZ-Y52 (Na-Y from Linde) in 1 L of distilled water and the pH is adjusted to 5 with nitric acid. A calculated amount of cadmium nitrate designed to give a specific exchange level is added to the slurry and the mixture is stirred at room temperature overnight. Collection of the exchanged zeolite by filtration and washing with distilled water is followed by drying and calcination. The powder is heated to 400° C at 3°/min in flowing dry oxygen (100 mL/min) then cooled in vacuo to 100° C. The zeolite is then exposed to flowing H_2S (40 mL/min) at 100° C for 30 min.

Finally, the still white zeolite is evacuated at 100° C for 30 mins, then sealed, and transferred to an inert atmosphere dry box for handling and storage. The zeolite turns pale yellow/cream during the final evacuation step. All zeolites prepared in this manner are moisture-sensitive becoming deep yellow (zeolite Y or X) or pale yellow (zeolite A) on prolonged exposure to atmospheric humidity. Chemical analysis confirms Cd and S are present in from 0 to 25 wt % depending on exchange conditions. XPS shows that there is no detectable Cd on the exterior surface of the zeolite crystallites. IR spectra show no SH groups but there are the expected OH groups attached to the zeolite framework. The exact nature of the CdS cluster units is revealed by a combined application of optical spectroscopies and X-ray techniques.

3.2.2 The Structure of the CdS Clusters in Zeolite Y and its Optical Consequences

Detailed analysis of the powder X-ray diffraction data on a series of CdS-loaded zeolite Y samples reveals the fundamental CdS cluster present consists of interlocking tetrahedra of Cd and S atoms (although some of the S atoms are occasionally replaced by O atoms) forming a distorted cube (Cd-S = 2.47 Å) (Figure 1). This structure, which is heavily dictated by the zeolite symmetry, is confirmed by EXAFS data at the Cd edge (which reveal the local symmetry and coordination environment of the Cd) and theoretical calculations [54] .

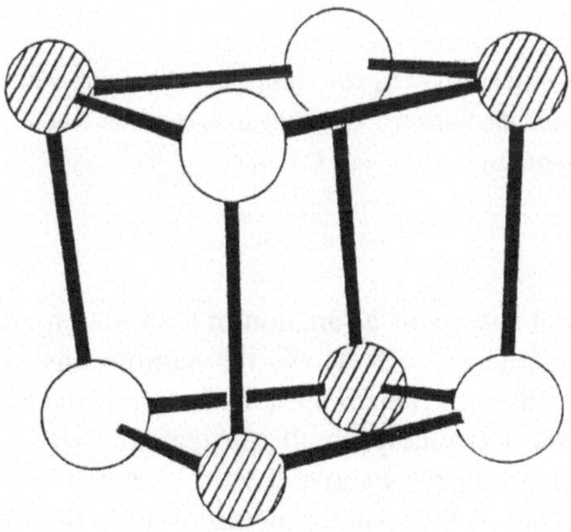

Figure 1. Structure of the $(CdS,O)_4$ unit found within the sodalite unit of zeolite Y (hatched circles = Cd; open circles = S).

To our initial surprise, these Cd_4S_4 cubes were not located in the supercages of the Y structure but were, instead, sited within the smaller sodalite cages. In retrospect, this location is entirely reasonable since these cages are the preferred sites for the original Cd ions upon exchange [57] and all that need occur upon exposure to H_2S is that the cube zips together. The Cd ions of the ideal cluster are in octahedral coordination to three sulfur atoms of the cube and three oxygen atoms of the zeolite framework six-ring window (Figure 2). The sodalite cage seems to have been made for this CdS cluster!

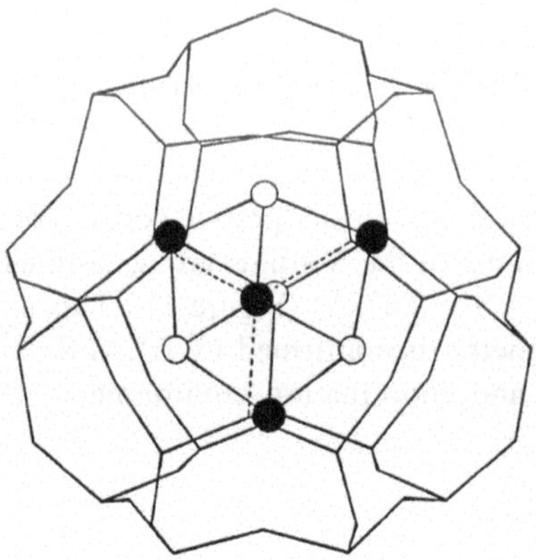

Figure 2. Representation of the $(CdS,O)_4$ cube inside a sodalite unit with dotted lines representing the bonds between the framework six-ring oxygen atoms (shown here as points connected to Si or Al atoms by sticks) and Cd atoms of the cluster (solid circles = Cd; open circles = S).

The evolution of the optical spectra as a function of CdS loading density is particularly revealing. At loading densities of < 5 wt %, the sample has an optical absorption spectrum with a shoulder at 280 nm. This represents a shift of the band edge by ~230 nm and correlates well with the extremely small clusters of Cd_4S_4 found in the X-ray analysis. However, as the loading rises above 5 wt %, a new absorption feature at 350 nm begins to appear and grows as the loading increases up to the maximum yet attained of ~25 wt %. While still quantum-confined CdS, the structural nature of the material responsible for this new absorption feature was unknown. The band edge observed is similar to that of colloidal CdS particles with 25-30 Å diameters [58], yet there are no void spaces of this kind of dimension within zeolite Y. The real clue as to the nature of this species comes from the threshold-loading density of 4±1 wt % - this corresponds to the concentration at which, statistically, CdS cubes must now populate adjacent sodalite units. Simple calculations based on the bulk density of CdS and the sodalite pore volume of Y reveal that the 4 wt % loading density corresponds to filling of ~14 % of the available sodalite volume. Percolation theory [59] predicts the percolation threshold in such systems to be at ~15 volume %, i.e., above 5 wt % of CdS, the individual clusters must begin to interact with one another in adjacent sodalite cages in a percolative fashion. In the limit, when all sodalite units are occupied (~28 wt % CdS), the structure would be such as that represented in Figure 3 and a hyperlattice of CdS clusters dictated by the zeolite Y topology results.

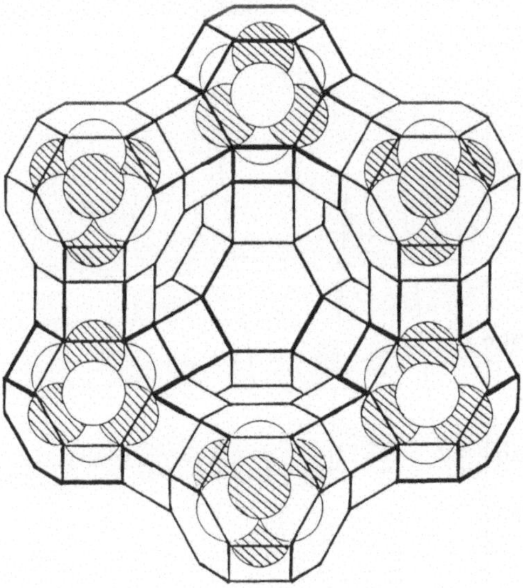

Figure 3. The ultimate, hyperlattice arrangement of CdS cubes in adjacent sodalite units of zeolite Y (hatched = Cd; open = S). The zeolite framework is represented as simple sticks connecting Si or Al atoms and the intervening O atoms are omitted for clarity.

This unique arrangement of individual clusters into an interconnected network gives rise to the luminescence and excitation spectra shown in Figure 4. As anticipated, the luminescence is not band-gap recombination in nature, but rather is dominated by Cd-related defects.

Since the novel optical behavior at higher loading densities of CdS is a consequence of the interaction between individual clusters as modulated by the zeolite structure, it is interesting to examine a zeolite with a different connectivity between the same kinds of Cd/S sodalite clusters. The spectral behavior should be very similar to zeolite Y at low loading densities but differ significantly as the interconnected CdS hyperlattice develops. Zeolite A is just such a zeolite system. In this case, the sodalite cages are connected via double 4-rings in a cubic arrangement and the distance between Cd_4S_4 cubes is projected at ~9 Å, i.e., the cluster-cluster interaction must be much weaker in A than in Y. The optical behavior follows these predictions precisely. While at low loading the absorption spectra of A and Y look very similar (isolated clusters), as the loading increases, there is no shift in the absorption edge of the A zeolite material indicating a much weaker interaction between clusters than in Y and so no significant development of an electronically cogent hyperlattice.

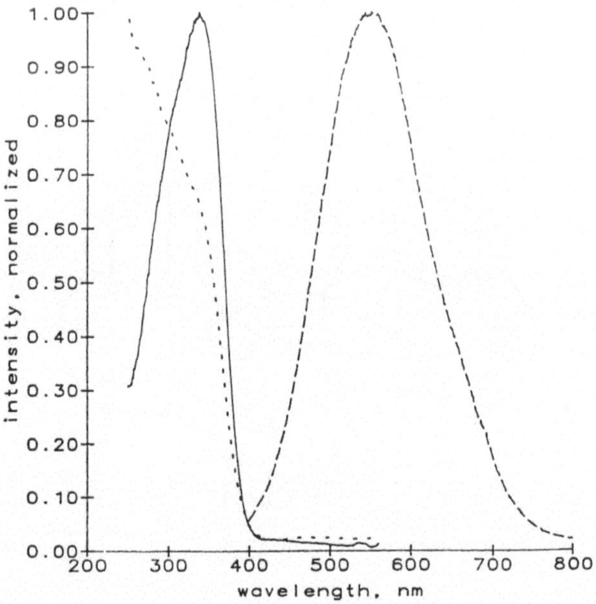

Figure 4. The absorption (dotted), excitation (solid), and emission (dashed) spectra of 6.5 wt % CdS in Y.

4. Conducting Polymers in Zeolites

One of the more intriguing developments in composite materials is the recent preparation of conductive polymeric materials within the channels of a zeolite host. Conjugated polymers of polypyrrole, polyaniline, and polythiophene have been introduced into zeolite channels using a strategy of monomer incorporation followed by polymerization in situ using oxidizing initiators [60-63]. Polyacetylene has also been generated using the acid sites of the zeolite host to initiate oligomerization of methylacetylene [64]. Conductive polythiophene was produced by oxidative doping of the polymer within the pores of pentasil zeolites [65]. All of these systems are examples of the concept of molecular wires - molecularly thin, conductive strands, electrically insulated from each other and aligned so as to allow a potentially high density of connections between tiny electronic devices. Bein's work on polyacrylonitrile (PAN) is a good example of the synthetic approach to these materials [66]. Acrylonitrile is absorbed from the vapor phase into zeolite Y to the level of ~7 molecules per supercage and then polymerized in situ using radical initiators (e.g., peroxodisulfate). The resultant polymer chains have been thoroughly characterized after removal from the zeolite pore system (via dissolution of the framework of the zeolite in HF) and proven to be true PAN with a molecular weight (by GPC) of ~19,000 (~360 monomer units) and approximate length of ~0.2 μ. After pyrolysis (400-650° C) of the PAN in the zeolite host, the material turns grey black and it is suggested to contain graphite-like conductive chains with a conductivity of $\sim 10^{-5}$ $S.cm^{-1}$.

By careful control of the synthesis and orientation of these kinds of conductive molecular wires, an intriguing (albeit distant) possibility exists for generation of electronic devices on the molecular scale. Semiconductor nanoparticulates in zeolite chambers (see section 3) may be connected using these molecular wires to set up an interconnected electronic network where information processing may be done at an unprecedented density. The zeolite framework acts essentially as the electrical insulation for the molecular wires, preventing short circuits and maintaining the spatial isolation of the active components. Input and read-out to such devices presents major difficulties, but may be able to take advantage of the spatial resolution of the Scanning Tunneling Microscope to control, program, and access the electronic network. An alternative approach to the same problem has been explored by Mallouk et al. who have created simple molecular diodes at the zeolite/solution interface and have used molecular species such as $Ru(bipy)_3^{2+}$ as a messenger molecule to both create and access the device [67].

5. Summary and Prospects

Zeolites provide novel hosts for the generation of both isolated and electronically connected metal and semiconductor hyperlattices within their pore volume. Furthermore, the dimensional confinement of conducting polymer filaments, "quantum wires", following their encapsulation as guests within zeolite frameworks produces fascinating novel composite materials. The control of the connectivity between these quantum dots and wires is unparalleled in any other host medium and so has allowed a detailed study of the optical consequences of such connectivity. One must always bear in mind, however, that these guest-host systems are intrinsically composites, by which I mean, that one should not think of the guest clusters as being an isolated electronic system, but rather a system which is intimately linked and even chemically bonded to the host zeolite framework. Of necessity, the small clusters discussed above have very high surface to volume ratios and so most of the cluster atoms are on the surface where their coordination needs are unfulfilled by the cluster lattice. The zeolite lattice then serves to passivate this unsatiated coordination demand with electronic consequences which are likely to be profound. This is strikingly evident in the case of the CdS/Y material described in section 3.2 above. These materials are best thought of as novel compositions of matter which may have unique electronic properties in their own right.

From a practical standpoint, such materials also have some severe drawbacks - most notably the lack of single crystals of sufficient size to produce viable, advanced optical devices such as optical transistors, spatial light modulators etc. Practical device applications of these kinds of novel electronic materials must await advances in the preparation of larger single crystals of the zeolite hosts coupled with an increased variety of pore and cavity sizes for the internal porosity of such systems. Once achieved, these materials offer the promise of an entire new field of quantum electronics where the skills of the device physicist and inclusion chemist merge [68].

References

1. L. E. Brus, *J. Phys. Chem.*, **90**, 2555 (1986) and references therein.
2. L. E. Brus, *J. Chem. Phys.*, **80**, 4403 (1984).
3. H. Weller, H. M. Schmidt, U. Koch, A. Fojtik, S. Baral, A. Henglein, W. Kunath, K. Weiss, and E. Dieman, *Chem. Phys. Lett.*, **124**, 557 (1986).
4. Y.-M. Tricot and J. H. Fendler, *J. Phys. Chem.*, **90**, 3369 (1986).
5. Y. Wang and W. Mahler, *Opt. Comm.*, **61**, 233 (1987).
6. N. F. Borelli, D. W. Hall, H. J. Holland, and D. W. Smith, *J. Appl. Phys.*, **61**, 5399 (1987).
7. N. Herron, J. C. Calabrese, W. E. Farneth, and Y. Wang, *Science,* **259**, 142 (1993).
8. T. Itoh, Y. Iwabuchi, and M. Kataoka, *Phys. Stat. Sol. B.*, **145,** 567 (1988).
9. N. Herron, *Inorg. Chem.*, **25,** 4714 (1986).
10. Y. Wang and N. Herron, *J. Phys. Chem.*, **91**, 257 (1987).
11. G. D. Stucky and J. E. MacDougall, *Science*, **247**, 669 (1990).
12. Y. Wang and N. Herron, *J. Phys. Chem.*, **95**, 525 (1991).
13. D. W. Breck, *Zeolite Molecular Sieves*, Wiley, New York (1974).
14. P. Gallezot, *Zeolites,* **2**, 103 (1982).
15. P. Gallezot, *Surface Science,* **106**, 459 (1981).
16. W. H. M. Sachtler, *Stud. Surf. Sci. Catal.*, **49,** 975 (1989).
17. Y. Kim, Y. W. Han, and K. Seff, *J. Phys. Chem.*, **97**, 12663 (1993).
18. P. H. Kasai, *J. Chem. Phys.*, **43,** 3322 (1965).
19. P. P. Edwards, L. J. Woodall, P. A. Anderson, A. R. Armstrong, and M. Slaski, *Chem. Soc. Rev.*, 305 (1993).
20. Y. Nozue, T. Kodaira, S. Ohwashi, T. Goto, and O. Terasaki, *Phys. Rev. B,* **48**, 12253 (1993).
21. V. N. Bogomolov and A. A. Kapanadze, *Sov. Phys.Solid State,* **20**(1), 57 (1978).
22. Y. A. Alekseev, V. N. Bogomolov, T. B. Zhukova, V. P. Petranovskii, and S. V. Kholodkevich, *Sov. Phys. Solid State,* **24**(8), 1384 (1982).
23. R. M. Barrer and J. L. Whiteman, *J. Chem. Soc. (A)*, 20 (1967).
24. V. N. Bogomolov, *Fiz. Tverd. Tela. (Leningrad)*, **15**, 1312 (1973).
25. K. Tsutsumi and H. Takahashi, *Bull. Chem. Soc. Jpn.*, **45**, 2332 (1972).
26. L. R. Gellens, W. J. Mortier, and J. B. Uytterhoeven, *Zeolites,* **1**, 11 (1981).
27. S. B. Jang, U. S. Kim, Y. Kim, and K. Seff, *J. Phys. Chem.*, **98,** 3796 (1994).
28. D. Exner, N. Jaeger, K. Moller, and G. Schulz-Ekloff, *J. Chem. Soc., Faraday Trans. 1*, **78**, 3537 (1982).
29. K. Moller, D. C. Koningsberger, and T. Bein, *J. Phys. Chem.*, **93**, 6116 (1989).

30. J. N. Miale and P. B. Weisz, *J. Catal.*, **20**, 288 (1971).
31. K. Seff, *J. Phys. Chem.*, **76**, 2601 (1972).
32. V. N. Bogomolov, A. I. Zadorozhnii, V. P. Petranovskii, A. V. Fokin, and S. V. Kholodkevich, *JETP Lett.*, **29**, 373 (1979).
33. V. N. Bogomolov, V. V. Poborchy, S. G. Romanov, and S. I. Shagin, *J. Phys. C: Solid State Phys.*, **18,** L313 (1985).
34. V. N. Bogomolov, S. V. Kholodkevich, S. G. Romanov, and L. S. Agroskin, *Solid State Commun.*, **47,** 181 (1983).
35. A. I. Zadorozhnii, L. K. Panina, V. F. Sakash, L. P. Strakhov, and S. V. Kholodkevich, *Sov. Phys. Solid State,* **22**, 1713 (1980).
36. J. B. Parise, J. E. MacDougall, N. Herron, R. Farlee, A. W. Sleight, Y. Wang, T. Bein, K. Moller, and L. M. Moroney, *Inorg. Chem.*, **27**, 221 (1988).
37. K. Tamura, S. Hosokawa, H. Endo, S. Yamasaki, and H. Oyanagi, *J. Phys. Soc. Jpn.*, **55**, 528 (1986).
38. V. N. Bogomolov, V. V. Poborchii, S. K. Kholodkevich, and S. I. Shagin, *J.E.T.P. Lett.*, **38**, 533 (1983).
39. V. N. Bogomolov, V. V. Poborchii, and S. V. Kholodkevich, *Pis'ma Zh. Eksp. Teor. Fiz.*, **42**, 419 (1985).
40. V. N. Bogomolov, E. L. Lutsenko, V. P. Petranovskii, and S. V. Kholodkevich, *J. E. T. P. Lett.*, **23**, 482 (1976).
41. V. N. Bogomolov, V. V. Poborchii, and S. V. Kholodkevich, *J. E. T. P. Lett.*, **41**, 435 (1980).
42. T. Hirono, A. Kawana, and T. Yamada, *J. Appl. Phys.*, **62**, 1984 (1987).
43. A. Stein, G. A. Ozin, and G. D. Stucky, *J. Photogr. Sci. Technol. Jpn.*, **53**, 322, (1990).
44. G. A. Ozin, J. Godber, and A. Stein, *U. S. Patent 4,942,119* (1990).
45. G. A. Ozin, S. Ozkar, and R. A. Prokopowicz, *Acc. Chem. Res.*, **25**, 553 (1992).
46. Y. Nozue, Z. K. Tang, and T. Goto, *Sol. State Commun.*, **73,** 531 (1990).
47. X. Liu, K. K. Iu and J. K. Thomas, *J. Chem. Soc., Faraday Trans.*, **89**, 1861 (1993).
48. R. D. Stramel, T. Nakamura, and J. K. Thomas, *J. Chem. Soc., Faraday Trans.*, **84,** 1287 (1988).
49. K. Moller, T. Bein, N. Herron, W. Mahler, and Y. Wang, *Inorg. Chem.*, **28**, 2914 (1989).
50. J. E. MacDougall, H. Eckert, G. D. Stucky, N. Herron, Y. Wang, K. Moller, and T. Bein, *J. Am. Chem. Soc.*, **111**, 8006 (1989).
51. T. Turk, F. Sabin, and A. Vogler, *Mat. Res. Bull.*, **27**, 1003 (1992).
52. X. Liu and J. K. Thomas, *Langmuir,* **5**, 58 (1989).
53. C. Luangdilok and D. Meisel, *Israel J. Chem.*, **33**, 53 (1993).
54. A. Jentys, R. W. Grimes, J. D. Gale, and C. R. A. Catlow, *J. Phys. Chem.*, **97**, 13535 (1993).
55. Y. Wang and N. Herron, *J. Phys. Chem.*, **92**, 4988 (1988).
56. N. Herron, Y. Wang, M. M. Eddy, G. D. Stucky, D. E. Cox, K. Moller, and T. Bein, *J. Am. Chem. Soc.*, **111**, 530 (1989).

57. M. Calligaris, G. Mardin, L. Randaccio, and E. Zangrando, *Zeolites,* **6**, 439 (1986).
58. R. Rossetti, R. Hull, J. M. Gibson, and L. E. Brus, *J. Chem. Phys.,* **83**, 1406 (1985).
59. S. Kirkpatrick, *Rev. Mod. Phys.,* **45**, 574 (1973) and references therein.
60. P. Enzel and T. Bein, *J. Phys. Chem.,* **93**, 6270 (1989).
61. P. Enzel and T. Bein, *J. Chem. Soc., Chem. Commun.,* 1326 (1989).
62. T. Bein and P. Enzel, *Angew. Chem. Int. Ed. Engl.,* **28**, 1692 (1989).
63. T. Bein and P. Enzel, *Mol. Cryst. Liq. Cryst.,* **181**, 315 (1990).
64. S. D. Cox and G. D. Stucky, *J. Phys. Chem.,* **95**, 710 (1991).
65. J. V. Caspar, V. Ramamurthy, and D. R. Corbin, *J. Am. Chem. Soc.,* **113**, 600 (1991).
66. T. Bein, *ACS Symp. Series,* **499**, 287 (1992).
67. Y. I. Kim, R. L. Riley, E. H. Yonemoto, D. Rong, and T. E. Mallouk, *ACS Symp. Series,* **499**, 333 (1992).
68. G. A. Ozin, *Adv. Mater.,* **4**, 612, (1992).

Abstract. A review of the use of zeolites as hosts for electronically active guest materials is presented. Metal, semiconductor and conductive polymer nanoparticles as electronic and optical material composites are reported with special emphasis on their synthesis.

Journal of Inclusion Phenomena and Molecular Recognition in Chemistry **21**: 299–324, 1995.

Alkali-Metal Clusters as Prototypes for Electron Solvation in Zeolites

NICK P. BLAKE and GALEN D. STUCKY

1. Introduction

In this section we describe, in some detail, work conducted on alkali-metal clusters in zeolites. Almost without exception, the clusters that we will discuss cannot exist in the gas-phase. As we shall demonstrate, the role of the zeolite is to provide an electrostatic containment field which gives rise to certain ionic sites within the zeolite. Alkali-metal ions which, of course, play the role of counter-ion in the aluminosilicate zeolites, only take on the character of a cluster through introduction of an alkali atom or electron to the zeolite. Under these conditions, the electron (either the valence electron of the alkali atom or the introduced electron) is solvated by the electron trap afforded by these counter-ions. The ions relax around the electron and the result is an alkali-metal cluster. The concept is more than merely notional, since many of the physical properties (EPR and absorption cross-section) of the associated electron can be explained in terms of an alkali-metal cluster. Further, circumstantial evidence that the valence electron of an introduced alkali-atom is auto-ionized comes from the fact that XRD reveals that the nucleus takes up one of the ionic sites in the zeolite.

There are a number of reasons why such systems are interesting:

1) Firstly, these clusters, when isolated, are color centers. They may be produced by exposure of the zeolite to γ-rays or UV light and can be bleached by exposure to a high-intensity white-light source. This cathodochromic and photochromic behavior has, in the past, found applications in CRTs and read/write devices [1].
2) Secondly, at higher cluster concentrations, aggregation of clusters leads to new types of materials. These are known to possess different electronic properties, most notably evident in the modified EPR signal and optical absorption spectrum. All of these changes point to the fact that the clusters are coupled and, as such, are highly suitable candidates for the study of metal-insulator transitions. The unique, channel-like structure of the zeolite ensures a highly directional coupling of neighboring clusters found only in these periodic porous structures. The magnetic and electronic properties of these coupled clusters are perhaps the most exciting aspects of the present alkali-metal cluster work.
3) Thirdly, the formation of alkali-metal clusters in photo-induced charge-transfer involving guests within zeolite hosts provides a mechanism for the stabilization of charge-transfer processes. Typically, charge-transfer processes occur at energies

1-4 eV lower than in the gas-phase. In this way, the zeolites exhibit great utility in the formation of new chemical species and stabilization of charge-separated intermediates. As we shall see, this may lead to selectivity in photo-oxidation reactions.

4) A full, quantitative understanding of the physical properties of these clusters is an indirect way of probing the all-important electrostatic fields within the cages and channels of the zeolites. This fundamental property of the zeolites is essential in the modeling of thermodynamic and transport processes; furthermore, with minor modifications, ab initio packages can be used to model electronic changes in guest molecules and introduction into the zeolite, as well as modifications in the magnetic properties of the system.

2. Methods for the Preparation of Alkali-metal Clusters.

To date, a great many alkali-metal clusters have been identified. Table 1 shows some of the alkali-metal clusters that involve exclusively Na; while in Figure 1, we show the structures of these zeolites and how they are related to the cubo-octahedral sodalite cage. The type of cluster observed, to a large extent, depends upon the method of preparation. We now discuss the primary techniques employed.

2.1 EXPOSURE OF THE ZEOLITE TO HIGH-ENERGY RADIATION

This technique is discussed first for historical reasons. In 1965, Kasai [2] found electron-capture induced through exposure of Na-Y to γ-radiation led to the formation of Na_4^{3+} color centers. Typically, in such studies, the zeolite sample is pressed into a semi-transparent disc which normally requires pressures of 5000 lb/in^2. In the case where the zeolite to be studied is dehydrated, it is necessary to mount the disc in a vacuum apparatus (10^{-3} Torr) and dehydrate at elevated temperatures (200-300° C) for periods of several hours. This disc can then be exposed to either a pulsed high-energy electron beam (10^6 rad/pulse, 2 ns pulse duration) [3] or if steady-state measurements are required, a ^{60}Co γ-ray source of dosage rates of 9.5 krad/min is sufficient.

In cathodochromic studies, air-sensitive zeolites are undesirable and so typically halo-sodalites are preferred. These can be prepared using traditional solid-state synthetic methods (make a gel of $NaAlO_2$, SiO_2, and NaBr in water and evaporate to dryness); however, to improve the cathodochromic properties of the solid, it is found that it is best to 'fire' the product for a few hours at temperatures of around 1000° C. The reasons as to why this procedure is required will become evident later. The resultant powder may then be deposited on an aluminum substrate to a thickness of 0.5 mm or for CRT applications, the powder may be settled in a water suspension and processed according to standard CRT manufacturing methods [1]. Coloration of the samples is effected using electron

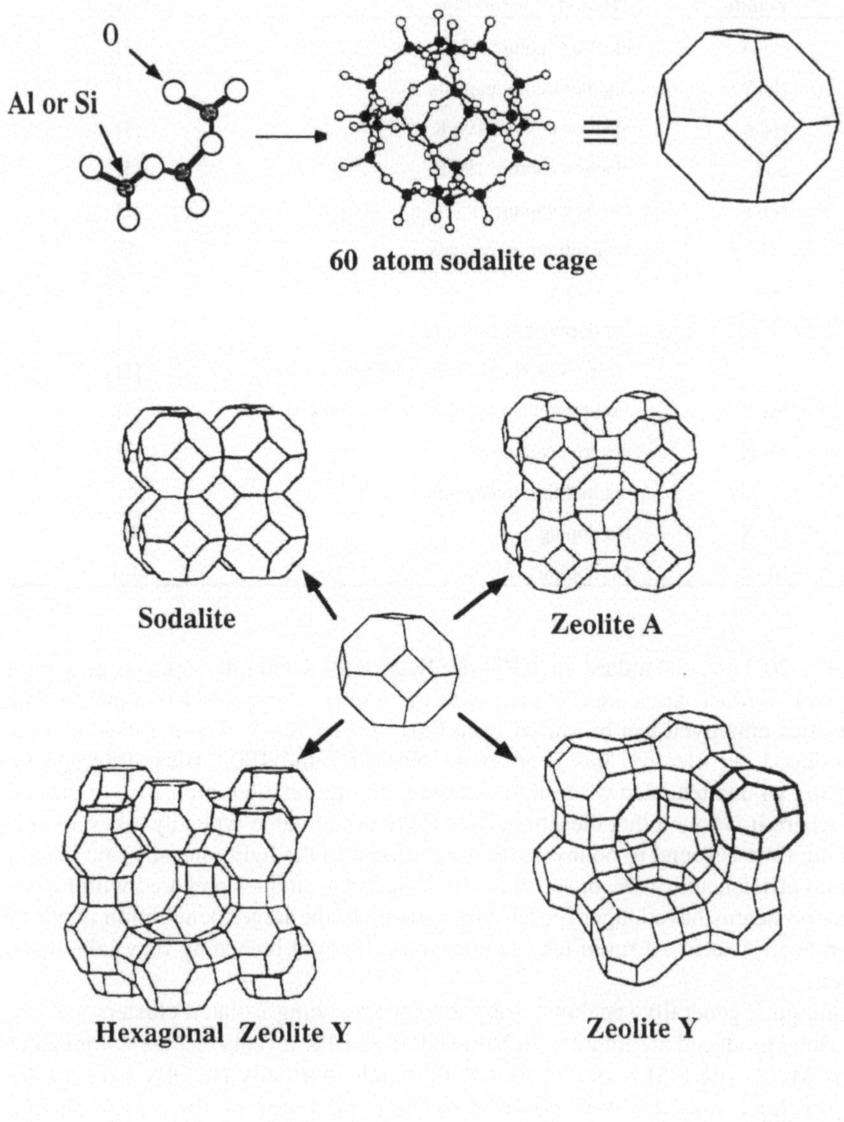

Figure 1. Structures of zeolites sodalite, Na-A, and Na-Y.

Table 1. $Na_{n+1}{}^{n+}$ observed to date and method of formation.

Cluster	Zeolite	Method of formation	Reference
$Na_2{}^+$	Na-A	electron capture 77 K	[3]
$Na_3{}^{2+}$	Na-Y	organolithium reagents	[4]
	Na-A	electron capture 196 K	[3]
	Sod	electron capture 196 K	[5]
$Na_4{}^{3+}$	Na-Y	electron capture	[2]
		organolithium reagents	[4a]
	Sod	Na-doping	[6,7]
		Na-doping halo-sodalites	[8]
		electron capture in halosodalites	[1]
	Na-Y	Europium metal exchange in liquid NH_3	[9]
	Na-X	electron capture	[3]
		organolithium reagents	[4a]
$Na_5{}^{4+}$	Na-X	Na-doping	[4b]
$Na_6{}^{5+}$	Na-X	Na-doping	[4b]

energies of 5-20 keV - standard in CRT applications. Generally speaking, higher sensitivity and contrast ratios are obtained with the higher energy electron beams. The charge densities employed can be varied from 0.01-100 mC/cm^2. The number of color centers produced per electron has been estimated at around 100. Bleaching can be achieved using an unattenuated white-light source with illumination intensities of around 10,000 lx, where it is found that bleaching to half the initial value typically takes 15 sec. The bleaching time is found to be inversely proportional to the light intensity and occurs with quantum efficiencies of 1% or so. It is also found that samples prepared with higher energy electron beams have longer bleach times owing to the larger penetration depth of the electron-beam itself. At room temperature dark, thermal bleaching may take up to several days.

Such techniques, generally speaking, are good for producing isolated clusters and the types of clusters produced are limited. In halo-sodalites, EPR reveals that the color center produced is $M_4{}^{3+}$ where M is one of the alkali-metals, normally Na. By varying the nature of the host sodalite, it is possible to vary the color of the alkali cluster. Empirically for the sodalites it is found that the wavelength of the absorption maximum is proportional to the lattice constant of the host - this is discussed in more detail in the penultimate section. In Na-X and Na-Y, it is found that the only color center produced is

Na_4^{3+}. In Na-A, at liquid nitrogen temperatures, EPR reveals that the preferred trapping site is Na_2^+ while at 196 K, Na_3^{2+} is the preferred site.

In dehydrated sodalite, the situation is more interesting however. The diffuse reflectance spectrum for the γ-irradiated sodalite gives two overlapping features of comparable intensity, one at 680 nm and the other at 550 nm. The distinct chemical identity of each absorption feature is confirmed by measuring the quenching of each feature in the presence of O_2. The assignment of these features comes from a comparison to the more unambiguous results for Na-A and Na-X where it is known that at 196 K irradiation gives Na_3^{2+} and Na_4^{3+}, respectively. Since Na-A has only one electronic feature centered around 650 nm and Na-X one feature at 560 nm, the two features in sodalite are attributed to Na_3^{2+} and Na_4^{3+}, respectively. Figure 2 shows the EPR signal of the γ-irradiated sodalite taken from [5]. The 10-line signal characteristic of Na_3^{2+} is clearly evident. It is, however, perplexing that the 13-line Na_4^{3+} signal is not *more* evident.

On the strength of the O_2 quenching data alone, Thomas et al. [11] have speculated that the Na_4^{3+} center in Na-A resides in the β-cage while the Na_3^{2+} must be produced in an α-cage. This conclusion arises from the premise that electron species in sodalite cages cannot be quenched by O_2 owing to the small pore size. If this conclusion is correct, it is

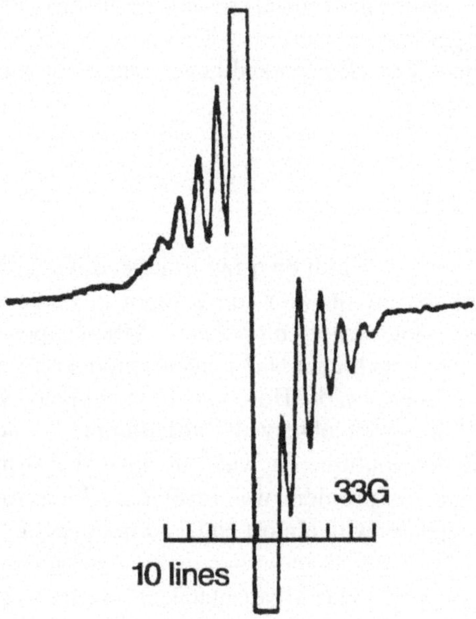

Figure 2. The observed EPR spectrum of the color center obtained upon irradiation of dehydrated sodalite at 196 K (taken from [5]).

puzzling that Na_3^{2+} exhibits the same absorption spectrum in both Na-A and sodalite. As we shall show in the theory section, the absorption spectrum for the color centers is *extremely* sensitive to the dimensions of the cluster; thus the fact that the Na_3^{2+} clusters absorb in the same spectral region in both Na-A and sodalite is more indicative that the species is present in the sodalite cage in both cases. In this case, it may be that the resilience of Na_4^{3+} to quenching is more a result of the fact that most windows into the sodalite cage are occupied by Na^+ ions themselves so that O_2 is unable to gain access to the cage. Bearing all this in mind, the inordinately larger quenching rate observed in Na-A may be simply a transport effect owing to the higher diffusion rates through the solid, coupled with the fact that the ionic sites within the sodalite cage accessible from the α-cage are not occupied by Na^+ ions; however, such an explanation cannot account for the fact that Na_2^+ is not quenched by O_2 .

In addition to the exposure of the zeolite to X-rays, one can, by exposing the zeolite to UV light with energies in excess of 5 eV, ionize the zeolite itself. These electrons are thought to originate from the valence band of the zeolite which Barr et al. report is comprised of the non-bonding O(2p) orbitals [10]. The electrons thus generated are then trapped at cationic sites with the subsequent formation of alkali-metal clusters [11]. Thomas et al. studied Na-X, Na-Y, Na-A, and sodalite after exposure to either 193 nm (6.4 eV) or 185 nm (6.7 eV) light. As with the studies conducted with γ-irradiated zeolites, they found that the same alkali-metal clusters are formed although the intensities are much weaker for UV generated electrons. This is to be expected, since each UV photon absorbed will produce one electron and hence one color center, whereas a γ-ray electron may produce 100 centers.

2.2 ALKALI-METAL DOPING

Shortly after the founding work of Kasai on γ-ray irradiated Na-Y, Kasai, Rabo, and co-workers [12] reported that exposure of Na-Y to Na vapor in vacuo acquired a bright red color attributable to the very same Na_4^{3+} color center. When exposed to Na vapor under the same conditions, Na-X, with its higher Na^+ content, produced a blue color which was interpreted in terms of the cluster Na_6^{5+}. However, EPR evidence was inconclusive and spectra with between 16-19 lines were observed; furthermore, the individual lines did not correspond to a binomial distribution. It was not until the work of Anderson and Edwards, in 1991 [4b], that this problem was resolved. They found that they could simulate the observed EPR spectra with almost equal admixtures of Na_5^{4+} and Na_6^{5+}, see Figure 3.

Thus, the introduction of additional alkali-metal atoms into the zeolite led to the formation of some different 'ion-rich' clusters not seen in the irradiated zeolites. Furthermore, through doping with different alkali-metal ions, it is possible to generate analogs of the Na clusters observed in the previous section. Edwards and co-workers have been able to elucidate the existence of three further centers - K_4^{3+} [13], K_3^{2+} (in K-A and K-X) [4c], and Na_3^{2+}, [4a] in zeolites X, Y, and A.

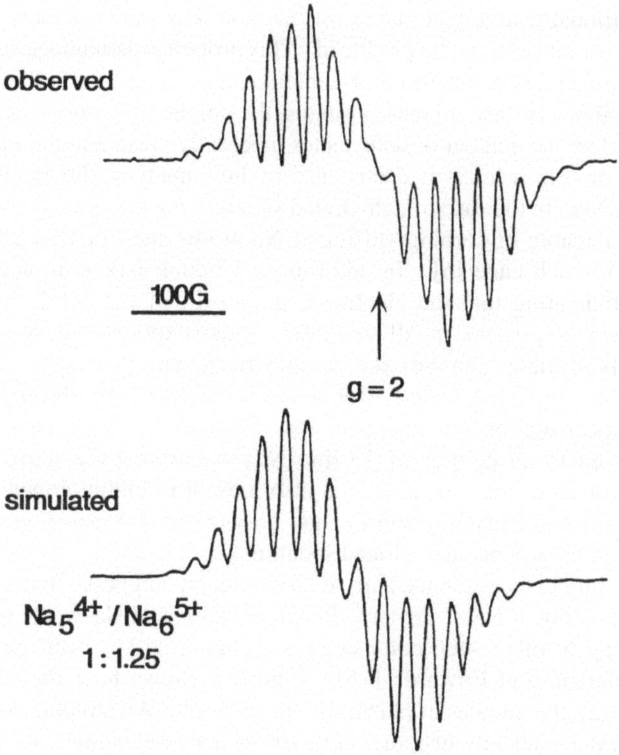

Figure 3. The top figure is the first derivative of the EPR spectrum obtained when Na-X is doped with Na at 298 K. The lower figure is the simulated spectrum for a mixture of the Na_5^{4+} and Na_6^{5+} color centers in the ratio 1:1.25 (taken from [4c]).

The doping amount can be controlled using methods outlined by Dye [14] whereby the alkali-metal is distilled into calibrated lengths of capillary. Doping in this way can have setbacks however. Generally, it is found that the initial reaction of the zeolite with the vapor is not uniform owing to the fact that certain regions of the zeolite have less of their surface exposed. Homogeneity can be achieved by annealing and frequent rotation of the reaction vessel. Owing to air sensitivity, it is necessary to keep the reaction chamber sealed at all times and, thus, EPR and NMR measurements must be performed in a spectrosil section of the reaction tube. Whilst reaction is most easily effected at higher temperatures, poorer crystallinity is the result. Anderson and Edwards have thus suggested that the best crystallinity is obtained at temperatures as low as 250° C (300° C for Na-A) with annealing times of several days. Combining these techniques, it is possible to obtain homogeneous samples of uniform color.

Doping has an additional feature that makes this method important: notably that much higher concentrations of clusters can be produced. This property distinguishes the zeolite systems from other materials in which color centers can be generated - the most well known being alkali-halide crystals. In these systems, the number of color centers that can be created is restricted by the number of defect sites inherently present in the material. In zeolites, the number of color centers is determined by how many of the zeolite cavities have available ionic sites. In the case of dehydrated sodalite, for example, Barrer showed that the crystal was capable of sorbing sufficient Na atoms such that an alkali-metal cluster was formed in each cage [6]. In addition, a gradual darkening accompanies continued sorption indicating that the electronic properties of the solid change with doping. This feature is present in all zeolites; thus, exposure of Na-Y to low concentrations of alkali-metal changes the sample from white to pink. Continued exposure then turns the sample red, which EPR shows is due to Na_4^{3+}, through purple to black. The EPR spectrum of the material also changes with increasing dopant concentrations, leading to an eclipse of 13-line Na_4^{3+} feature by a narrow singlet. Anderson and Edwards show that this singlet is also present in highly doped Na-A and Na-X, while Barrer [6] and Srdanov, Stucky, and co-workers [7] (see Figure 4) have revealed its presence in fully doped dehydrated sodalite.

The fact that this singlet is evidenced in sodalite clearly suggests that its origin is somehow linked to coupling of Na_4^{3+} centers. In Na-X, Na-Y, and Na-A, the singlet may be due to the coupling of other such color centers. This issue has been meticulously investigated by Anderson and Edwards [15]. Figure 5 shows how the EPR signal changes upon increasing the number of Na atoms per unit cell. All structure attributable to the Na_4^{3+} color center gradually becomes eclipsed by a singlet signal. Anderson and Edwards found that *the singlet signal becomes important at concentrations as low as 3 additional Na atoms per unit cell where the probability that two neighboring cages both possess Na_4^{3+} centers becomes significant.* When 8 additional atoms are added, all

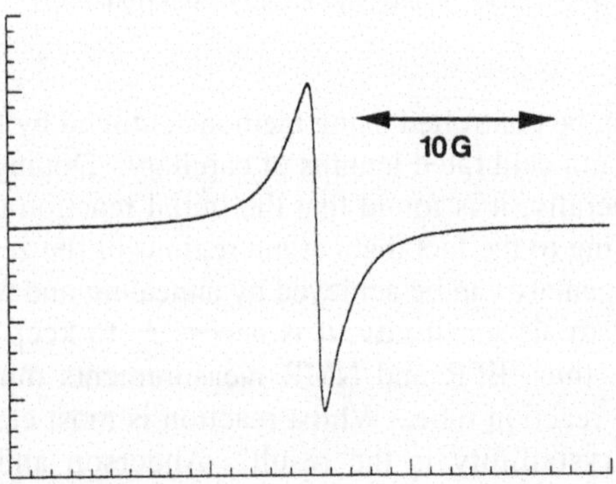

Figure 4. First derivative EPR signal observed in the fully Na-doped system (reproduced by kind permission of Dr. V. I. Srdanov).

β-cage sites are occupied and the singlet eclipses the multiplet structure. This observation adds force to the conclusion that the singlet signal is a result of coupling color centers.

One complication that arises in the cases of Na-Y, Na-X, and Na-A is that exposure to high concentrations of Na leads to a gradual destruction of the crystallinity of the solid. Thus, even though in Na-Y up to 150 Na atoms can be sorbed per unit cell, damage to the framework is observed with concentrations equivalent to 50-60 atoms per unit cell; in Na-X, this threshold is 40-50 atoms per unit cell, while in Na-A damage to the crystallinity is observed at all doping levels.

The K variants of the Na clusters can be prepared by K doping of K-exchanged zeolites. At low doping amounts, and at low temperatures (77 K), K_3^{2+} is the preferred trapping site [4c] in both K-A and K-X; both lead to the characteristic 10-line EPR signals with

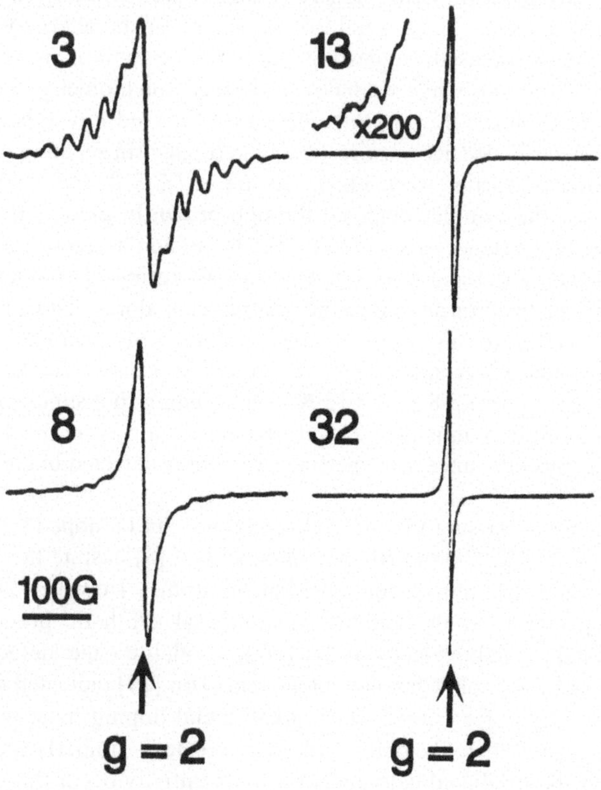

Figure 5. Dependence of the EPR signal of Na/Na_{55}-Y for 3, 8, 13 and 32 additional Na atoms per unit cell (taken from [15]).

hyperfine coupling constants of 12.8 G. This can be viewed as evidence that the cluster must reside in the sodalite cage in both systems. It is unclear whether loss of hyperfine coupling at higher temperatures occurs because of thermal instability or motional averaging.

Sun and Seff [16] recently reported a single crystal XRD study of $K_{12.5}$-A. They found that the occupancy of the sodalite cages in the crystal increases from 1.5 to 2.5 on doping with K vapor and a concomitant K-O bond length increase on K sorption from 2.6 Å to 2.8 Å. The K ions are thus recessed further inside the sodalite cavity. The formation of K_3^{2+} inside the sodalite cavity is thus supported by both XRD and EPR. Furthermore, it is consistent with the work of Thomas' group on cluster formation in dehydrated sodalite and dehydrated Na-A where the similarity in the UV-Vis spectra attributable to Na_3^{2+} suggests that the species both originate within the sodalite cavity.

More recently Armstrong, Anderson, and Edwards [17] have analyzed the K-A system at higher doping levels using XRD. In the K_1/K_{12}-A doping regime, the solid turned an intense blue color. Results from XRD suggest that the K preferentially enters the sodalite cage in the so-called K3 site. They found that the thermal parameters for occupation of the possible ionic sites decreased, suggesting that tumbling motion present in the empty K_{12}-A had ceased. However, while we have the correct stoichiometry to form a K_4^{3+} in the sodalite cage, XRD shows this not to be the case. This led Edwards and co-workers to suggest that the electron density must reside in the α-cage (in seeming contradiction to Anderson and Edwards' earlier work [4c]). In the K_3/K_{12}-A and K_5/K_{12}-A doping regimes, the color of the samples changed through brown to green. In addition, they found evidence for the existence of supercell structure arising as a result of K ordering in neighboring α-cages. The ionic sites K1, K3, and K4 refined to values of 0.5 which turns out can be rationalized through proximity arguments alone. Thus, neighboring K3 and K1 sites are too close to one another to be simultaneously occupied, as are K1 and K4 sites which are only 3.3 Å apart. To overcome this possible source of instability, alternate α-cages either contain 8 K1 or 12 K4 sites leading to a supercell consisting of two conventional zeolite A unit cells. Unfortunately, however, the location of the electrons remains a mystery, since no hyperfine structure is detected in the EPR spectrum down to 77 K.

It is, perhaps, rather surprising that the halo-sodalites can be doped with alkali-metal atoms. Nonetheless Na, K, Cs, and Rb can all be sorbed [8] despite the fact that in the perfect crystal no ionic sites remain for sorbed alkali atoms. However, NMR studies of Stucky et al. [18], clearly show that the conventional synthetic procedures for the production of the halo-sodalites leads to dehydrated sodalite cage 'defect sites'. Later DORS NMR studies by Jelinek, Chmelka, Stein, and Ozin [19] indicated that 10% of the sodalite cages were dry-defect sites. Thus, alkali-metal doping must proceed through uptake into defect cages, where the color centers are created - similarly the production of the color centers in the irradiated halo-sodalites must initially occur through trapping at the dry defect sites, as observed in the work of Thomas' group, followed by ion migration to the cluster site. Smeulders et al. [8] studied the effects of alkali-metal (Li, Na, K, Rb, and Cs) doping in the chloro-, bromo-, and iodo-sodalites. The EPR results obtained suggested that, in all cases, the Na_4^{3+} color center was formed. This result suggests one of three things:

1) The Na_4^{3+} cluster is more stable in the halo-sodalites of type $Na_4[AlSiO_4^-]_3X_2$ than the possible mixed-ion clusters. This certainly makes sense for the K, Cs, and Rb experiments where their electron affinities are lower. It would not necessarily explain the Li results; however, Li is known to be anomalous in many such situations.
2) If the mixed-ion clusters were of comparable stability to the Na cluster, then the result may be due to Na ion transport. If, for example, we have a cluster Na_3K^{3+}, then even if K and Na ions have equal diffusion rates through the zeolite (in fact, it is most probable that Na has a higher diffusion rate than K ion), it is 3 times as likely that a Na ion will leave as a K ion. When the ion leaves, it will take the electron with it leaving a Na_4^{3+} cluster with a K ion in a neighboring 'empty cage'. Thus, if the observations of the Smeulders group are correct, then both thermodynamic and kinetic effects may be responsible for the exclusive formation of Na_4^{3+} clusters. The third possibility takes another tack:

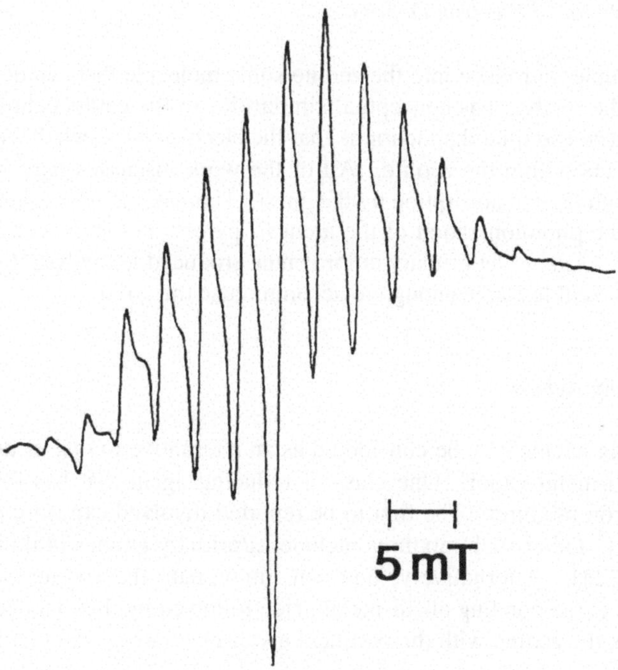

Figure 6. Observed EPR spectrum of Li-doped $Na_4[AlSiO_4]_3Cl$ [8].

3) Mixed ion clusters do exist but could not be discerned in the EPR spectrum. This last alternative is given credence by virtue of the fact that in some cases, i.e., Li/Cl-sodalite (see Figure 6), and Cs/Cl-sodalite systems, non-binomial distributions of the multiplets are observed. This observation is usually synonymous with multiple clusters being present. Furthermore, some of the multiplet lines exhibit shoulders suggesting hidden features in the spectrum.

It seems likely that the behavior of the heavier alkali-metals in such systems may be quite different - as one sees a decrease in the ionicity of the guests. These systems are generally prepared first by ion-exchange, followed by doping with the alkali-metal vapor. Seff and co-workers [20], for example, report the existence of linear Rb_4^{3+}, Cs_3^{2+}, and Cs_4^{3+} clusters along with a triangular Rb_3^+ cluster in sodalite cages of zeolite-A. XRD of $Rb_{14}[AlSiO_4]_{12}$ [21] reveals that a cluster of the type Rb_6^{4+} may exist. This species involves 3 Rb ions in the sodalite cage connected to a further 3 Rb ions in the supercage connected via the six-ring window. Similar clusters have also been seen in $Rb_{12}Ag_x$-A (x=3,7) [22].

2.3 INTRODUCTION OF MOLECULAR ELECTRON DONORS

2.3.1 Photo-Induced Electron Donors

The idea is simple, introduce into the zeolite some molecule that can donate electrons to the zeolite. The method has conceptual similarities to the cathodochromic work of the first section. The essential distinction is that the electron is 'injected' without creation of structural defects within the zeolite. All of the work discussed here involves electron donation through photon absorption by the guest. Thomas and co-workers at Notre Dame have studied the photoionization of the arenes - pyrene and anthracene sorbed in Na-Y and Na-X [23]. Again, the primary color center produced is the Na_4^{3+} and studies have to be limited to zeolites large enough to accommodate the guest.

2.3.2 Reducing Agents

Use of reducing agents may be considered as an alternative to that of doping the zeolite with the alkali-metal vapor. One class of reducing agent that has found favor is the solvated electron mixtures. The first to be reported involved exposure of Na-X to Eu in liquid NH_3 [9]. Other such mixtures include Li/primary amines [4a] and M/THF (M = alkali-metal) [24]. Alternatively, one can impregnate the zeolite with an alcoholic mixture of the corresponding alkali-metal azide followed by thermal decomposition [25] or directly mix the zeolite with the pertinent alkali-metal azide and thermally decompose [26].

3. Spectroscopic Properties of Alkali-metal Clusters and What They Tell Us About Them

3.1 STUDIES OF ISOLATED ALKALI-METAL (F-CENTER) CLUSTERS

3.1.1 Electron Paramagnetic Resonance

From the bulk of the work discussed above, it is clear that the main technique used in the identification of the type of cluster present has been EPR. At low cluster densities and low temperatures, the electron is observed to couple with the spin $^3/_2$ alkali-metal nuclei of the cluster. It does this democratically, i.e., all nuclei are seen as equivalent to the electron leading to 3N+1 lines with a binomial intensity distribution. This has led workers to the notion of an alkali-metal cluster with an electron 'solvated' in the field supplied by the alkali ions themselves either in simple 'electrons-in-a-box' models [3,6] or more sophisticated one-electron models [3,7,27,28]. In addition, the hyperfine coupling constants are characteristically small, becoming smaller with size of cluster. Since the hyperfine coupling gives information regarding the electron density at the nuclei themselves (i.e., the degree of S character of the valence electron), this indicates that the larger the cluster, the more delocalised the electron - again consistent with the picture of an electron solvated in the ionic-field of the cluster. For the Na_4^{3+} cluster for example, it is estimated that the spin-density at each nucleus is around 10% of that of the valence electron in Na, while in Na_6^{5+} the spin density is estimated at only 7.5%. In both cases, these figures are less than the 100%/N expected for an electron residing in the valence orbital of the free atom and therefore suggests that the electron occupies the space *between* the ions. These arguments are equally valid for the K clusters as well, thus, for K_3^{2+}, for example, the spin density is 16% of the free atom. In the case of the Na_4^{3+} color center, further spectroscopic evidence points to the fact that the cluster is tetrahedral and subject to a central forcefield. This comes from the only Raman study of the cluster reported to date by Sen et al. [29]. They find the characteristic normal modes A, E, and T_2 for the cluster with the correct intensity ratios expected for a tetrahedral cluster subject to a central forcefield - adding weight to the interpretation of an electron delocalised in the space between the ions of the cluster. In cases where several types of cluster are present, interpretation of EPR becomes difficult without modeling of the data. Unfortunately, all clusters exhibit similar hyperfine coupling constants and, as such, the observed spectrum may be a complicated superposition of signals. The best example is that of the Na-doped Na-X work of Anderson and Edwards discussed in the previous section. The hallmarks for admixtures of different clusters appear to be i) non-binomial intensity ratios and ii) non-constant numbered multiplet features from sample to sample.

3.1.2 *Optical Properties of the F-centers*

In addition to the EPR data, perhaps the second most extensively used technique for characterization has been electronic absorption spectroscopy. This seems natural considering that these clusters are also color centers. In all cases, the absorption bands are broad and featureless containing no vibronic structure. Thus, obtaining any structural information from this technique is difficult without comparison to theoretical calculation. However, in principle, this method reveals information not only about the ground electronic state but also the excited electronic surfaces, which in turn contain information regarding the electrostatic fields inside the zeolite. This led Haug, Srdanov, Stucky, and Metiu [7,27] and later Blake, Srdanov, Stucky, and Metiu [28] to use 3-d wavepacket calculations for the simulation of the electronic absorption spectrum for an electron solvated in the Coulombic field supplied by the sodalite. In this model, a pseudo-potential description of the electron-frame interaction was used. In the first papers by Haug and co-workers, a Shaw-type pseudo-potential description was used. The sensitivity of the obtained spectrum to both charges of the framework atoms and the tetrahedral radius of the color center were investigated. They found, in all charge schemes, that the electron did delocalize at the center of the alkali cluster in the electronic ground state and that the spectrum was mostly sensitive to the tetrahedral radius of the cluster. With this in mind, later calculations by Blake et al. included calibration of a Born-Mayer Na^+-ion framework potential:

$$V_{Na\text{-}ion}(r) = A_{ion} \exp(-r/\rho_{ion}) + Q_{Na}Q_{ion}/r \qquad (1)$$

where Q_{ion} is the partial charge assigned to a given framework atom. In addition, an improved pseudo-potential based on the notion of Gaussian charge distribution was adopted:

$$V_{electron\text{-}ion}(r) = - Q_{ion} \,\mathrm{erf}(r/2^{1/2}\Delta_{ion})/r \qquad (2)$$

where Δ_{ion} is the crystal radius of the atom as catalogued by Shannon [30]. The parameters that describe these potentials are given in Table 2.

Using this approach, both Na-ion positions and the electronic ground-state energy were minimized. These calculations were successful in reproducing the absorption spectra of not only the Na-doped sodalite but also all of the color center spectra in the halo-sodalites (Figure 7). The model successfully predicts the apparent linear relationship observed in the wavelength of the absorption maxima vs. lattice constant of the sodalite host (Figure 8) while the calculated distance for the nearest Na-ion neighbors to the color center agree to within 0.02 Å with those inferred from the EPR work of Smeulders and co-workers (Table 3). Thus, indicating that a picture of electron solvation within the sodalite is consistent with the observed absorption spectra for the color centers in sodalites. In the

Table 2. Parameters that define the electron-sodalite pseudo-potential and the Na-framework forcefield.

Atom	A (eV)	ρ (Å)	Q (e)	$2^{1/2}\Delta$ (Å)
Na	0	0	+1.0	1.65
O	1224.5	0.284	-0.7	0.82
Si	0	0	+1.0	0.75
Al	0	0	+0.8	0.57
Cl	1551.7	0.334	-1.0	2.56
Br	1594.6	0.347	-1.0	2.77
I	1595.4	0.373	-1.0	3.10

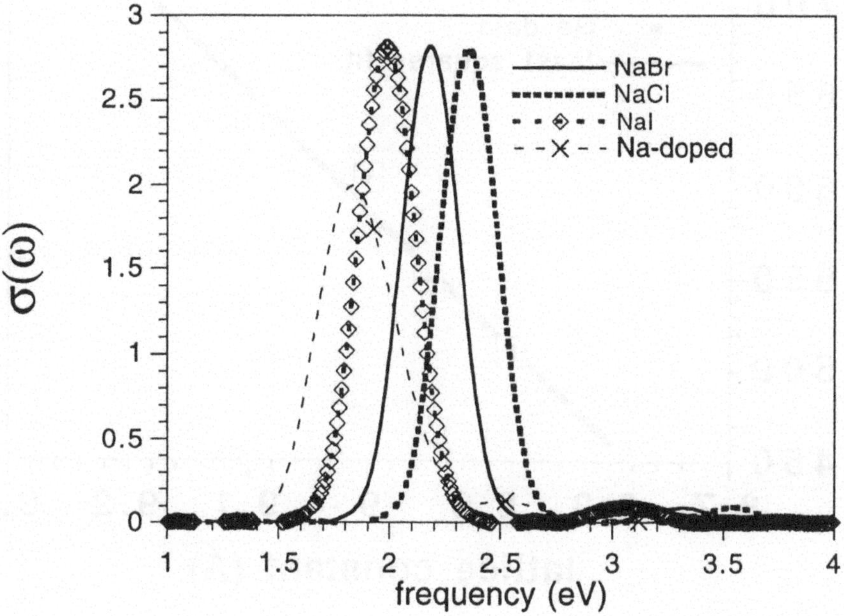

Figure 7. Calculated absorption spectra for the Na_4^{3+} color center in various sodalite hosts.

Table 3. Radial displacement of the Na-ions from the cage center. The values given in parentheses for shell 1 are those for the crystalline host.

Host Lattice	Shell 1 (Å)	Shell 2 (Å)
dry sodalite	3.32 (3.71)	4.14
Cl-sodalite	2.61 (2.73)	4.97
Br-sodalite	2.73 (2.88)	4.89
I-sodalite	2.89 (3.12)	4.72

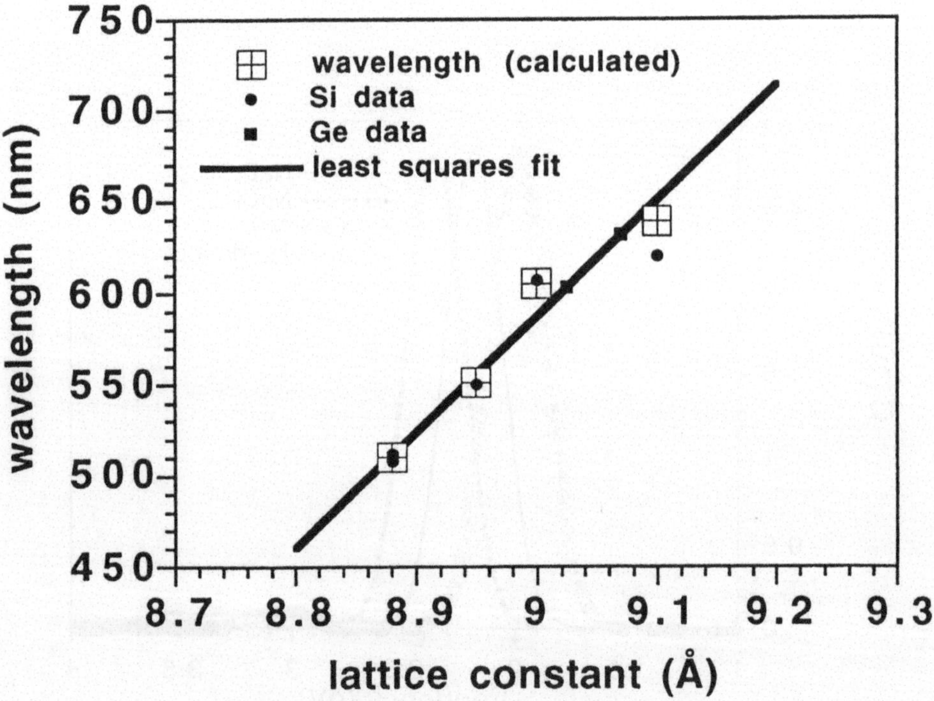

Figure 8. A plot of both the experimentally and theoretically determined wavelengths of maximum absorption for the Na_4^{3+} color center as a function of the lattice constant of the sodalite host.

ground state, the electron does indeed delocalize around the center of the alkali-metal cluster (Figure 9) in the space between the ions. This is in accord with the interpretations from the EPR coupling constants and confirms the geometry of a perfect tetrahedral cluster (in agreement with the Raman spectrum). The electronic states accessed through photon absorption are characterized by their t_2 symmetry (triply degenerate irreducible representation of the T_d point group) with two lobes (one of positive phase and the other negative) pointing along the *{100}* axes (Figure 10). The node is located at the center of the cluster, thus ensuring orthogonality with the ground state (Figure 9). On optical excitation, the resultant deshielding of the ions from one another forces the Na-ions into motion and an effective breathing of the cluster [31]. In part, it is this motion that leads to the broad absorption profile. Recent calculations by us suggest that this motion in itself is not enough to explain the width observed in experiment for the halo-sodalites. Widths of the order of 0.15 eV FWHM are expected while widths from single crystal work are of the order 0.3 eV. It is therefore likely that both inhomogeneity (maybe coupling with the electron hole in the cathodochromic work) and coupling with O is important in determining the final width.

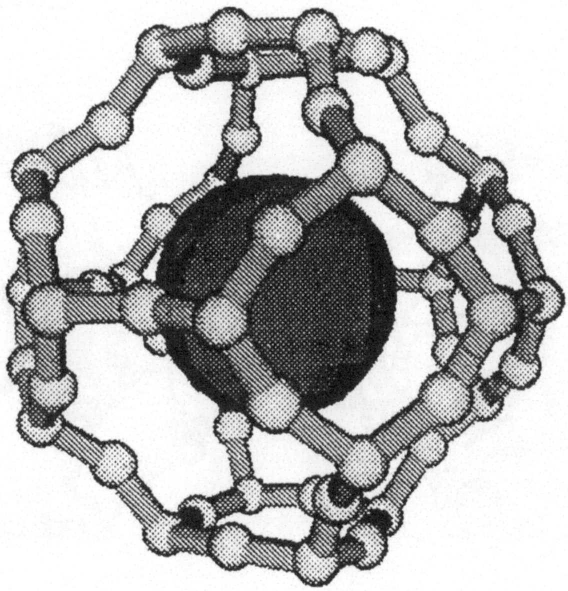

Figure 9. A plot of the electron density for the solvated electron in the ground electronic state in the Na-doped chloro-sodalite ($Na_4[AlSiO_4]_3Cl$) .

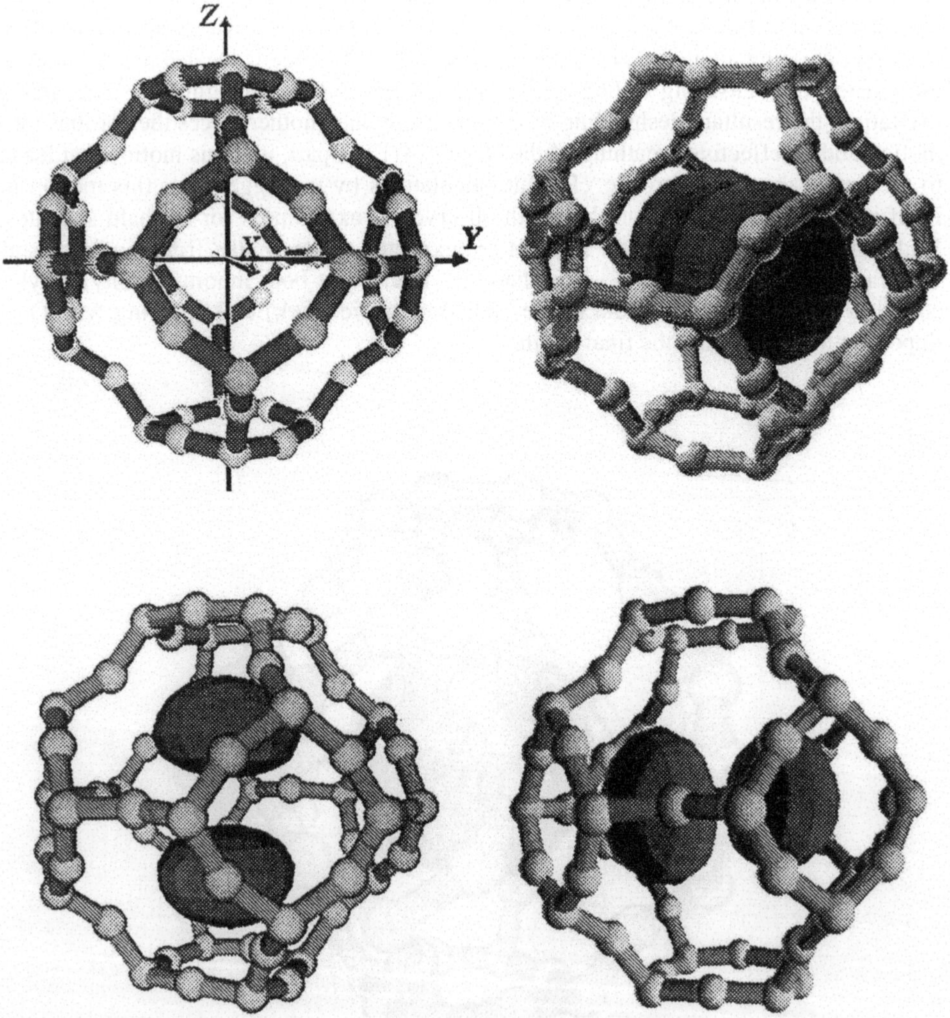

Figure 10. Plots of the electron density for the solvated electron in the first excited, triply degenerate t_2 states in the Na-doped chloro-sodalite ($Na_4[AlSiO_4]_3Cl$).

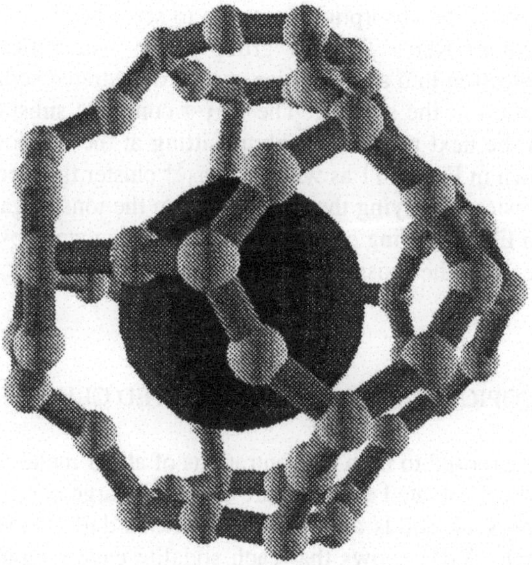

Figure 11. A plot of the electron density for the solvated electron in the ground state in the electron injected dehydrated sodalite ($Na_3[AlSiO_4]_3$).

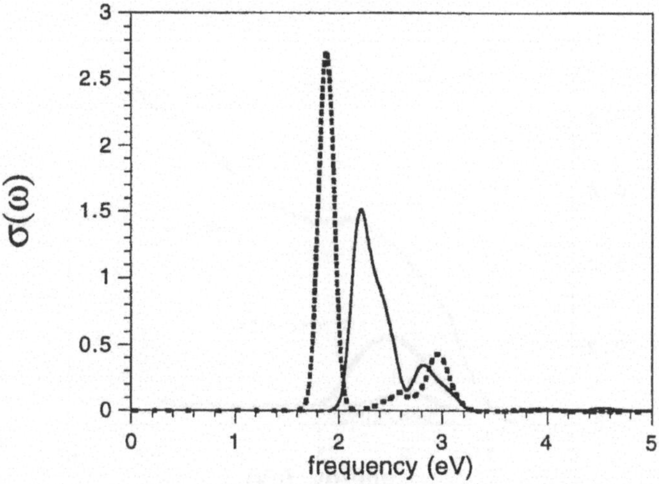

Figure 12. Calculated absorption spectra for Na_3^{2+} in dehydrated sodalite. The dotted line is the spectrum for the cluster in the dipole-aligned crystal, while the solid line is the spectrum for the dipole-misaligned sample.

Recently, this pseudo-potential description of electron solvation has been applied to the Na_3^{2+} cluster within dry sodalite [32]. In particular, we have been interested in the theoretical prediction of the absorption spectrum to see whether of not this is consistent with the assignment of Kerry Thomas' group [3,5]. Theoretically, it is found that introduction of an electron into an otherwise perfect dehydrated sodalite lattice leads to a significant polarization of the system. The Na_3^{2+} contracts substantially on formation of the cluster with the next nearest neighbors sitting at the windows of the cage. The ground state is shown in Figure 11 as with the Na_4^{3+} cluster the electron is delocalised at the center of the cluster occupying the space between the ions - again in agreement with the inferences from EPR coupling constant data. The calculated absorption cross-section is shown in Figure 12. The cross-section exhibits a maximum at 1.7 eV in agreement with experiment.

3.2 SPECTROSCOPIC STUDIES OF AGGREGATED CLUSTERS

When the zeolite is exposed to high concentrations of alkali-metal vapor, it is possible to produce samples where solvated electrons are present in large percentages of the cavities. As we noted above, such solids are typically black or dark brown in color. In black sodalite, for example, XRD shows that each sodalite cage contains an excess Na-ion filling all ionic sites within the sodalite (Srdanov, Monnier et al.). The cell constant contracts considerably from the dehydrated sodalite starting material, changing from 9.1 Å to 8.9 Å, and the Na ions move toward the center of the cage. The optical absorption spectrum lineshape broadens to a continuum with a threshold for absorption at around 0.8 eV.

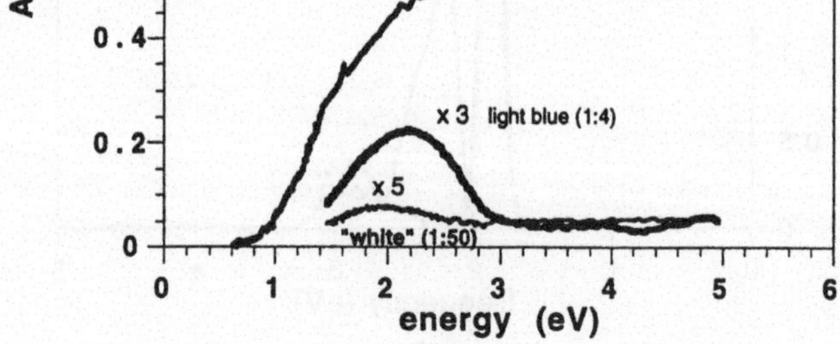

Figure 13. The evolution of the optical absorption spectrum of the Na-doped dehydrated sodalite, (reproduced by kind permission of Dr. V. I. Srdanov).

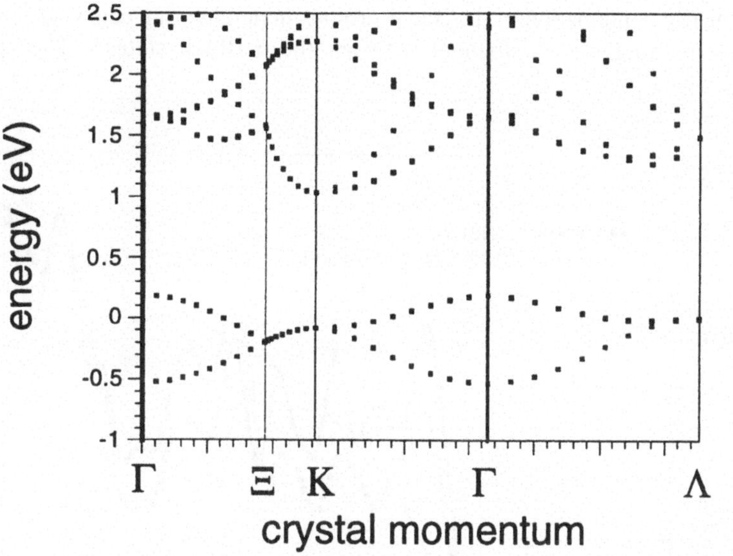

Figure 14. The calculated band structure of black sodalite in the independent electron approximation. There is a total of two electrons per unit cell in black sodalite; therefore, at 0 K, only the lowest band is occupied.

Figure 13 shows how the absorption profile changes as a function of the doping level. From these reflectance studies, it appears that, at high sorption levels, the material is a semiconductor with a band gap of 0.8 eV or so. Recently, improved doping methods lead to a black sodalite phase which exhibits a very narrow EPR singlet resonance (see Figure 4) the width and intensity of which is apparently temperature independent down to 4 K. SQUID measurements of this same material confirm its metallic nature.

In an attempt to rationalize these apparently contradictory observations, Blake et al. [33] have repeated the earlier band-structure calculations of Monnier, Stucky, and Metiu [34], with the same pseudo-potentials used to model the isolated color-center systems as Blake et al. [28]. The band structure is shown in Figure 14. In the absence of band degeneracies, the 2 electrons per unit cell would imply that the material was an insulator at 0 K. However, degeneracy of this valence band with the lowest lying excited band at the edges of the Brillioun zone, as well as along *{100}↔{110}* , indicate that the crystal is a semi-metal and as such will be weakly ferromagnetic. The joint density of states and the transition probability are shown in Figure 15 along with the experimental band structure. Firstly, note the reasonably large contribution to the joint density of states at 0 eV indicating a large region of degeneracy between the lowest two bands. Secondly, transitions from the valence band (this band can be considered as a symmetric combination of the Na_4^{3+} orbitals in neighboring cages) to the lowest unoccupied band (the antisymmetric combination of the Na_4^{3+} orbitals in neighboring cages) are forbidden owing to a vanishing matrix element between these Bloch states. The spectrum

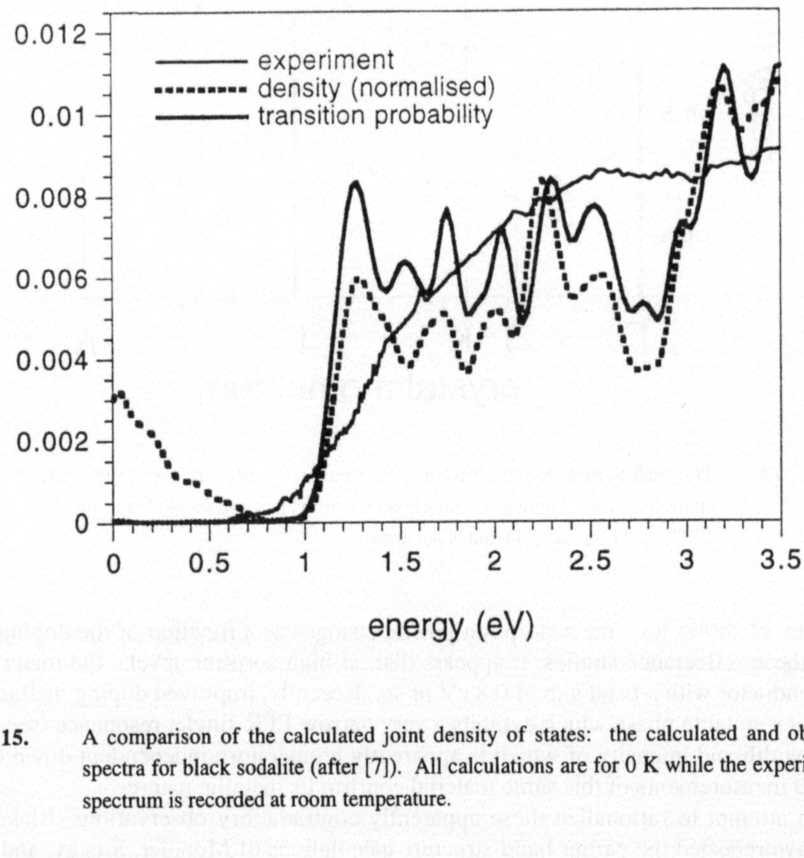

Figure 15. A comparison of the calculated joint density of states: the calculated and observed spectra for black sodalite (after [7]). All calculations are for 0 K while the experimental spectrum is recorded at room temperature.

therefore exhibits an onset for absorption of approximately 1 eV - in reasonable agreement with experiment. The structure in the spectrum corresponds to transitions at the edges of the Brillioun zone. The strongest feature corresponds to vertical transitions at the K and Γ points to the T_2 states at around 2.4 eV, in agreement with experiment. The point of inflexion at around 3 eV corresponds to the gap between the states comprised of the T_2 ("p") states and states comprised of the "d" orbitals of the color center. The calculations, however, predict a much sharper onset for absorption, suggesting that perhaps the preliminary absorption data are not for 100% black sodalite. This has been confirmed recently through NMR, which estimates the material to be 68% black sodalite. Band structure calculations of $Na_7(AlSiO_4)_6{\cdot}e$ reveal that this onset is lower (0.8 eV) than the fully-doped material [33]. Thus, at this point, it seems likely that the "black" sodalite spectrum is a superposition of these two materials and perhaps even some of the isolated clusters.

EPR of the black phase shows, that as one increases the Na doping levels, the EPR spectrum collapses from the 13-line pattern to a singlet. This observation was first noted by Edwards and Harrison in 1984 [13] in conjunction with Na-X and Na-Y, where the similarity to the spectrum obtained in colloidal Na metal led to a metallic cluster assignment. In the years following this influential work, all observations of such clusters were attributed to metallic Na - even in Na-doped halo-sodalites, where there is no room for such a species! Later, Anderson and Edwards [15] re-assessed the validity of such an assignment - realizing that in all other respects this EPR signal bore little resemblance to a conduction-electron EPR signal, (i.e., no Knight shift, wrong temperature dependence for the line-width, etc.) Careful analysis of the lineshape as a function of the dopant levels showed that the collapse to a singlet occurred at levels where it was statistically likely that solvated electron clusters were present in neighboring cages. They postulated that the singlet signal arose as a result of an exchange mechanism and showed by modeling of the EPR spectrum how reasonable values for the exchange integral led to collapse to a singlet structure with even one neighboring cluster. Thus, the EPR spectrum shows evidence that color centers in neighboring cages are coupled - dashing hopes of using alkali-metal clusters in zeolites to study quantum size effects - where such interactions are normally ignored. These observations are also consistent with the magnetic properties of the less than fully doped black sodalite phase, which exhibits an antiferromagnetic transition at around 60 K [35].

4. Prospects for the Future

In the preceding pages we have brought the reader to the present state of alkali-cluster research in zeolites. To some extent, we have concentrated on the theory of the optical properties of these color centers - reflecting simply our present research interests. We have shown how electron solvation theory is successful in predicting the observed optical behavior of these color centers in sodalites and, at the other extreme how the theory, when used in conjunction with Bloch theory, can qualitatively reproduce the optical behavior of the black phase. The present level of theory is thus helping us to qualify some of the interesting properties exhibited by such alkali-metal clusters. Preliminary results for the K derivatives indicate that these solids may have very different electronic and magnetic properties to the Na derivatives and the understanding of the K-doped sodalites is one of the main thrusts of our group. The characterization of the highly doped solids is, however, the object of many groups' research. In particular, interest is focusing on the possibility of effecting an insulator-metal transition in these systems [36].

Estimates for the faujasites indicate that they are close to obtaining the required 10^{21} electrons cm^{-3} required by the Mott criterion [37] for transition to the metallic phase [36]. Interestingly, the fully doped sodalite fulfills this condition and indications are that, if carefully made, it is metallic.

For the future, it seems certain that careful, temperature-dependent, EPR and SQUID measurements will uncover the mysteries of the highly doped materials and the unique properties conferred on them by the zeolite.

Acknowledgments

We gratefully acknowledge partial funding from O.N.R. and the Center for QUantised Electronic STructures (QUEST). Furthermore, we would like to thank Dr. Vojislav I. Srdanov for both the helpful discussions and for the use of both his EPR and optical absorption data for black sodalite. Finally, we would like to thank Prof. J. Kerry Thomas and Prof. Peter Edwards for kindly granting permission for the reproduction of EPR data.

References

1. P. T. Bolwijn, D. J. Schipper, and C. Z. van Doorn, *J. Appl. Phys.*, **43**, 132 (1972).
2. P. H. Kasai, *J. Chem. Phys.*, **43,** 3322 (1965).
3. K. K. Iu, X. Liu and J. K. Thomas, *J. Phys. Chem.*, **97**, 8165 (1993).
4. (a) P. A. Anderson, D. Barr, and P. P. Edwards, *Angew. Chem. Int. Ed. Engl.*, **30**, 1501 (1991);
   (b) P. A. Anderson and P. P. Edwards, *J. Chem. Soc., Chem. Commun.*, 915 (1991);
   (c) P. A. Anderson, R. J. Dancer, and P. P. Edwards, *J. Chem. Soc., Chem. Commun.*, 914 (1991).
5. X. Liu and J. K. Thomas, *Langmuir*, **8**, 1750 (1992).
6. R. M. Barrer and J. F. Cole, *J. Phys.Chem. Solids*, **29**, 1755 (1968).
7. V. I. Srdanov, K. Haug, H. Metiu, and G.D. Stucky, *J. Phys. Chem.*, **96**, 9039 (1992).
8. J. B. A. F. Smeulders, M. A. Hefni, A. A. K. Klassen, E. de Boer, U. Westphal, and G. Geismar, *Zeolites*, **7**, 347 (1987)
9. G. D. Stucky, L. Iton, T. Morrison, G. Shenoy, S. Suib, and R. P. Zerger, *J. Mol. Catal.*, **27**, 71-80 (1980); *J. Chem. Phys.*, **80**, 2203 (1984).
10. T. L. Barr, L. M. Chen, M. Mohsenian, and M. A. Lishka, *J. Am. Chem. Soc.*, **110**, 7962 (1988).
11. X. Liu, K. K. Iu, and J. K. Thomas, *Chem. Phys. Lett.*, **224**, 31 (1994).
12. J. A. Rabo, C. L. Angell, P. H. Kasai and V. Schomaker, *Discuss. Faraday Soc.*, **41**, 328 (1966)

13. P. P. Edwards, M. R. Harrison, J. Klinowski, S. Ramdas, J. M. Thomas, D. C. Johnson, and C. J. Page, *J. Chem. Soc., Chem. Commun.*, 982 (1984)
14. J. Dye, *J. Phys. Chem.*, **84**, 1084 (1980)
15. P. A. Anderson and P. P. Edwards, *J. Am. Chem. Soc.*, **114**, 10608 (1992).
16. T. Sun and K. Seff, *J. Phys. Chem.*, **98**, 10156 (1994).
17. P. A. Anderson, R. J. Singer, and P. P. Edwards, *J. Chem. Soc., Chem. Commun.*, 914 (1994).
18. A. Stein, P. M. Macdonald, G. A. Ozin, and G. D. Stucky, *J. Phys. Chem.*, **94**, 6943 (1990).
19. R. Jelinek, B. F. Chmelka, A. Stein, and G. A. Ozin, *J. Phys. Chem.*, **96**, 6744-6752 (1992)
20. S. H. Song, Y. Kim, and K. Seff *J. Am Chem. Soc.*, **97**, 10139 (1993);
N. H. Heo and K. Seff, *J. Am. Chem. Soc.*, **109**, 7986 (1987); *J. Chem. Soc., Chem. Commun.*, 1225 (1987); *Zeolites*, **12**, 819 (1992);
N. H. Heo, C. Dejsupa, and K. Seff, *J. Phys. Chem.*, **91**, 3943 (1987); *Zeolites*, **9**, 146 (1989).
21. S. H. Song, U. S. Kim, Y. Kim, and K. Seff, *J. Phys. Chem.*, **96**, 10937 (1992).
22. S. H. Song , Y. Kim, and K. Seff, *J. Phys. Chem.*, **95**, 9919 (1991).
23. K. K. Iu and J. K. Thomas, *J. Phys. Chem.*, **95**, 506 (1991); *Colloids Surf.*, **63**, 39 (1992).
24. K. B. Yoon and J. K. Kochi, *J. Chem. Soc., Chem. Commun.*, 510 (1988).
25. L. R. M. Matens, P. J. Grobet, and P. A. Jacobs, *Nature*, **315**, 568 (1985).
26. B. Xu, X. Chen, and L. Kevan , *J. Chem. Soc., Faraday Trans.*, **87**, 3157 (1991).
27. K. Haug, V. I. Srdanov, G. D. Stucky, and H. Metiu, *J. Chem. Phys.*, **96**, 3495 (1992).
28. N. P. Blake, V. Srdanov, G. D. Stucky, and H. Metiu, *J. Phys. Chem.*, *in press.*
29. P. Sen, C. N. R. Rao, and J. M. Thomas, *J. Mol. Struct.*, **146**, 1711 (1986).
30. R. D. Shannon, *Acta Crystallogr.*, **A32**, 751 (1976).
31. N. P. Blake and H. Metiu, *J. Chem. Phys.*, *submitted.*
32. N. P. Blake and H. Metiu, *J. Phys. Chem*, *to be submitted.*
33. N. P. Blake, V. I. Srdanov, G. D. Stucky, and H. Metiu, *Phys. Rev. Lett.*, *to be submitted.*
34. A. Monnier, G. D. Stucky, and H. Metiu, *J. Chem. Phys.*, **100**, 6944 (1994).
35. V. I. Srdanov and G. D. Stucky, *in preparation.*
36. P. P. Edwards, L. J. Woodhall, P. A. Anderson, A. R. Armstrong, and M. Slaski, *Chem. Revs.*, 305 (1993).
37. P. P. Edwards and M. J. Sienko, *J. Am. Chem. Soc.*, **103**, 2697 (1981).

Abstract. Zeolites are unique in that they can play host to a large number of alkali-metal clusters of the type M_n^{+p} hitherto unseen in any other system. When isolated, these clusters behave as color centers. The alkali ions reside in ionic sites within the cavities and so the nature of the cluster is very much a function of the zeolite host, the Si:Al ratio, and the method chosen to prepare the cluster. Since these centers are created within zeolite cages rather than as structural defects (as is the case with the alkali halides) high cluster concentrations can be achieved at which point the optical and magnetic properties of the zeolite change profoundly. We review experimental work in this area, as well as our own attempts to understand both the electronic and optical properties of these systems in terms of an electron solvation model.

Subject Index